Topics in Applied Physics   Volume 51

Topics in Applied Physics Founded by Helmut K. V. Lotsch

Light Scattering in Solids III

Recent Results

Edited by M. Cardona and G. Güntherodt

With Contributions by
M. Cardona   G. Dresselhaus   M. S. Dresselhaus
G. Güntherodt   W. Hayes   M. V. Klein   D. J. Lockwood
J. R. Sandercock   R. G. Ulbrich   C. Weisbuch

With 128 Figures

Springer-Verlag Berlin Heidelberg GmbH 1982

Professor Dr. *Manuel Cardona*

Max-Planck-Institut für Festkörperforschung, Heisenbergstraße 1,
D-7000 Stuttgart 80, Fed. Rep. of Germany

Professor Dr. *Gernot Güntherodt*

Universität zu Köln, II. Physikalisches Institut, Zülpicher Straße 77,
D-5000 Köln 41, Fed. Rep. of Germany

ISBN 978-3-662-31175-2   ISBN 978-3-540-39208-8 (eBook)
DOI 10.1007/978-3-540-39208-8

2153/3130-543210

Preface

This book is the third of a series of four volumes devoted to the scattering of light by solids (Raman and Brillouin spectroscopy and related phenomena). The first volume appeared in 1975 as Vol. 8 of the springer series "Topics in Applied Physics." It contains a succinct description of the general principles of spontaneous and stimulated light scattering by solids and treats a selection of a few topics of potential growth, such as scattering by amorphous solids, scattering by electronic excitations, and resonant scattering. A second revised and updated printing of this volume in soft cover is in preparation.

The second volume (Topics in Applied Physics, Vol. 50) contains a detailed treatment of the theory of light scattering in molecules and solids, starting with a macroscopic or phenomenological description and followed by the microscopic treatment. It also includes a chapter on instrumental aspects (in particular multichannel detection systems) and another on high-power stimulated scattering phenomena (CARS, stimulated Raman effect, hyper-Raman effect).

Recent developments have been so numerous that two more volumes have become necessary to do justice to this field of endeavour. In the present volume we collect case studies involving light scattering in solids by a few important families of materials of current research interest. They include intercalated layer compounds, transition-metal compounds, superionic conductors, Brillouin scattering by surface excitations, and Brillouin scattering resonance at electronic polaritons. A forthcoming volume will contain work on electronic Raman scattering, scattering by magnetic and spin excitations, surface-enhanced Raman scattering, and Raman scattering under mechanical stress.

The range of information obtained by means of light scattering spectroscopy techniques is very wide and touches many branches of the natural sciences, from the life sciences to physics, chemistry, and ecology. These books are primarily intended for solid-state physicists. A number of chapters, however, should find interest among all practitioners of light scattering spectroscopy. Most of them should be useful to other spectroscopists interested in looking beyond the narrow range of their immediate concerns.

The editors would like to thank again all contributors for keeping their deadlines and for cooperation and understanding in considering their suggestions.

Stuttgart and Köln,
March 1982

Manuel Cardona
Gernot Güntherodt

Contents

Contributors

Cardona, Manuel
Max-Planck-Institut für Festkörperforschung, Heisenbergstraße 1,
D-7000 Stuttgart 80, Fed. Rep. Germany

Dresselhaus, Gene
Massachusetts Institute of Technology, Francis Bitter National
Magnet Laboratory, Cambridge, MA 02139, USA

Dresselhaus, Mildred S.
Massachusetts Institute of Technology, Center for Materials Science
and Engineering, and Department of Electrical Engineering and Computer
Science, Cambridge, MA 02139, USA

Güntherodt, Gernot
Universität zu Köln, II. Physikalisches Institut, Zülpicher Straße 77,
D-5000 Köln 41, Fed. Rep. Germany

Hayes, William
University of Oxford, Department of Physics, Clarendon Laboratory,
Parks Road, GB-Oxford OX1 3PU

Klein, Miles V.
University of Illinois at Urbana-Champaign, Dept. of Physics,
Urbana, IL 61801, USA

Lockwood, David J.
National Research Council, Division of Physics,
Ottawa, Canada K1A OR6

Sandercock, John R.
RCA Laboratories Ltd., Badener Straße 569, CH-8048 Zürich, Switzerland

Ulbrich, Rainer G.
Universität Dortmund, Institut für Physik,
D-4600 Dortmund 50, Fed. Rep. Germany

Weisbuch, Claude
Ecole Polytechnique, Laboratoire de Physique de la Matière Condensée,
F-91120 Palaiseau, France

1. Introduction

M. Cardona and G. Güntherodt

Ahora ya no soy más que luz y clamor;
palabra de palabras, sin tacto, resplandor...
Atravieso la hondura de un universo turbio
y huyo de mí al azar.

Angeles Cardona

This volume is the second in a series of three which reviews developments and significant advances in the field of light scattering in solids. These three volumes are a continuation of the 1975 volume, "Light Scattering in Solids," edited by *Cardona* [1.1]. The first of these three volumes [1.2] includes in its first chapter, written by the editors, an extended description of its contents and those of all other volumes in the series. Consequently the present description will be kept as brief as possible. In [Ref. 1.2, Chap. 2], *Cardona* reviewed in detail the macroscopic and microscopic treatment of light scattering, with emphasis on resonance phenomena in semiconductors. An effort was made to keep the theoretical side simple and self-contained. It should serve as a general background for most of the treatments in the series. The technical aspects of optical multichannel detection and its application to time- and space-resolved measurements were discussed by *Chang* and *Long* [Ref. 1.2, Chap. 3]. Finally, *Vogt* discussed nonlinear phenomena in the scattering of light by solids and the various spectroscopic techniques which result from them [Ref. 1.2, Chap. 4].

While [1.1 and 1.2] focused on general principles, the present volume focuses mainly on case studies of light scattering in several classes of materials. Its aim is to demonstrate the versatility of the light scattering technique as a modern solid-state physicist's tool in investigating materials of current interest to research. In this respect, the present volume goes beyond the basic principles outlined in [1.1] and in [Ref. 1.2, Chap. 2], and applies these principles to obtain information on particular material- or phenomena-oriented questions.

Chapter 2, written by M.S. Dresselhaus and G. Dresselhaus, deals with Raman scattering in graphite intercalation compounds. It describes the contributions of Raman scattering to the understanding of lattice dynamics, i.e., superlattice formation due to staging, of graphite-intercalant interactions, and of systematic trends in this vast variety of materials.

Chapter 3, by D.J. Lockwood, summarizes extensive work on various transition-metal halides, for which Raman scattering is shown to be very useful in studying magnetic, electronic, and coupled electron–phonon excitations. Emphasis is placed on scattering by one- and two-magnons in pure and mixed antiferromagnets as well as on electronic Raman scattering of iron-group transition-metal ions.

Chapter 4, by W. Hayes, describes the significance of Raman scattering in understanding disorder and dynamical processes in superionic conductors. Characteristic features, such as vibrational excitations due to oscillatory motion of ions at lattice sites and quasielastic scattering arising from diffusive motion of the ions, will be illustrated by specific examples of the different classes of superionic conductors.

Raman scattering by phonon anomalies in transition-metal compounds, such as carbides, nitrides, dichalcogenides, and A15 compounds, is the subject of Chap. 5, written by M. V. Klein. As a preface to theoretical understanding, a rather general theory of light scattering by electrons, phonons, and coupled electron–phonon excitations is included. Such theory, based on many-body perturbation techniques and Green's functions, was omitted in [Ref. 1.2, Chap. 2] for the sake of simplicity. It is a welcome complement to the theoretical background contained in the series. A theoretical description of the relationship between strong electron–phonon processes, which are responsible for phonon anomalies and phase transitions, and strong Raman scattering is provided. Particular emphasis is placed on Raman scattering from charge-density-wave phonons in superconducting systems and on their coupling to electronic excitations across the superconducting gap.

The increasing interest in applying light scattering to opaque and in particular metallic solids is reflected in Chap. 6, written by J. Sandercock. It outlines the most recent developments in multipass and synchronously scanned tandem Fabry–Perot interferometers and their application to Brillouin spectroscopy. These advances have made possible investigations of surface phonons and magnons on opaque materials, as well as quasielastic, diffusive-type excitations.

Finally, Chap. 7, by C. Weisbuch and R. G. Ulbrich, describes the experimental verification of the theoretically predicted resonant light scattering by excitonic polaritons, one of the rare and most beautiful recent examples of the *predictive* nature of solid-state theory. The elusive role of the additional boundary conditions, the resonant scattering vs. hot luminescence controversy in multiphonon processes, and exciton polariton-mediated electronic scattering are discussed.

Contrary to the above described case studies, the forthcoming volume [1.3] will focus mainly on basic investigations of new scattering mechanisms in solids. By this are meant free-carrier excitations in doped semiconductors (G. Abstreiter, M. Cardona, and A. Pinczuk), spin-dependent scattering in CdS (S. Geschwind and R. Romestain), and in magnetic semiconductors (G. Güntherodt and R. Zeyher), and a microscopic description of light scattering in valence-fluctuating rare earth compounds (G. Güntherodt, R. Merlin), and of surface-enhanced Raman scattering (K. Arya, A. Otto, R. Zeyher). This following volume concludes with a summary of morphic effects (e.g., effects of hydrostatic and uniaxial pressure on phonons and electrons as observed by light scattering by B. A. Weinstein and R. Zallen).

References

1.1 M. Cardona (ed.): *Light Scattering in Solids*, Topics Appl. Phys., Vol. 8 (Springer, Berlin, Heidelberg, New York 1975)
1.2 M. Cardona, G. Güntherodt (eds.): *Light Scattering in Solids II*, Topics Appl. Phys., Vol. 50 (Springer, Berlin, Heidelberg, New York 1982)
1.3 M. Cardona, G. Güntherodt (eds.): *Light Scattering in Solids IV*, Topics Appl. Phys., Vol. 54 (Springer, Berlin, Heidelberg, New York 1983)

2. Light Scattering in Graphite Intercalation Compounds

M. S. Dresselhaus and G. Dresselhaus

With 26 Figures

Abstract

Light scattering in graphite intercalation compounds gives key insights into the physics of these layered structures. In this chapter a review is presented of experimental Raman scattering studies and their interpretation based on models of the lattice dynamics of pristine and intercalated graphite. The periodic layer structure of intercalation compounds makes it possible to model the dynamical matrix by a Brillouin zone folding of the pristine graphite matrix. The stage dependence of the Raman-active modes is reported which correlates with a stage dependent strain. Resonant enhancement of the scattering cross-section permits observation of modes related to the intercalate layer. Explicit results are obtained for the internal modes of Br_2 molecules in the graphite-Br_2 system. Stage 1 alkali metal compounds show a lineshape of the Breit-Wigner-Fano form which implies a coupling between a sharp vibrational mode and a Ramanactive continuum. Second-order Raman scattering results for intercalated graphite are reported. A brief summary is also given on Raman scattering studies of intercalated graphite fibers, adsorbed molecules on graphite surfaces and ion-implanted graphite.

2.1 Background

2.1.1 Introductory Comments

Raman scattering has proven to be an important technique for the characterization of layered materials [2.1]. These materials generally exhibit strong intralayer and weak interlayer bonding, thus giving rise to large anisotropies in their properties generally and in their phonon modes in particular. The layered compounds often include materials exhibiting superlattice structures either through polytype formations, growth procedures such as molecular beam epitaxy or chemical procedures such as intercalation.

Because of the small wave vector of light relative to Brillouin zone dimensions in solids, first-order light scattering is confined to the excitation of phonon modes close to the zone center. Superlattice structure resulting from polytype formation folds nonzone-center modes of the parent structure into zone-center modes for the polytype, thus resulting in the excitation of additional

Ramanactive modes. Such modes associated with the zone folding of five polytype structures have been studied extensively in SiC [2.2] and for various transition metal dichalcogenides such as $2H$-MoS_2, $MoSe_2$, $MoTe_2$, $NbSe_2$, WS_2, WSe_2 [2.1] and related materials such as PbI_2 [2.3, 4]. Raman scattering has also been used to study the superlattice structures prepared by molecular beam epitaxy [2.5].

Graphite intercalation compounds offer a unique system for the generation of large varieties of superlattices by proper choice of more than 100 intercalate species and a wide range of intercalate concentrations. For the graphite intercalation compounds, superlattice structures perpendicular to the layer planes are generated by the staging mechanism, and in some cases, in-plane superlattices are also found.

The calculation of the lattice dynamics of graphite intercalation compounds is simplified by their very strong intralayer and weak interlayer binding, so that as a first approximation, the lattice modes in the graphite layers are related to the graphite host and in the intercalate layers to the bulk intercalant. Thus, information on the coupling between the graphite and intercalant is provided by studies of departures from the lattice mode structure of the unperturbed constituent layers. In this context, we distinguish between graphite *bounding* layers adjacent to an intercalate layer, where the perturbation is largest, and the graphite *interior* layers which are adjacent only to graphite layers. Raman scattering, being a spectroscopic tool, allows the independent investigation of lattice modes associated with the graphite interior, graphite bounding and intercalate layers, because each of these lattice modes occurs at different mode frequencies. The layer structure of the intercalation compounds further allows the separation of the lattice modes into in-plane and c-axis modes. Experimental studies have largely focused on the in-plane lattice modes because the cleavage plane is a c-face, and good optical surfaces can be prepared by cleavage.

Three complementary experimental techniques are especially important for the study of lattice mode structure: inelastic neutron, Raman and infrared spectroscopy. All three techniques have been successfully applied to graphite and its intercalation compounds, and general reviews of this work have appeared [2.6, 7]. Infrared and Raman spectroscopy probe predominantly zone-center modes, while inelastic neutron scattering is capable of exploring the entire Brillouin zone. Information on the phonon dispersion relations for other points in the Brillouin zone can be obtained from second-order Raman spectra, where contributions are made by pairs of phonons with vectors q and $-q$, the major contribution coming from those regions in the Brillouin zone having high densities of states.

The high symmetry of the pristine graphite lattice is an approximate symmetry for the larger real space unit cell for the intercalation compounds. This approximate symmetry gives rise to zone-folding phenomena which map certain nonzone-center q-vectors for graphite into the zone center for the intercalation compound, thereby turning on new Raman-active modes.

Because of the high optical absorption of graphite and its intercalation compounds, Raman scattering experiments on in-plane modes are conveniently performed on cleaved c-faces using a Brewster angle backscattering geometry. Incident photons are typically provided by a cw argon-ion laser, though other excitation sources (e.g. dye lasers) are preferred in certain cases, such as for resonant enhancement studies. The scattered radiation is analyzed by a conventional double grating monochromator. To prevent sample instabilities and intercalate desorption, Raman measurements are conveniently made using low laser power (< 50 mW) and low temperatures (e.g. < 77 K). Encapsulation of the samples in ampoules fitted with suitable optical windows is necessary for maintaining stage fidelity of the samples during the spectroscopic measurements. To excite c-axis modes, light with the E vector along the c-axis is introduced onto an optical a-face which can be prepared by sputtering with argon ions [2.8, 9].

2.1.2 General Properties of Graphite Intercalation Compounds

Graphite intercalation compounds consist of highly ordered graphite planes into which the insertion of atomic or molecular layers of a different chemical species has been made [2.10]. The intercalation process generally occurs in highly anisotropic layered structures where the intraplanar binding forces are large in comparison with the interplanar binding forces. Some common examples of host materials for intercalation compounds are graphite and transition metal dichalcogenides (e.g. TaS_2 and $TaSe_2$ [2.11]). The graphite compounds are of particular physical interest because of their high degree of structural ordering. The dominant superlattice symmetry is associated with the *staging* phenomenon, whereby the intercalate layers are periodically arranged in the matrix of graphite layers and are characterized by a stage index n denoting the number of graphite layers between adjacent intercalate layers, as illustrated in Fig. 2.1.

Intercalation provides to the host material a means for controlled variation of many physical properties over wide ranges [2.10]. Because the free carrier concentration of the graphite host is very low [2.7] ($\lesssim 10^{-4}$ free carriers/atom), intercalation with different chemical species and concentrations permits a wide variation of the electrical properties of the host material. Though the effect of intercalation on the electrical conductivity has attracted the most attention, other transport and structural properties are also strongly modified [2.6, 7, 10, 12]. The intercalation process yields synthetic metals where the dominant carriers are either electrons (donor compounds) or holes (acceptor compounds). Also striking is the range of electrical conductivity behavior, ranging from almost insulating behavior for the c-axis conductivity in certain acceptor compounds to superconducting behavior below 1.0 K for stage 1 alkali metal donor compounds (C_8K) where neither of the parent chemical species individually exhibit superconductivity [2.13].

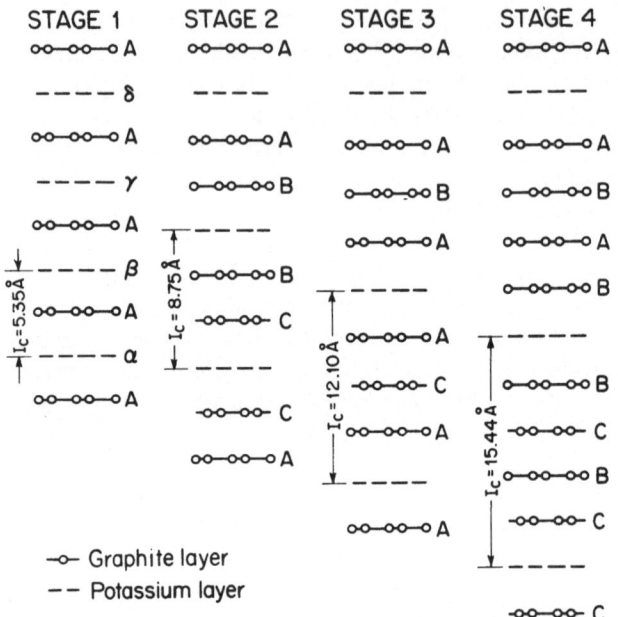

Fig. 2.1. Schematic diagram illustrating the staging phenomenon in graphite-potassium compounds for stages $1 \leq n \leq 4$. The potassium layers are indicated by dashed lines and the graphite layers by solid lines connecting open circles, and schematically indicating a projection of the carbon atom positions. The ...$ABAB$... graphite layer stacking for stages $n \geq 2$ is maintained between intercalate layers, although a rhombohedral stacking arrangement appears across intercalate layers. The stacking ordering is well-confirmed by x ray diffraction (hkl) patterns. For each stage, the distance I_c between adjacent intercalate layers is indicated

Intercalation can be achieved starting from a solid, liquid or gaseous reagent, though preparation from the vapor is the most common for the controlled preparation of well-staged materials [2.10]. The intercalation rate and resulting stage are strongly dependent on the reaction conditions such as pressure and temperature of the intercalate reagent, the temperature, the degree of crystalline order and defect density of the graphite host material, and its physical dimensions. Intercalation occurs for many types of reagents ($> 10^2$) ranging from simple ionic species such as alkali metals and diatomic molecules such as the halogens to larger organic molecules such as benzene. The simpler binary and ternary compounds are usually prepared by direct synthesis, and the more complicated materials by a variety of stepwise intercalation procedures. A special set of growth procedures has been developed for each of the common intercalants [2.10].

It is convenient to use highly oriented pyrolytic graphite (HOPG) as the host material for graphite intercalation compounds [2.14] because of the ease with which large-sized samples can be fabricated. Because of the high reactivity of most graphite intercalation compounds, they are commonly prepared in

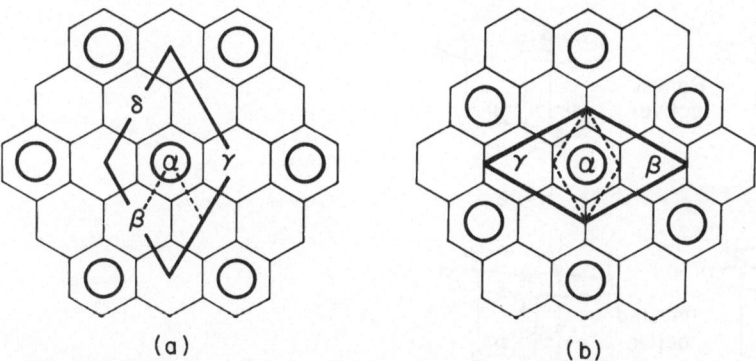

Fig. 2.2a, b. Schematic diagram of an in-plane unit cell for the (a) $p(2 \times 2)R0°$ and (b) $p(\sqrt{3} \times \sqrt{3})R30°$ structures showing the carbon atoms at the corners of the hexagons and the intercalants as the open circles. In these diagrams, the intercalate and graphite layers are projected onto a single plane. For structure (a) there are four equivalent intercalate sites labeled α, β, γ, δ and on a given intercalate layer, only one of these sites is occupied, while for structure (b) there are three equivalent intercalate sites labeled α, β, γ. The dashed lines represent the in-plane unit cell for pristine graphite. For C_8K and C_8Rb, the intercalant is arranged in sequential layers on α, β, γ, δ sites while for C_8Cs, the intercalant in (a) is arranged on alternating α, β, γ sites. For C_6Li, the intercalate ordering is given by (b)

glass ampoules for convenience in handling. Since the measured properties are stage dependent, the principal technique for the characterization of graphite intercalation compounds is x ray diffraction using $(00l)$ reflections, since such measurements provide a sensitive measure of stage fidelity and stage index.

Graphite intercalation compounds are of particular physical interest because of their high degree of ordering. The most striking type of ordering is the staging phenomenon which gives rise to the dominant c-axis superlattice symmetry. With careful sample preparation techniques, it is possible to prepare single-staged material (admixtures of secondary-staged regions in the range $\lesssim 1\%$). The staging phenomenon is long range, and it is possible with present techniques to prepare and characterize single-staged samples up to stage ≈ 10. The intercalation process has a negligible effect on the graphite layers with respect to the in-plane ordering, the lattice constant a_0 of the graphite layers, the c-axis graphite interplanar spacing c_0 and the $ABAB$ stacking of uninterrupted graphite layers [2.6, 10].

In addition to the staging periodicity, in-plane ordering of the intercalant commonly occurs. An especially high degree of intercalate ordering is found in the first stage alkali metal compounds which exhibit a commensurate (2×2) superlattice for the K, Rb, and Cs intercalants and a $(\sqrt{3} \times \sqrt{3})$ superlattice for Li (Fig. 2.2). The lattice constants in the intercalate layer are typically large compared with a_0 because the intercalate ionic radii and intramolecular bond lengths tend to be large compared with the in-plane carbon-carbon separation. The ordering on the intercalate layer tends to be similar to that of the parent material. Therefore, in most cases, the intercalate and graphite layers are

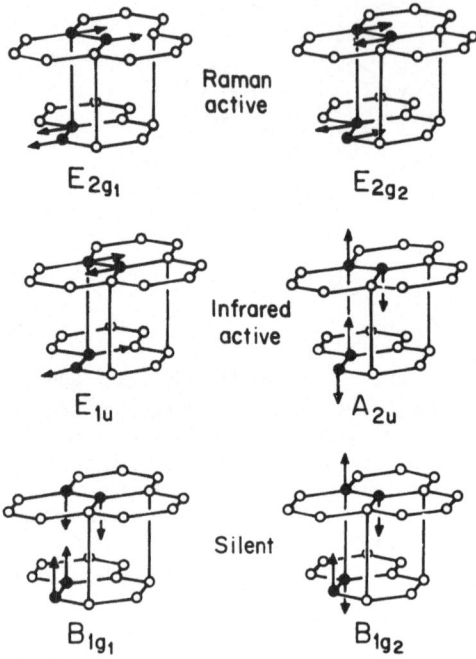

Fig. 2.3. Zone-center optical lattice modes in pristine graphite. For the in-plane modes (E_{1u}, E_{2g_1}, E_{2g_2}), only one of the degenerate pair of modes is shown. The c-axis modes (A_{2u}, B_{1g_1}, B_{1g_2}) are nondegenerate. Infrared and Raman activity are indicated. The zero frequency acoustic modes (E_{1u}, A_{2u}) corresponding to pure translations are not shown

incommensurate or not in registry, though local correlation between atoms on the intercalate and graphite bounding layers commonly occur. An arrangement of the intercalant in a distribution of ordered island structures surrounded by disordered regions commonly occurs [2.15].

2.1.3 Raman Scattering and Lattice Modes in Pristine Graphite

Because of the strong connection between the lattice mode structure in the intercalation compounds and the graphite host material, the lattice mode structure of pristine graphite is summarized. Pristine graphite crystallizes according to the D_{6h}^4 space group and has twelve vibrational modes at $q = 0$. These modes, shown in Fig. 2.3, are classified as three acoustic modes ($A_{2u} + E_{1u}$), three infrared-active modes ($A_{2u} + E_{1u}$), four Raman-active modes ($2E_{2g}$), and two silent modes ($2B_{1g}$).

The frequencies of the two in-plane Raman-active E_{2g} modes have been measured directly by Raman spectroscopy [2.16–18] yielding $\omega(E_{2g_2}) = 1582 \pm 1$ cm^{-1} for the high frequency mode and $\omega(E_{2g_1}) = 42$ cm^{-1} for the low frequency mode. In addition, infrared and inelastic neutron scattering experiments have been performed [2.17, 19, 20] yielding values for the E_{1u}, A_{2u}, B_{1g_1}, and E_{2g_1} modes, as summarized in Table 2.1: Because of the strong intralayer force constants relative to the interlayer force constants, the vibrational frequencies of the E_{2g_2} and E_{1u} modes are nearly degenerate. The small frequency difference of ~ 6 cm^{-1} between $\omega(E_{2g_2})$ and $\omega(E_{1u})$ is thus associated

Table 2.1. Experimental lattice mode frequencies [cm^{-1}] and elastic constants [10^{11} dyn/cm^2] in pristine graphite

Parameter	Value	Measurement technique
$\omega_{TO}(E_{2g_2})$	1582.0 ± 1.0	Raman spectroscopy[a, b]
$\omega_{TO}(E_{1u})$	1588.0 ± 2.0	Infrared spectroscopy[b, c]
$\omega_{LO}(A_{2u})$	868.0 ± 1.0	Infrared spectroscopy[c]
$\omega_{TA}(E_{2g_1})$	42.0	Raman spectroscopy[d]
$\omega_{TA}(E_{2g_1})$	48.0	
$\omega_{LA}(B_{1g_1})$	128.0	
$\omega_{TA}(M\ pt)$	~ 466	Inelastic neutron spectroscopy[e]
$\omega_{TA}(A\ pt)$	~ 30	
$\omega_{LA}(A\ pt)$	~ 90	
$C_{11} = C_{22}$	106.0	
$C_{66} = (C_{11} - C_{12})/2$	44.0	
C_{33}	3.65	Velocity of sound[f, g]
C_{13}	1.50	
$C_{44} = C_{55}$	0.40	

[a] From *Tuinstra* and *Koenig* [2.16].
[b] From *Brillson* et al. [2.17].
[c] From *Nemanich* et al. [2.19].
[d] From *Nemanich* et al. [2.18].
[e] From *Nicklow* et al. [2.20].
[f] From *Blakslee* et al. [2.28].
[g] From *Seldin* and *Nezbeda* [2.29].

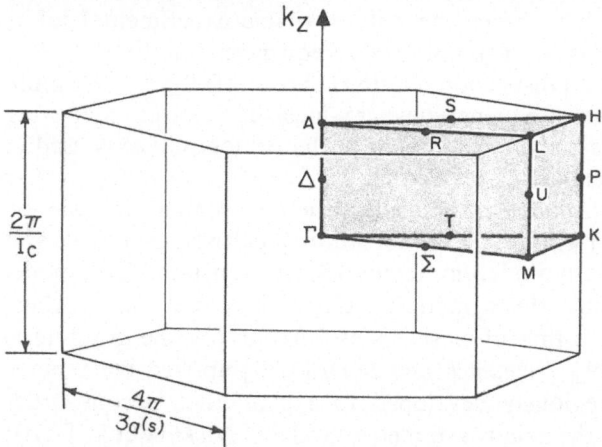

Fig. 2.4. Hexagonal Brillouin zone showing several high symmetry points and axes. For the intercalation compounds, I_c is the c-axis repeat distance and $a^{(s)}$ is the in-plane lattice constant for the superlattice. For pristine graphite $I_c = 2c_0 = 6.70\ \text{Å}$ and $a^{(s)} = a_0 = 2.46\ \text{Å}$

with interlayer force constants arising from differences with respect to interplanar displacements (see Fig. 2.3). *Nemanich* et al. [2.18, 19] have discussed the physical mechanisms needed to explain the ordering of these modes and the magnitude of their splitting. Analysis of the inelastic neutron scattering data for graphite by *Nicklow* et al. [2.20] has provided phonon dispersion curves for the two lowest frequency phonon branches along the high symmetry direction ΓM (see the Brillouin zone in Fig. 2.4). Near the M point, the two lowest frequency modes are nearly degenerate at $\sim 466\,\mathrm{cm}^{-1}$ and correspond to z-axis displacements.

On the basis of these measurements, detailed force constant models for the lattice dynamics of graphite have been developed by a number of workers [2.21]. Early calculations of the phonon dispersion relations for pristine graphite were carried out by *Yoshimori* and *Kitano* [2.22] using a valence force model in the Born-von Kármán formalism. These authors used 4 force constants with both bond-bending and bond-stretching forces, estimating their in-plane vibrational force constants from those of the benzene molecule and the out-of-plane force constants from compressibility data [2.23] and low temperature specific heat measurements [2.24]. *Young* and *Koppel* [2.25] used the basic lattice dynamic model of *Yoshimori* and *Kitano* [2.22] but evaluated their force constants more accurately using specific heat data over a wide temperature range.

The availability of inelastic-neutron scattering measurements by *Nicklow* et al. [2.20] of the low frequency phonon dispersion relations and by *Brillson* et al. [2.17] of the high frequency Raman and infrared-active zone-center modes made a more detailed calculation of the phonon dispersion relations possible. In this work, *Nicklow* et al. [2.20], using an axially symmetric force constant model involving 8 force constants, were able to fit the experimental information available at that time on the phonon modes. Later measurements by *Nemanich* et al. [2.19, 26] of the infrared-active out-of-plane A_{2u} mode frequency indicated a major discrepancy (by $\sim 500\,\mathrm{cm}^{-1}$) with the Nicklow model and stimulated further calculations. Since the only available experimental information above $\sim 500\,\mathrm{cm}^{-1}$ is the Γ-point mode frequencies for the E_{2g_2}, E_{1u}, and A_{2u} modes, the calculated dispersion relations above $\sim 500\,\mathrm{cm}^{-1}$ are more tentative than at the lower frequencies where inelastic neutron scattering information along several symmetry directions in the Brillouin zone is used in obtaining the dispersion relations.

Another force constant model introduced in the intervening time was by *Nicholson* and *Bacon* [2.27] who used a combination of pairwise potentials and a many-body valence constant model employing 13 force constants. This model allowed for a nonvanishing shear modulus C_{44} which previous workers assumed had vanished. A listing of elastic constants for pristine graphite is given in Table 2.1 [2.28, 29]. *Nemanich* and *Solin* [2.30] applied the various models that had been previously developed to explain their second-order Raman spectra, and found the axially symmetric model of *Nicklow* et al. [2.20] to give the best fit to the second-order spectra, but the discrepancy with the measured A_{2u} mode frequency remained.

A fit to the A_{2u} mode frequency was, however, provided by the model of *Maeda* et al. [2.31] who took explicit account of the anisotropic coupling in graphite and employed 8 force constants. A good fit was obtained for most of the currently available experimental data, except that the calculated low frequency phonon branches out to the M point were significantly below the inelastic neutron scattering measurements of *Nicklow* et al. [2.20].

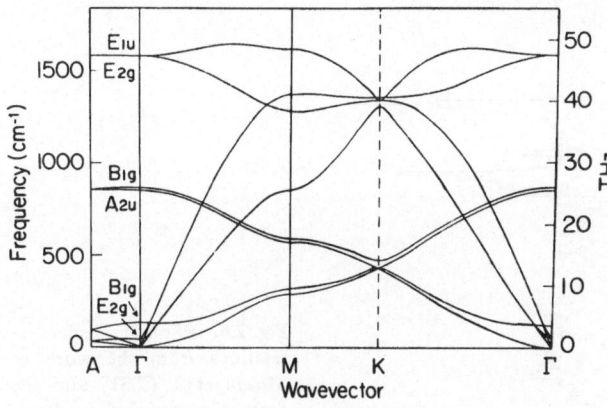

Fig. 2.5. Graphite phonon dispersion curves along several high symmetry axes from *Maeda* et al. [2.31]. The Γ-point symmetries for the graphite structure are indicated

Table 2.2. Force constants of hexagonal graphite[a]

Notation[b]	Significance	Value [10^5 dyn/cm]
$\phi_r^{(1)}$	Radial force constant	3.066
$\phi_r^{(2)}$	Radial force constant	1.363
$\phi_{ti}^{(1)}$	Tangential force constant responsible for change in bond angle	2.810
$\phi_{ti}^{(2)}$	Tangential force constant responsible for change in bond angle	−0.5269
$\phi_{to}^{(1)}$	Tangential force constant responsible for c-axis displacement	0.8561
$\phi_{to}^{(2)}$	Tangential force constant responsible for c-axis displacement	−0.1421
ϕ_r^{int}	Radial interplanar force constant	0.0579
ϕ_t^{int}	Tangential interplanar force constant	0.00718

[a] From *Maeda* et al. [2.31].
[b] For the intraplanar force constants $\phi_r^{(n)}$, $\phi_{ti}^{(n)}$, $\phi_{to}^{(n)}$, $n=1$ denotes nearest-neighbor and $n=2$ next-nearest-neighbor interactions. For the interplanar force constants ϕ_r^{int}, ϕ_t^{int}, only nearest-neighbor interactions are considered.

Furthermore, the Maeda calculation did not attempt to fit the measured elastic constants. The phonon dispersion relations obtained by *Maeda* et al. [2.31] are given in Fig. 2.5 along the $A\Gamma$, ΓM, MK, and $K\Gamma$ directions and are applied in Sect. 2.2.2 to the intercalation compounds. In the Maeda model, interactions are treated up to second-nearest-neighbor in-plane and first-neighbor out-of-plane force constants. The anisotropic properties are taken into account by two tangential force constants: one responsible for changing the bond angle within a layer and the other responsible for displacements perpendicular to the layer. The force constants used in this model are highly anisotropic and their values are listed in Table 2.2.

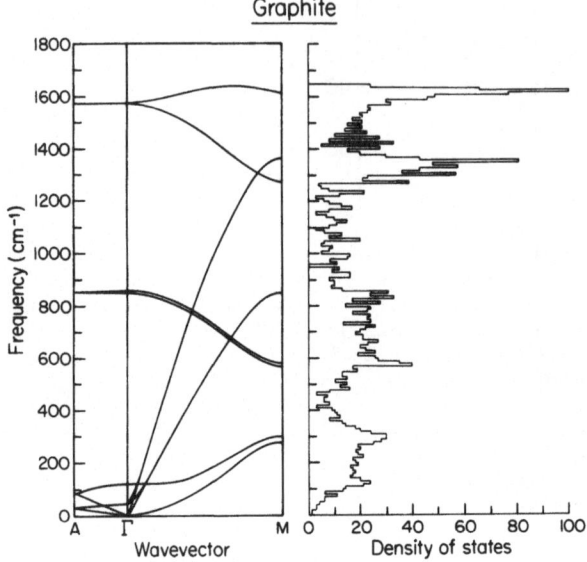

Fig. 2.6. Phonon dispersion relations from the work of *Maeda* et al. [2.31] and the corresponding phonon density of states from the work of *Leung* et al. [2.32] for pristine graphite. The highest peak in the density of states is arbitrarily scaled to 100

Using the phonon dispersion relations of Fig. 2.5, the phonon density of states has been calculated by *Leung* et al. [2.32] and the results are given in Fig. 2.6. A high density of states is associated with the highest lying phonon branch with a large contribution arising from modes near the Brillouin zone boundary. For this highest lying branch, we note that the maximum phonon frequency does not occur at the zone center but at a lower symmetry point Σ in the zone.

The other dominant peak in the phonon density of states occurs near $1355 \, \text{cm}^{-1}$, consistent with the occurrence of the dominant peak in the second-order Raman spectra near $2730 \, \text{cm}^{-1}$. The major contribution to the $1355 \, \text{cm}^{-1}$ peak in the density of states comes from the phonon modes near the Brillouin zone boundary, especially near the M point and in the vicinity of the MK axis. Since there are no zone-center modes near this peak in the density of states, the observation of Raman intensity near $1355 \, \text{cm}^{-1}$ has been identified with various types of disorder in the graphite structure [2.16].

In this regard, highly oriented pyrolytic graphite (HOPG), the synthetic graphite material commonly used as a host for the intercalation compounds, shows no Raman intensity near $1355 \, \text{cm}^{-1}$, consistent with the long range in-plane ordering in this material. The synthetic HOPG material is formed by the cracking of hydrocarbons at high temperature and subsequent heat treatment, often combined with the application of pressure. The resulting material, which has been characterized extensively by *Moore* [2.14], is highly oriented along the c-axis (orientational deviations less than $1°$), but in the layer planes it consists of a randomly ordered collection of crystallites of $\sim 1 \, \mu\text{m}$ average diameter. More poorly ordered graphite materials such as activated charcoal [2.16] or glassy

Table 2.3. Characteristics of Raman spectra for different forms of carbon

Graphitic form	L_a	First order Raman shift	Linewidth (FWHM)
		$[cm^{-1}]$	
Natural crystals[a]	3 mm	1582	7
Pyrolytic (HOPG)[a]	1–2 μm	1582	11
Fisher powder[a]	130 Å	1583	–
Carbon rod[a]	25 Å	1585	–
Vitreous carbon[b]	30 Å	1590	–
GY70[c]	200 Å	1585	28
UC4104B[c]	400 Å	1583	26
B–D Fiber[d]	500 Å	1582	16

[a] From *Nemanich* et al. [2.26].
[b] From *Nathan* et al. [2.33].
[c] From *Kwizera* et al. [2.35].
[d] Benzene-derived fiber heat treated to 2900 °C. From *Chien* et al. [2.110].

carbon [2.33] exhibit two high frequency Raman lines, a broadened line peaking near 1580 cm^{-1} and an additional line near ~1355 cm^{-1} where the phonon density of states has a strong maximum. Using L_a as a measure of the in-plane crystallite size, the intensity and linewidth of the 1355 cm^{-1} line increases with decreasing L_a [2.16, 26, 34]. In addition, the linewidth of the 1580 cm^{-1} line increases as L_a decreases, and the peak frequency of the 1580 cm^{-1} line upshifts since the breakdown in the wave vector selection rule brings in a range of modes away from the zone center which (Fig. 2.5) are upshifted relative to the zone-center mode. A list of the characteristics of the Raman spectra for various disordered graphites is given in Table 2.3. One commercially significant category of disordered graphite is the carbon fiber, and characteristics of the Raman spectra for some carbon fibers [2.35] are also included in Table 2.3.

Other features in the phonon density of states are found at lower frequencies, such as the two-peaked structure near 850 cm^{-1} and 830 cm^{-1} with contributions from both zone-center and zone-edge modes, and at ~580 cm^{-1} and ~300 cm^{-1} with contributions primarily from modes near the Brillouin zone boundary (M point). No features in the Raman spectra of disordered graphites have, however, been attributed to these lower frequency peaks in the phonon density of states.

2.2 Lattice Dynamics of Intercalated Graphite

2.2.1 Introductory Comments

Models for the phonon dispersion relations provide an important basis for the interpretation of Raman scattering experiments. Thus, the recent availability of models of the phonon dispersion relations for intercalated graphite has made a significant impact on our understanding of the previously observed Raman spectra and has predicted new and interesting areas for future study.

The lattice dynamical models for ordered and disordered solids have been traditionally formulated in terms of Born-von Kármán models for ordered structures and localized bonding models for disordered solids. Intercalation compounds represent a blend of these two types of systems, on the one hand exhibiting highly ordered structures in the graphite layers and the long range staging superlattice periodicity of the intercalant, and on the other hand, exhibiting less-ordered intercalate layers. Because of these dual characteristics, the calculation of the lattice dynamics for these systems presents a challenging theoretical problem.

2.2.2 Model Calculations for Phonon Dispersion Relations

The first calculation of the phonon dispersion relations for an intercalation compound was carried out by *Horie* et al. [2.36] who applied a Born-von Kármán model to a stage 1 alkali metal compound, where the intercalant was considered to occupy the commensurate (2×2) in-plane superlattice structure with the $\alpha\beta\gamma\delta$ intercalant interlayer stacking. This structure is observed in the C_8K and the C_8Rb compounds. The set of 8 force constants previously applied by *Maeda* et al. [2.31] to describe the lattice dynamics of pristine graphite was used by *Horie* et al. to model the force constants in and between graphite layers in the intercalation compounds. In addition, they introduced nearest-neighbor radial and tangential intercalant-graphite interlayer force constants ϕ_r^{XG} and ϕ_t^{XG}, and estimated their values by fitting them to the inelastic neutron scattering results of *Ellenson* et al. [2.37] on C_8Rb. Nearest-neighbor radial and tangential intercalate intralayer force constants ϕ_r^{XX} and ϕ_t^{XX} were also introduced to take into account vibrations on the intercalate layer. Because of the high degree of symmetry of this lattice, a large unit cell in real space is obtained, thereby predicting a large number of zone-center lattice modes as shown in Fig. 2.7. This approach has not been more widely pursued because of (a) the difficulty of extending this type of calculation to higher stage compounds for which the unit cells are larger, (b) the ambiguity of the ordering in the intercalate layer and in the interlayer intercalant stacking for higher stage compounds, and (c) the lack of experimental observation of the large number of zone-center modes predicted by the calculation. The Raman spectra observed for the stage 1 alkali metal compounds are anomalous, and the Horie model has in fact been applied to the interpretation of the observed spectra, as discussed in Sect. 2.3.3.

Another approach to the lattice dynamics of intercalated graphite based on symmetry considerations has been presented recently by *Leung* et al. [2.38]. Some attractive features of the Leung model are its applicability to any intercalant and any stage and to various orderings (or total disorder) within the intercalate layer. The model is based on a zone folding of the basic Born-von Kármán model for pristine graphite which takes into account explicitly the

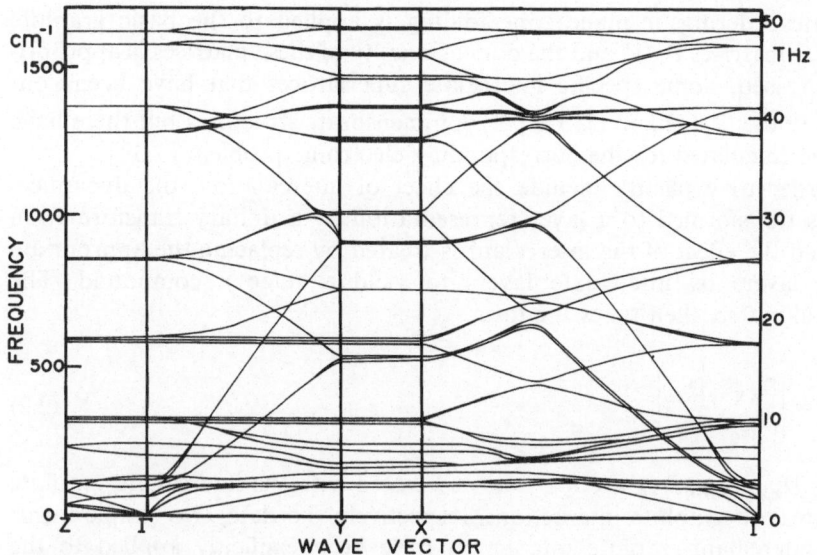

Fig. 2.7. Phonon dispersion relations for C_8K from the work of *Horie* et al. [2.36]. The symmetry points are for the Brillouin zone appropriate to $\alpha\beta\gamma\delta$ stacking of the (2×2) intercalate layers

superlattice periodicity due to staging, as well as any additional in-plane periodicity that may occur in the intercalate layer. One limiting form of the model corresponding to the jellium approximation allows the intercalate layer to be completely structureless. Because of the wide applicability of this model to graphite intercalation compounds, a summary of the major features of the model is given below.

The model is based on the dynamical matrix for pristine graphite on to which the dominant staging periodicity is introduced by k_z axis zone-folding. For the dynamical matrix of graphite, the best currently available model by *Maeda* et al. [2.31] is used which includes force constants up to second-neighbor in-plane and first-neighbor out-of-plane interactions. The k_z-axis zone folding is performed on the graphite dynamical matrix. The resulting dynamical matrix for the intercalation compounds takes the form

$$D_{ZF} = \begin{vmatrix} D_0(k) & 0 & \cdots & \\ 0 & D_0\left(k+\dfrac{\pi}{lc_0}\hat{z}\right) & & \\ \vdots & & \ddots & \\ 0 & & & D_0\left(k+\dfrac{l-1}{l}\dfrac{\pi}{c_0}\hat{z}\right) \end{vmatrix}, \tag{2.1}$$

where $D_0(k)$ is the 12×12 graphite dynamical matrix, $l=n+1$ for even stages, $l=(n+1)/2$ for odd stages and $c_0 = 3.35\,\text{Å}$. To consider explicitly any in-plane intercalate ordering, in-plane zone folding is applied to the basic graphite dynamical matrices $D_0(\boldsymbol{k})$ and the dimensionality of these matrices is appropriately increased. Some specific intercalate superlattices that have been considered are the (2×2) and $(\sqrt{3} \times \sqrt{3})$ commensurate structures but these have only been calculated for the corresponding electronic problem [2.39].

In order to explicitly include the effect of intercalation, the dynamical matrix is transformed to a layer representation by a unitary transformation [2.40] and the effect of the intercalant is treated by replacing the appropriate graphite layers by intercalate layers to yield a stage n compound. The dynamical matrix then takes the form

$$
D_{GIC} = \begin{vmatrix} D_{XX} & D_{XG} \\ D_{XG}^{\dagger} & D_{GG} \end{vmatrix} \tag{2.2}
$$

in which D_{XX}, D_{GG}, D_{XG} are matrices associated with the intercalant, graphite and intercalant-graphite interactions, respectively. To date, two simple forms for the intercalant-graphite interaction have been explicitly applied to the calculation of the phonon dispersion relations: (a) an elastic continuum model which introduced a single force constant coupling of the intercalant to the graphite bounding layer [2.41], and (b) an ordered commensurate (1×1) superlattice [2.32] in which the intercalant is placed over the centers of the graphite hexagons (the $C_{2n}X$ structure). In the site representation of (2.2) the masses and force constants of graphite are modified to model the intercalate layer and the intercalant-graphite bounding layer interaction. The placement of the intercalant over the centers of the graphite hexagons requires the introduction into the basic Maeda model of second-neighbor out-of-plane radial and tangential force constants. These force constants are also present in graphite because of the AB planar stacking but are of secondary importance, while for the $C_{2n}X$ structure, these second-neighbor out-of-plane force constants are dominant in the intercalant-bounding layer interaction. In addition, for the intercalate layer the masses are appropriately modified and the nearest-neighbor intraplanar radial and tangential force constants are suitably modified to conform to the occupation of only half of the graphite sites.

The force constants for this model were evaluated using the following experimental observations. The experimentally observed splitting of the Raman-active E_{2g_2} mode for the graphite interior and graphite bounding layers (Sect. 2.3.2) was used to evaluate the intercalant-graphite bounding layer interaction. As discussed in Sect. 2.3.2, the bounding layer mode in the dilute intercalate limit is *upshifted* relative to the interior layer mode by 19 ± 2 cm^{-1} for all acceptors that have been studied and by 31 ± 3 cm^{-1} for donors [2.41]. In addition, a stage dependence of the 8 graphite force constants of *Maeda* et al. [2.31] of the form

$$
\phi = \phi_0(1 + \alpha/n) \tag{2.3}
$$

Table 2.4. Additional parameters for a generalized donor and acceptor compound used in the lattice dynamical calculations[a]

	Donor	Acceptor
ϕ_r^{XX} [10^3 dyn/cm]	1.0	1.0
ϕ_t^{XX} [10^2 dyn/cm]	1.0	1.0
ϕ_r^{XG} [10^4 dyn/cm]	2.66	1.70
ϕ_t^{XG} [10^4 dyn/cm]	2.66	1.70
ϱ	1.0	1.0
α	0.0388	0.0378

[a] From the work of *Leung* et al. [2.38]. In this calculation, the graphitic force constants in Table 2.2 are used, α is defined by (2.3) and ϱ is the square root of the in-plane intercalate mass density relative to that on the graphite layers.

is introduced, based on the observation of an upshift of these mode frequencies with reciprocal stage for acceptor compounds and a downshift in frequency for donors. These stage-dependent shifts in mode frequencies are associated with a contraction of the in-plane lattice constant for acceptors [2.42, 43] and an expansion for donors [2.44, 45]. A model calculation by *Pietronero* and *Strässler* [2.46] for the magnitude of the electron-phonon coupling has yielded a quantitative relationship between the change of the in-plane lattice constant δa and the magnitude and sign of the charge transfer f_c per carbon atom:

$$\delta a = 0.157 f_c + 0.146 |f_c|^{3/2} + 0.236 f_c^2, \tag{2.4}$$

with δa being positive for donor intercalants and negative for acceptors. In their model calculation for the phonon dispersion relation, *Leung* et al. [2.38] used the magnitude of α in (2.3) as given by the stage dependence of the Raman mode frequencies, and consistent with that reported by *Nixon* and *Parry* [2.44] for the graphite-potassium compounds. The sign of α was taken to be positive for acceptors and negative for donors. A summary of the force constants used in the Leung model is given in Table 2.4.

The dispersion relations thus obtained by *Leung* et al. [2.38] are shown in Fig. 2.8 for a model stage 3 acceptor compound along various high symmetry axes. For the high-lying levels, the phonon branches are arranged in pairs with the lower component associated with in-plane vibrations in the graphite interior layers and the upper component with the graphite bounding layers, adjacent to the intercalant where the intercalant-graphite interaction is strong. With increasing stage index, the number of graphite interior layers per unit cell increases, thereby increasing the number of lattice mode branches. However, model calculations for higher stage compounds [2.38] show the additional modes to be essentially degenerate (on the scale of Fig. 2.8) with each other,

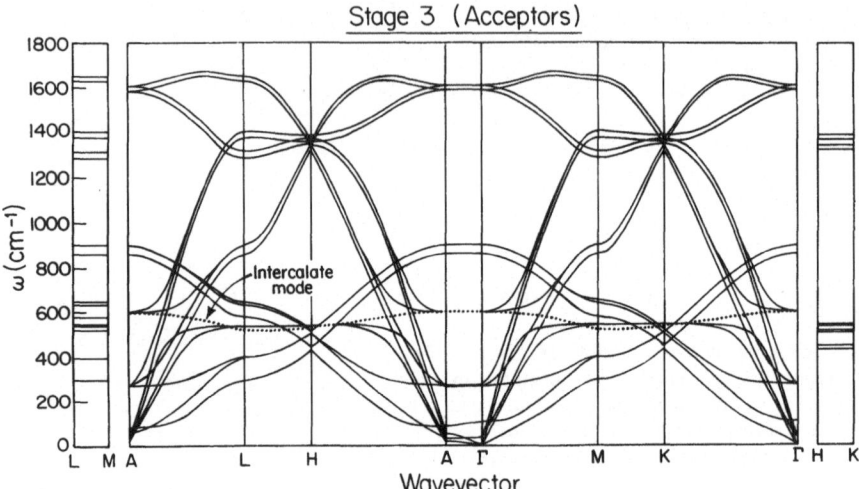

Fig. 2.8. Phonon dispersion relations calculated for a third stage $C_{2n}X$ acceptor compound for $n=3$ along several high symmetry axes. The force constants used in this calculation are given in Table 2.4. (From *Leung* et al. [2.38])

thereby effectively giving rise to a single slightly broadened Raman line for the graphite interior layers. Except for the lowest-lying modes along k_z (i.e., in the ΓA direction), the modes propagating along k_z show very little dispersion. Recent experimental evidence for k_z dispersion of low frequency modes provides valuable information for the evaluation of several force constants, as discussed in Sect. 2.3.4. Also shown in Fig. 2.8 is the dispersion for modes associated with a model intercalate layer (dotted curve). Application of the Leung model to a specific intercalant would involve the evaluation of force constants ϕ_r^{XX} and ϕ_t^{XX} to fit some aspects of the observed intercalate mode, as for example, its behavior at and near the zone center.

This model gives a reasonably consistent picture for the graphite-derived modes, as discussed in Sects. 2.3.2 and 2.3.3. However, the modes associated with the intercalate layer (especially for large molecular intercalants) have not been quantitatively modeled, nor have they been extensively studied experimentally. This is an area in which light scattering experiments can be expected to make important advances during the next few years. Preliminary results on the low frequency modes have recently been reported (see Sect. 2.3.4).

The model of *Leung* et al. [2.38] suggests that Brillouin scattering measurements could be carried out on the acoustic mode of intercalated graphite. From the model, the longitudinal sound velocity is inferred to increase upon intercalation for the acceptors. However, the intercalate modes (whether molecular or atomic) are generally in the low frequency range, and a strong coupling between the longitudinal graphite and intercalate modes is expected to strongly perturb the low frequency modes and the longitudinal sound velocity in intercalated graphite. For the case of the alkali metal compounds, the model of

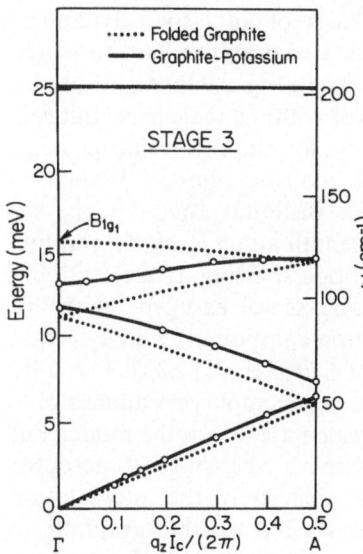

Fig. 2.9. Phonon dispersion relations measured by *Magerl* and *Zabel* [2.50] for third stage graphite-potassium using inelastic neutron scattering (points). The solid curves are a calculated fit to their data using a simple linear chain model and the dashed curves are for k_z axis zone-folded phonon modes for pristine graphite [2.20, 52]

Horie et al. [2.36] gives in-plane *intercalate* modes with ω in the range $76 < \omega < 106$ cm^{-1} for C_8Rb and $101 < \omega < 120$ cm^{-1} for C_8K. However, Raman-active alkali metal modes have only recently been reported for C_8K and C_8Rb, and are not incorporated in the model, which was developed before the measurements were reported (see Sect. 2.3.3). The frequencies for these intercalate modes are comparable to $k\theta_D$ for the parent metals, where the Debye energies for the alkali metals are ~ 63 cm^{-1}, 39 cm^{-1}, and 26 cm^{-1} for metallic K, Rb, and Cs, respectively. Some evidence for a soft intercalate mode is also provided by the observation of anomalies in the low temperature heat capacity in the low stage compounds with Cs [2.47–49] and Rb [2.49]. *Alexander* et al. [2.49] carried out their measurements on HOPG-based samples in the temperature range $0.48 < T < 90$ K while *Mizutani* et al. [2.47, 48] made measurements on intercalated spectroscopic graphite powders in the temperature range $1.5 < T < 5$ K. By using a wide temperature range, *Alexander* et al. [2.49] were able to show that the low temperature anomaly was of the Einstein form, and they obtained values for the Einstein temperature of 48 K (36 cm^{-1}) for first stage C_8Cs and 65 K (49 cm^{-1}) for C_8Rb. These values for the Einstein temperature are approximately proportional to the Debye temperatures for the corresponding pristine alkali metals.

Inelastic neutron scattering experiments measure directly the longitudinal acoustic branch in the k_z direction, and such experiments have been carried out on a limited number of graphite intercalation compounds: by *Ellenson* et al. [2.37] on first stage C_8K and C_8Rb, by *Magerl* and *Zabel* [2.50] on the stage 3 compound $C_{36}K$ and by *Axe* et al. [2.51] on stage 2 graphite-FeCl$_3$. These experiments all give strong support for k_z-axis zone folding associated with the staging phenomenon, which is the basis of all lattice dynamical calculations that have been carried out for intercalated graphite. The measured low frequency longitudinal phonon branches shown in Fig. 2.9 for $C_{36}K$ are for a compound with 3 graphite and 1 potassium layer within a real space unit cell. The 3 lowest observed phonon branches (points) for the intercalation compound are closely related to the longitudinal acoustic phonon branch of graphite (dashed curve), folded 4 times into a Brillouin zone for the intercalation compound of approximately 1/4 the length along k_z of the graphite zone [2.20, 52]. The results of Fig. 2.9 show a mode splitting at the Brillouin zone boundaries and zone center of the smaller zone, as well as a general upshift of the phonon mode frequencies in the intercalation compound relative to that of graphite, in agreement with the calculations of *Leung* et al. [2.38]. The solid curves represent a fit to the experimental data using a simple one-dimensional chain model [2.50]. The highest-lying branch is calculated from the model, but was not observed experimentally. For the case of the stage 2 acceptor compounds with $FeCl_3$, the observed mode frequencies in the intercalation compound are *downshifted* with respect to zone-folded pristine graphite, in contrast to the observations in the stage 3 donor compound with K where the modes in the intercalation compound are *upshifted* (Fig. 2.9). Mode splittings at the zone boundary and zone center are also observed for the stage 2 acceptor compound with $FeCl_3$, similar to the behavior in Fig. 2.9. Zone folding along k_z could lead to Raman-active modes, and it remains an experimental challenge to observe such modes by light scattering techniques; such experiments would require use of *a*-face samples.

2.2.3 Phonon Density of States

The availability of a model for the phonon dispersion relations in the intercalation compounds allows calculation of the phonon density of states. In addition to the use of the density of states in model calculations involving lattice modes such as the electron-phonon interaction and the lattice specific heat, there are also important ramifications of the phonon density of states on such aspects of light scattering as disorder-induced Raman scattering (Sects. 2.1.3, 2.3.8) and second-order Raman scattering (Sects. 2.1.3, 2.3.5).

In calculating the electronic dispersion relations for intercalated graphite, many workers have used a single layer (two-dimensional) graphite model [2.53, 54]. A calculation of the phonon dispersion relations for single layer graphite based on the Maeda model has also been carried out [2.40] and the results are shown in Fig. 2.10 together with the associated density of states. The main difference between a single layer model and that for three-dimensional graphite is the elimination of certain phonon branches and the elimination of k_z dispersion in the remaining branches (compare Figs. 2.6 and 2.10). Except for the absence of a well-defined structure in the single layer graphite phonon density of states in the low frequency region, the density of states for single layer graphite is very similar to that for three-dimensional graphite. The behavior of single layer

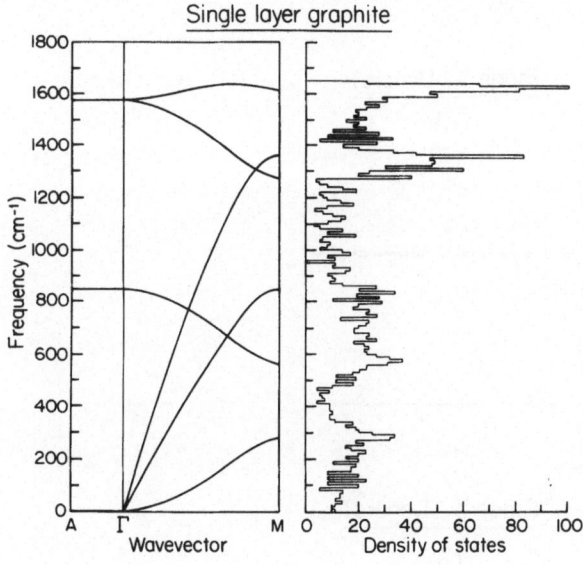

Fig. 2.10. Phonon dispersion relations and phonon density of states for single layer graphite from the work by *Leung* [2.40]. The highest peak in the density of states is arbitrarily scaled to 100

graphite is also pertinent to the results for first stage intercalated graphite where the graphite layers are isolated from each other by an intercalate layer.

For the case of intercalated graphite, model calculations of the phonon density of states based on the phonon dispersion relations obtained by *Leung* et al. [2.32, 38, 40] have been made for stage 1, 3, 5, 7, 9, and 11 donor and acceptor compounds and results for the dispersion relations are shown in Fig. 2.11 for model stage 1 and 3 donor compounds together with their corresponding density of states. Calculations of the density of states for the higher stage donor compounds yield results similar to that for stage 3, consistent with the behavior of the phonon dispersion relations for the higher stage compounds discussed above. For both the stage 1 and stage 3 compounds, peaks in the density of states are found in the vicinity of 1650, 1410, and 1360 cm^{-1}, similar to what is found in pristine and single layer graphite. A very strong and sharp peak in the density of states is found at ~ 650 cm^{-1} for the stage 1 compound. This peak is also found in the stage 3 compound but with a lower intensity and a greater linewidth, as well as in pristine and single layer graphite with still lower intensities and greater linewidths. For the stage 1 compound, the density of states of the graphitic modes at very low frequencies is relatively small compared with the stage 3 compound and with three-dimensional graphite, but is more similar to that for the single layer graphite. The density of phonon states calculated for the acceptor compounds is qualitatively similar to that for donor compounds of a comparable stage [2.40], and is not shown here explicitly.

As discussed in Sect. 2.1.3, Raman studies of disordered graphite show that disorder-induced Raman scattering gives rise to structure at ~ 1355 cm^{-1}. The general absence of Raman intensity in the region of 1355 cm^{-1} for single crystal

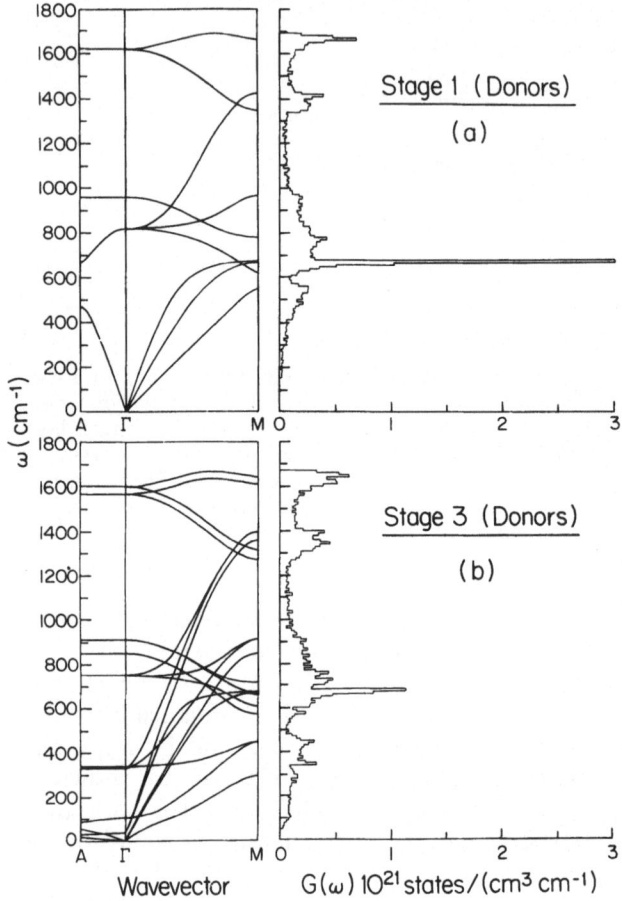

Stage 1 (Donors)

(a)

Stage 3 (Donors)

(b)

Wavevector $G(\omega)\,10^{21}$ states/(cm^3 cm^{-1})

Fig. 2.11a, b. Phonon dispersion curves and density of states $G(\omega)$ calculated by *Leung* et al. [2.38] for a first and third stage $C_{2n}X$ donor compound based on the parameters given in Tables 2.2, 2.4

or HOPG-based intercalation compounds indicates that the dominant disorder in graphite intercalation compounds is associated with the intercalate layers and not with the graphite layers. Disorder in the intercalate layer is expected to induce non q-conserving modes in the graphite bounding layers, though q-conservation should still apply to the graphite interior layers. This argument implies that the disorder-induced density of states scattering should be most pronounced for stage 1 compounds.

The calculated densities of states will be utilized in discussions of second-order and disorder-induced Raman scattering in intercalated graphite. The curves shown in Figs. 2.10, 11 do not distinguish between in-plane modes and out-of-plane modes. Since all modes above ~ 870 cm^{-1} are in-plane modes and it is these modes which correlate well with the second-order Raman spectra, it would seem appropriate to calculate an in-plane density of states for c-face Raman experiments and an out-of-plane density of states associated with a-face Raman experiments. Such calculations for the density of states have not yet been carried out.

The models for intercalation compounds discussed in this section are very general and must be specialized to each compound if quantitative results are to be obtained. These models do, however, prove to be extremely useful in obtaining a qualitative understanding of the experimental results presented in Sect. 2.3.

2.3 Raman Spectra of Graphite Intercalation Compounds

2.3.1 Introductory Comments

The Raman spectra in graphite intercalation compounds are dominated by features identified with the graphite layers. This experimental fact, consistent with other findings relevant to the electronic structure, provides strong evidence that the graphite layers are not strongly perturbed by the intercalant. The detailed behavior of the Raman spectra further indicates that the interaction between the intercalant and the graphite is largely confined to the graphite bounding layers adjacent to the intercalant. For conducting materials (e.g. metals and semimetals), light scattering is generally limited to a Brewster-angle backscattering geometry. Intercalation gives rise to a large increase in electrical conductivity resulting in a decrease in optical skin depth and therefore a decrease in scattering volume. Experimentally, it is, however, found that the full range of Raman studies that can be carried out in pristine graphite can also be done in intercalated graphite.

In particular, light scattering allows examination of the Raman-active modes in the graphite interior layers, the graphite bounding layers and the intercalate layers, and these modes can each be studied separately because their Raman peaks occur at well-separated frequencies. The behavior of these modes has been studied in detail for a large number of intercalants and stages, and the general features of the graphitic modes have now been well established (Sect. 2.3.2). A unique behavior is found for the graphitic modes in the stage 1 alkali metal compounds and this is described in Sect. 2.3.3. These sections are followed by a review of studies of the intercalate modes and resonant enhancement effects associated with these modes (Sect. 2.3.4). The relationship between the second-order Raman scattering in intercalated and pristine graphite is treated in Sect. 2.3.5, while Raman scattering from intercalated graphite fibers is described in Sect. 2.3.6. Scattering from surface molecules is briefly discussed in Sect. 2.3.7 and finally, the disorder-induced scattering associated with ion implantation is described in Sect. 2.3.8.

Though the number of Raman lines observed in graphite intercalation compounds is small, their observation has been extremely important in establishing the physical structure of the intercalation compounds and in determining the form and magnitude of the force constants used in the lattice dynamics model presented in Sect. 2.2.2. The resulting model in turn makes predictions about the effect of intercalation on graphite mode frequencies, thereby suggesting new light scattering experiments in these materials.

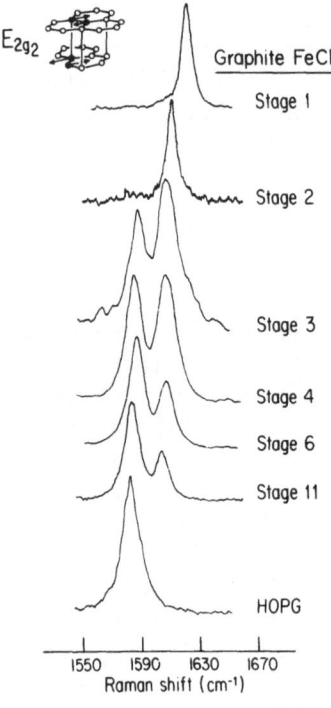

Fig. 2.12. Unpolarized room temperature Raman spectra taken by *Underhill* et al. [2.59] in the backscattering geometry ($E \perp c$) for stage $n = 1, 2, 3, 4, 6$, and 11 graphite-ferric chloride compounds and for pristine graphite (HOPG). Laser excitation at 4880 Å and a power level < 50 mW were used to excite in-plane Raman-active E_{2g_2} modes (see inset). The upper frequency component ($\hat{E}_{2g_2}$) is identified with the graphite bounding layer mode and the lower component ($E_{2g_2}^0$) with the graphite interior layer mode

One finding of special interest is the observation that the parity selection rule is preserved under intercalation, even though intercalation formally destroys the inversion symmetry of the compound. This finding also follows from the calculation of the phonon dispersion relations by *Leung* et al. [2.38]. Thus, even-parity modes of graphite remain Raman-active in the intercalation compounds. Furthermore, odd-parity graphitic modes seen in the infrared spectra have not been identified in the reported Raman spectra for the intercalation compounds.

2.3.2 Raman Scattering from Graphitic Modes in Donor and Acceptor Intercalation Compounds for Stages $n \geq 2$

In discussing the Raman spectra for graphite intercalation compounds, it is advantageous to distinguish between the behavior for stages $n \geq 2$ which is the characteristic behavior for these materials, and that for $n = 1$ which tends to be anomalous, especially for the donor compounds. For this reason, Raman scattering for stage 1 compounds is discussed separately (Sect. 2.3.3), while the behavior for $n \geq 2$ is presented in this section.

Many authors [2.55–58] have found that the Raman spectra for graphite intercalation compounds with stage $n > 2$ characteristically exhibit a two-peak structure at frequencies close to the single E_{2g_2} peak found in pristine graphite. This two-peak structure for the intercalation compounds is illustrated in

Fig. 2.12 for the graphite-FeCl$_3$ system, along with the spectrum for the pristine HOPG graphite host material, as reported by *Underhill* et al. [2.59]. A similar two-peak structure with a frequency separation of ~ 20 cm^{-1} is found for $n > 2$ for all intercalants studied, including the acceptors Br$_2$, IBr, and ICl [2.55], FeCl$_3$ [2.59], AlCl$_3$ [2.60], SbF$_5$ [2.57], SbCl$_5$ [2.61] and the donors K, Rb, and Cs [2.57, 58, 62, 63]. The calculated phonon dispersion relations for intercalated graphite (Fig. 2.8) are also consistent with such a two-peak or doublet structure.

The lower frequency component of the doublet is attributed to a superposition of the E_{2g_2} carbon atom vibrations in interior graphite layer planes and the unresolved lattice modes associated with these layers are denoted by $E_{2g_2}^0$. This identification is supported by the proximity of the $E_{2g_2}^0$ mode to the E_{2g_2} mode of pristine graphite, by the decrease in intensity of the $E_{2g_2}^0$ with decreasing stage (increasing intercalate concentration) and by the vanishing of the $E_{2g_2}^0$ line in stage 1 and stage 2 compounds where there are no graphite interior layers [2.56, 57]. It is also significant that for $n \geq 2$, the lineshapes are Lorentzian and the linewidth of the $E_{2g_2}^0$ component is only slightly sensitive to intercalate concentration. This result suggests that all interior graphite layers have approximately the same set of in-plane force constants and have nearly degenerate modes, also consistent with calculations of the phonon dispersion relations for the high stage compounds [2.38]. The small stage dependent shift of the $E_{2g_2}^0$ mode frequency suggests that the intercalant transfers a small charge density which is spread over the graphite interior layers. This stage dependence is consistent with a stage-dependent strain occurring in both graphite bounding and interior layers.

The upper frequency component of the doublet structure (Fig. 2.12) is identified with an E_{2g_2}-type graphitic mode occurring in a bounding graphite layer, and this mode is denoted by $\hat{E}_{2g_2}$. Support for this identification comes from the absence of the $\hat{E}_{2g_2}$ line in pristine graphite, the increase of its intensity with increasing intercalate concentration and the occurrence of a single $\hat{E}_{2g_2}$ line in stage 1 and stage 2 compounds, where all graphite layers are bounding layers. The upshift in frequency of the $\hat{E}_{2g_2}$ mode relative to the $E_{2g_2}^0$ mode is due to the difference in force constants arising from the different environment of the carbon atoms in the graphite bounding and interior layers. The dependence on reciprocal stage of the relative intensities of the $E_{2g_2}^0$ and $\hat{E}_{2g_2}$ modes indicates that a single bounding layer on either side of the intercalate layer effectively screens the intercalate layer from the graphite interior layers. This behavior is consistent with the z-dependence of the charge distribution calculated by *Pietronero* et al. [2.64, 65] on the basis of a Thomas-Fermi model, yielding a high charge density in the graphite bounding layers and a rapid decrease in charge density with distance from the intercalate layer.

The frequencies for the $E_{2g_2}^0$ and $\hat{E}_{2g_2}$ modes both exhibit a distinctive dependence on reciprocal stage. For the acceptor compounds such as with intercalants FeCl$_3$, AlCl$_3$, NiCl$_2$, Br$_2$, AsF$_5$, SbCl$_5$, ICl, and HNO$_3$, both mode frequencies $\omega(E_{2g_2}^0)$ and $\omega(\hat{E}_{2g_2})$ exhibit approximately the same frequency

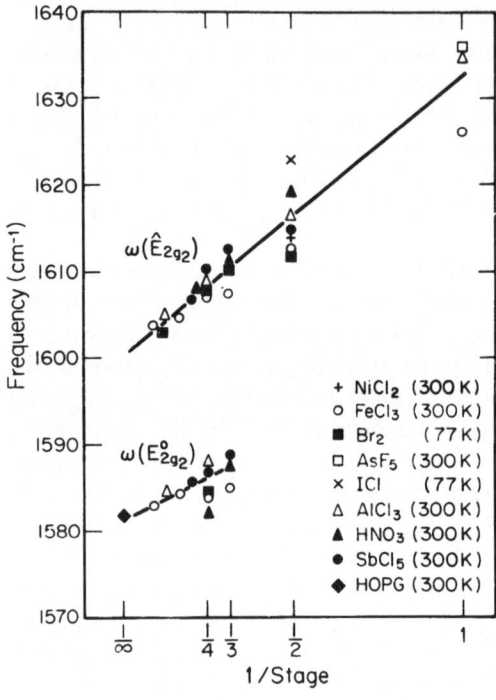

Fig. 2.13. Reciprocal stage $(1/n)$ dependence of the Raman frequencies associated with the graphite interior layers $(E_{2g_2}^0)$ and the graphite bounding layers $(\hat{E}_{2g_2})$ for various *acceptor* intercalants. An *upshift* is found for both $\omega(E_{2g_2}^0)$ and $\omega(\hat{E}_{2g_2})$ vs $(1/n)$ and the behavior is similar for all acceptor intercalants. (See text for references)

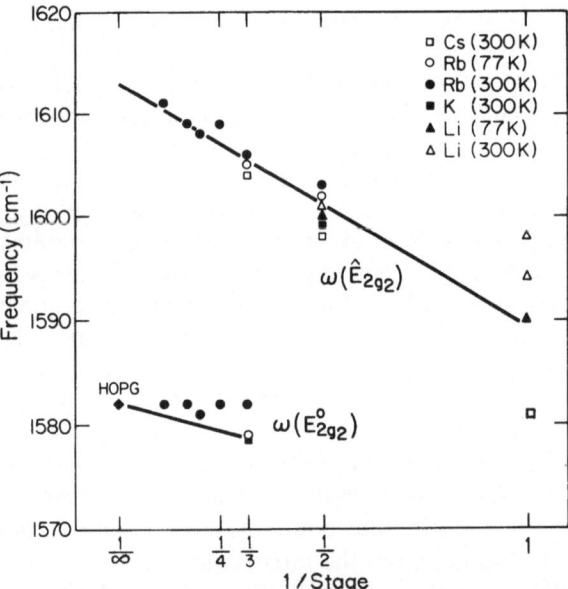

Fig. 2.14. Reciprocal stage $(1/n)$ dependence of the Raman frequencies associated with the graphite interior layers $(E_{2g_2}^0)$ and the graphite bounding layers $(\hat{E}_{2g_2})$ for various *donor* intercalants. A *downshift* is found for both $\omega(E_{2g_2}^0)$ and $\omega(\hat{E}_{2g_2})$ vs $(1/n)$ (see text for references). Also included is the point for $\omega(\hat{E}_{2g_2})$ determined for first stage C_8Cs (300 K) by a calculation of the Breit-Wigner-Fano lineshape in terms of a coupling to a frequency dependent multiphonon continuum [2.79]

upshift as a function of reciprocal stage $(1/n)$ as shown in Fig. 2.13. A similar stage-dependent upshift of the two modes gives rise to an average doublet separation of 22 ± 2 cm^{-1} for these acceptor compounds and an extrapolated value of 19 ± 2 cm^{-1} in the limit $(1/n)\rightarrow0$ (Fig. 2.13). In contrast, the donor alkali metal compounds (with K, Rb, Cs, and Li intercalants) exhibit a downshift in these mode frequencies with increasing reciprocal stage [2.57, 58, 62, 63, 66] as shown in Fig. 2.14. In this case, the downshift of the $E_{2g_2}^0$ mode is less pronounced than for the $\hat{E}_{2g_2}$ mode. Extrapolation to the dilute limit $(1/n)\rightarrow0$ yields 31 ± 3 cm^{-1} for the mode splitting. The mode splittings $[\omega(\hat{E}_{2g_2})-\omega(E_{2g_2}^0)]$ in the limit $(1/n)\rightarrow0$ for both donor and acceptor compounds are utilized in evaluating the force constants for the intercalant-graphite bounding layer interaction (Sect. 2.2.2).

The mode softening observed for the donor compounds is consistent with the lattice expansion with increasing $(1/n)$ reported by *Nixon* and *Parry* [2.44] and *Guérard* et al. [2.45]. The stiffening of these lattice modes in the acceptor compounds and their softening in the donor compounds has been attributed to lattice strain associated with stage dependent changes in the lattice parameters by *Underhill* et al. [2.59]. On this basis, the hardening of the lattice modes as a function of reciprocal stage for the acceptor compounds suggested an in-plane lattice contraction in acceptor compounds [2.59] which was subsequently observed by *Flandrois* et al. [2.42] in stage 2 $C_{11.3}NiCl_{2.13}$ and by *Krapchev* [2.43] in stage 2 $FeCl_3$. For the case of graphite-$FeCl_3$ compounds, the approximate constancy of the linewidth for the $E_{2g_2}^0$ interior layer mode as a function of intercalate concentration and the identical frequency upshift for both interior and bounding Raman modes for a sample of given stage has been interpreted by *Underhill* et al. [2.59] in terms of a similar strain within both the bounding and interior graphite layers for a given compound, a condition also necessary to prevent sample fracture upon intercalation. The magnitude of the strain increases as the number of interior graphite layers decreases, consistent with the idea that the stress introduced by intercalation is shared by fewer layers as the stage index n decreases. A summary of mode frequencies and linewidths for the Lorentzian $E_{2g_2}^0$ and $\hat{E}_{2g_2}$ modes for some typical intercalants is given in Table 2.5, in which we note that the linewidths for acceptors are smaller than for donors of comparable stage.

A strong dependence of the relative intensities of the $E_{2g_2}^0$ and $\hat{E}_{2g_2}$ components on intercalate concentration is also observed. Increasing the intercalate concentration causes the intensity of the lower frequency component to decrease and the intensity of the upper frequency component to increase. In this connection, it is of interest to note in Fig. 2.12 that for stage 4 where the number of graphite bounding and interior layers is equal, the two peaks have approximately equal intensity. A simple expression relating the relative intensities of the graphite interior to bounding layer modes is written down from geometrical arguments as

$$\frac{I(E_{2g_2}^0)}{I(\hat{E}_{2g_2})} = \left(\frac{f_0}{1-f_0}\right)\frac{\sigma_0}{\hat{\sigma}} = \left(\frac{n-2}{2}\right)\frac{\sigma_0}{\hat{\sigma}}, \tag{2.5}$$

Table 2.5. Lorentzian Raman lineshape parameters for some typical intercalants[a]

Intercalant	Stage	T [K]	$\omega(\hat{E}_{2g_2})$ [cm^{-1}]	$\Gamma(\hat{E}_{2g_2})$	$\omega(E^0_{2g_2})$	$\Gamma(E^0_{2g_2})$
–	∞[b]	300	–	–	1582	11
Rb	2[c]	77	1602±1	7±0.5	–	–
	3[c]	77	1605±1	5±0.5	1579±1	6±0.5
	2[d]	300	1603	22.0	–	–
	3[d]	300	1606	14.9	1582	8.3
	4[d]	300	1609	15.9	1582	8.3
	5[d]	300	1608	18.2	1581	12.1
	6[d]	300	1609	18.4	1582	10.6
Cs	2[e]	300	1598	22	–	–
	3[e]	300	1604	~ 9	1579	~ 5
K	2[e]	300	1599	13	–	–
Li	2[f]	300	1604±4	33±4	–	–
	2[f]	77	1601±4	27±4	–	–
	2[f]	4	1602±4	28±4	–	–
AsF$_5$	1[g]	300	1636	5.1	–	–
SbCl$_5$	2[g]	300	1615	–	–	–
HNO$_3$	2[g]	300	1619	–	–	–
FeCl$_3$	1[h]	300	1626	3	–	–
	2[h]	300	1613	3	–	–
	3[h]	300	1608	13	1585	13
	4[h]	300	1607	12	1584	12
	6[h]	300	1605	9	1584	10
AlCl$_3$	1[i]	300	1635	3	–	–
	2[i]	300	1616	5	–	–
	4[i]	300	1608	6	1588	6
	8[i]	300	1605	3	1584	6
SbCl$_5$	2[j]	300	1616	9	–	–
	3[j]	300	1611	9	1591	9
	4[j]	300	1607	9	1589	9
	5[j]	300	1607	9	1588	9

[a] For some of the low stage alkali metal compounds, the lineshapes have been fit to Breit-Wigner-Fano lines with $\Gamma/q \ll 1$.
[b] From *Brillson* et al. [2.17].
[c] From *Eklund* et al. [2.62].
[d] From *Solin* [2.58].
[e] From *Nemanich* et al. [2.57].
[f] From *Eklund* et al. [2.66].
[g] From *Eklund* et al. [2.70].
[h] From *Underhill* et al. [2.59].
[i] From *Gualberto* et al. [2.60].
[j] From *Eklund* et al. [2.61].

where σ_0 and $\hat{\sigma}$ are, respectively, the Raman scattering cross sections for the graphite interior and bounding layers and f_0 is the fraction of carbon atoms on interior graphite layers in which f_0 has a stage dependence $f_0 = 1 - 2/n$ for $n \geqq 2$. The validity of relation (2.5) has been verified by *Solin* in the alkali metal compounds with the intercalants Rb [2.58] and K [2.63], and in the acceptor FeCl$_3$ compounds [2.63], by obtaining a straight line in a plot of $[I(E_{2g_2}^0)/I(\hat{E}_{2g_2})]$ vs n with a zero intercept at $n = 2$. These straight line plots support the idea that the Raman cross sections for the graphite bounding and interior layers $\hat{\sigma}$ and σ_0 are independent of stage with a value of $(\sigma_0/\hat{\sigma}) = 0.3 \pm 0.1$ for both the K and Rb intercalants and $(\sigma_0/\hat{\sigma}) = 1.2$ for the graphite-FeCl$_3$ system [2.63]. Thus, a similar behavior is found for the two alkali metal intercalants but showing a large difference in Raman cross section between graphite bounding and interior layers, whereas for the FeCl$_3$ acceptor compounds, this difference is much smaller. This is consistent with the much smaller charge transfer from the intercalant to the graphite bounding layers in the case of acceptors (such as FeCl$_3$) as compared with the large transfer occurring in the alkali metal donor compounds. For the case of the alkali metal compounds, these results imply that the graphite bounding layers are more polarizable than the graphite interior layers which are expected to have approximately the same polarizability as pristine graphite. In this connection, it should be mentioned that the Raman efficiency S has been shown by *Wada* and *Solin* [2.67] to be much larger for graphite than for diamond, with $(S_{\text{graphite}}/S_{\text{diamond}}) \approx 55$.

In principle, the intensity of these graphitic modes can also be dependent on laser excitation energy through a resonant enhancement effect involving the electronic levels of the system. However, no resonant enhancement effects have been reported for the graphitic modes though such effects have been observed for intercalate modes (Sect. 2.3.4).

The temperature dependence of the observed Raman lines in graphite intercalation compounds has been most extensively studied in those materials which are known to have structural phase transitions. For example, the intercalation compounds with FeCl$_3$ and NiCl$_2$ have been found to order magnetically: at ~ 4 K for stage 1 graphite-FeCl$_3$, at ~ 4 K for stage 1 graphite-FeCl$_3$, at ~ 8 K for stage graphite-CoCl$_2$ [2.68] and at ~ 18 K for stage 2 graphite-NiCl$_2$ [2.69]. No discontinuous changes in mode frequency or linewidth have, however, been reported in crossing these transition temperatures. Although graphite-AsF$_5$ is known to exhibit structural phase transitions for the stage 1 compound at 150 K and 220 K, no anomalous change in mode frequency or linewidth was observed in the temperature dependent Raman spectra [2.70]. Recently a change in line intensity associated with structural phase transitions in graphite-Br$_2$ has been reported (see Sect. 2.3.4). The alkali metal compounds with intercalants K, Rb, and Cs show phase transitions at temperatures T_l and T_u which occur in the various compounds between ~ 100 K and ~ 200 K [2.71]. Anomalies associated with these transitions have recently been reported for low frequency modes (see Sect. 2.3.4).

The Raman lines do, however, shift to higher frequency with decreasing temperature, as one would expect at least from thermal contraction considerations, and this effect has been explicitly investigated and verified by *Eklund* et al. [2.70] for the case of first stage graphite-AsF_5.

The earliest report of a significant change in the Raman spectrum on passing through a known phase transition involves the low frequency lines in first stage C_8Cs where *Caswell* and *Solin* [2.72] reported a discontinuous downshift of the Raman line near 590 cm^{-1} at the intercalant "melting" transition ($T_m \simeq 608$ K). This particular transition is discussed further in Sect. 2.3.3.

These results for the temperature dependence of the Raman spectra for intercalated graphite indicate that the intercalate layers behave essentially independently of the graphite layers. The mechanism of charge transfer changes the lattice constant of the graphite [2.46], thus giving rise to a shift in the Raman $E^0_{2g_2}$ and $\hat{E}_{2g_2}$ modes. However, the internal structure of the intercalate layer seems to have little effect on the mode frequencies for these graphitic modes, as can be seen from Figs. 2.13, 14 where the dependence of mode frequencies on reciprocal stage seems to be independent of stoichiometric ratio ξ in $C_{n\xi}X$ for stage n or whether the intercalate layer is commensurate or incommensurate with the graphite layer.

The absence of a significant temperature dependence of the linewidth suggests that most of the linewidth arises from inhomogenous broadening due to a variety of different crystalline environments. Indeed, the graphite interior layer modes are generally broader than the bounding layer modes because in high stage compounds, each crystallographically distinct interior layer contributes in a somewhat different frequency range, but unresolved from the other graphite interior layers, whereas all graphite bounding layers are crystallographically indentical and contribute equivalently to the observed line.

The model presented in Sect. 2.2.2 shows the low frequency and out-of-plane graphitic modes to be very sensitive to the intercalant-graphite bounding layer interaction parameters, indicating that these graphitic modes would be much more sensitive than the E_{2g_2} mode to structural phase changes associated with the intercalant. One reason why these effects might be difficult to observe is related to inhomogeneities in the intercalate layers [2.15, 73], giving rise to a range of values for the intercalant-graphite bounding layer interaction parameters and hence a broadened Raman line.

In addition to temperature dependent studies, the effect of hydrostatic pressure on the Raman spectra of graphite intercalation compounds has been investigated up to 10 kbar by *Wada* [2.74] for the case of the alkali metal intercalants Rb and Cs for stages 2, 3, and 4. This work gives new insight into the staging mechanism through observations of pressure-induced phase transitions involving stage changes. Using x ray diffraction techniques, *Wada* was able to identify pressure-induced reversible phase transitions for stage 2 compounds with Rb and Cs at a pressure of ~ 5 kbar, from $C_{24}X$ (stage 2) to $C_{24}X$ (stage 3), indicating a 50% increase in the in-plane intercalate density with a (2×2) commensurate in-plane superlattice observed in the high pressure phase. The Raman spectra showed the appearance of the $E^0_{2g_2}$ interior layer mode as pressure was applied, with a variation in the intensity of the interior to bounding layer modes in general agreement with (2.5) for pressures above the stage transition. The similarity between this intensity ratio for the conventional stage 3 compounds (with Rb) and for the pressure-induced stage 3 compounds indicates that the Raman cross sections are approximately independent of in-plane intercalate density.

For the stage 3 compounds, pressure-induced transitions to mixed stage 4 and 5 compounds were observed, while the stage 4 compounds exhibited pressure-induced transitions to stage 6 compounds for which (2×2) in-plane superlattice structures were identified in the high pressure phases. The increase in intensity $I(E^0_{2g_2})$ relative to $I(\hat{E}_{2g_2})$ was again found to be consistent with (2.5) in terms of the stage transitions identified by x ray characterization.

For all the compounds that were investigated, both the graphite interior and bounding layer modes were upshifted with increasing pressure by an amount roughly similar to that of graphite which was measured to be $[\partial\omega(E_{2g_2})/\partial p] = (0.50 \pm 0.08)\,\mathrm{cm}^{-1}/\mathrm{kbar}$. In particular, values for the pressure dependent upshifts for the bounding and interior layers were reported to be $[\partial\omega(\hat{E}_{2g_2})/\partial p] = (0.70 \pm 0.12)$, (0.51 ± 0.09), (0.44 ± 0.05), $(0.54 \pm 0.08)\,\mathrm{cm}^{-1}/\mathrm{kbar}$ for $C_{24}Cs$ and $C_{12n}Rb$ ($n = 2, 3, 4$) and $[\partial\omega(E^0_{2g_2})/\partial p] = (0.78 \pm 0.08)$, (0.32 ± 0.12) $\mathrm{cm}^{-1}/\mathrm{kbar}$ for $C_{12n}Rb$ ($n = 3, 4$). Pressure results in a lattice contraction and hence a lattice stiffening or an upshift in mode frequency, similar to the lattice contraction associated with acceptor intercalation (Fig. 2.13). The pressure-induced transition to a *higher* stage donor compound is expected from Fig. 2.14 to result in an additional frequency upshift, consistent with *Wada's* observations. The effect of pressure on the Raman spectra for the stage 1 alkali metal compounds is discussed in Sect. 2.3.3.

2.3.3 Raman Scattering in Stage 1 Compounds

Two calculations of the phonon dispersion relations have been carried out for stage 1 compounds (Sect. 2.2.2). In the calculation by *Horie* et al. [2.36] for C_8K and C_8Rb, the in-plane (2×2) intercalate superlattice and the $\alpha\beta\gamma\delta$ intercalate stacking sequence for these compounds are treated explicitly and a large number of phonon branches are obtained (Fig. 2.7). In the calculation by *Leung* et al. [2.38] which is applicable to any stage and intercalant (Fig. 2.11), it is necessary to fit experimental data for a given compound to determine explicit values for the intercalant-graphite bounding layer interaction parameters. To date, neither of these models have been applied to the detailed analysis of the observed Raman spectra for a specific compound.

For all stage 1 compounds, each graphite layer is sandwiched between two intercalate layers, unlike the situation for $n \geq 2$ where each graphite layer is adjacent to at least one other graphite layer. For the case of the acceptor compounds where the intercalant-graphite bounding layer interaction is weak, the observed Raman spectra for the stage 1 compounds are qualitatively

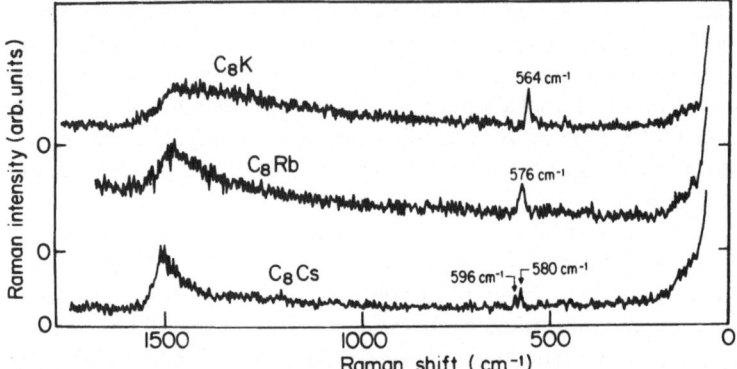

Fig. 2.15. Raman intensity for the stage 1 alkali metal compounds C_8K, C_8Rb, and C_8Cs at $T = 4\,K$ over a wide frequency range, as measured by *Eklund* et al. [2.62]

similar to those for the higher stages, insofar as the lineshape remains Lorentzian, the linewidth remains comparable to that for the higher stages, and the mode frequency upshift fits the plot of $\omega(\hat{E}_{2g_2})$ vs $1/n$ in Fig. 2.13. On the other hand, the behavior of the Raman spectra for the alkali metal donor compounds is anomalous, particularly for the intercalants K, Rb, and Cs, and requires special consideration.

In contrast to most intercalation compounds, these alkali metal compounds can form well-ordered (2×2) superlattices at low temperatures, commensurate with the graphite layers. Some effects in the Raman spectra associated with this superlattice ordering have in fact been observed. We here review the experimental results for stage 1 compounds, comment on some of the proposed explanations, and suggest further work to clarify those issues where the correct model is not yet firmly established.

Although the Raman spectra for the alkali metal stage 1 compounds C_8K, C_8Rb, and C_8Cs are qualitatively different from spectra that have been reported for stage 1 acceptors or for higher stage donor and acceptor compounds, these spectra are all very similar to each other (Fig. 2.15), as shown by the work of *Nemanich* et al. [2.57] and *Eklund* et al. [2.62]. These spectra are, however, in sharp contrast with spectra observed for higher stage compounds with the same intercalant, as illustrated in Fig. 2.16 for the graphite-Rb system [2.62].

The spectra for the stage 1 alkali metal compounds C_8K, C_8Rb, and C_8Cs in Fig. 2.15 show a broad asymmetric line at high ($\sim 1500\,cm^{-1}$) frequencies and sharp structure at lower ($\sim 600\,cm^{-1}$) frequencies. For each of the compounds, the broad lines peaking near $1500\,cm^{-1}$ have the asymmetric Breit-Wigner-Fano lineshape [2.75]

$$I(\omega) = I_0 \left[1 + \left(\frac{\omega - \omega_0}{q\Gamma} \right) \right]^2 \left[1 + \left(\frac{\omega - \omega_0}{\Gamma} \right)^2 \right]^{-1} \qquad (2.6)$$

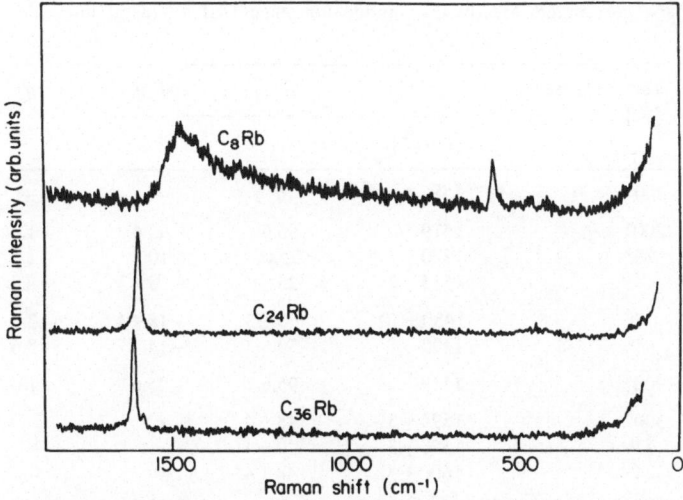

Fig. 2.16. Raman spectra of C_8Rb (stage 1), $C_{24}Rb$ (stage 2), and $C_{36}Rb$ (stage 3) at 77 K and over a wide frequency range. The data were taken by *Eklund* et al. [2.62] in the Brewster angle, backscattering geometry

in which I_0 is a normalization factor, ω_0 is the renormalized resonant frequency corresponding to I_0, Γ is a measure of the linewidth and q^{-1} is the Breit-Wigner-Fano coupling coefficient. In the limit $q^{-1} \rightarrow 0$, the Lorentzian lineshape is recovered. Breit-Wigner-Fano lineshapes arise in Raman scattering through the interaction between discrete Raman-active phonon modes and a continuum of Raman-active excitations, as has been discussed in [2.75, 76]. Values for the Breit-Wigner-Fano parameters ω_0, Γ, and Γ/q for first stage C_8K, C_8Rb, and C_8Cs are summarized in Table 2.6. It is of interest to note that as the temperature is reduced from 300 K to 77 K, the linewidth Γ and the coupling coefficient q^{-1} decrease sharply with little further change in going to 4 K, while the resonant frequency ω_0 is almost temperature-independent.

In contrast to the behavior shown in Fig. 2.15, the Raman spectra for the stage 1 acceptor compounds, such as with the intercalants $FeCl_3$ [2.59]), $AlCl_3$ [2.60], and AsF_5 [2.70], characteristically exhibit a single Lorentzian line (Fig. 2.12) associated with the graphite bounding layer. The linewidth is comparable to that for $\hat{E}_{2g_2}$ modes for higher stage compounds and the mode frequency is significantly upshifted, in accordance with the dependence of the $\hat{E}_{2g_2}$ mode frequency on reciprocal stage as shown in Fig. 2.13. This difference in behavior between stage 1 donor and acceptor compounds has been attributed by *Dresselhaus* and *Dresselhaus* [2.6] to the greater coupling between intercalant and graphite boundary layers for the case of donor compounds. This greater coupling is manifested by a greater overlap in the $s-p$ alkali metal electronic orbitals with the π orbitals in the graphite bounding layers, by a larger charge transfer between the alkali metal and graphite bounding layers, by an increase in the c-axis conductivity, by the relatively high electronic in-

Table 2.6. Raman lineshape parameters for the $\hat{E}_{2g_2}$ mode for stage 1 alkali metal intercalation compounds[a]

Material	Temperature [K]	ω_0	Γ	Γ/q	q
		$[\text{cm}^{-1}]$			
Graphite (HOPG)	300[b]	1582	6	–	–
C_8Cs	300[c]	1519	95.6	−95.6	−1.0
	77[d]	1520	25.0	−10	−2.5
	4[d]	1515	25	−9	−2.8
C_8Rb	77[d]	1480±10	60±5	−18±1	−3.3
	4[d]	1490	50	−14	−3.6
C_8K	300[c]	1519	95.6	−95.6	−1.0
C_6Li	300[e]	1594±3	48±4	–	–
	77[e]	1590±3	40±4	–	–
	4[e]	1600±3	48±4	–	–

[a] The parameters for the Lorentzian lineshapes are Γ and ω_0, and for the Breit-Wigner-Fano lineshapes are ω_0, Γ, and q^{-1} [see (2.6)], where the background continuum is assumed to be frequency-independent.
[b] From *Underhill* et al. [2.59].   [d] From *Eklund* et al. [2.62].
[c] From *Nemanich* et al. [2.57].   [e] From *Eklund* et al. [2.66].

plane scattering in the graphite bounding layers for the stage 1 donors, and finally by the large magnitude of the asymmetric Breit-Wigner-Fano parameter $(1/q)$ characterizing the broad Raman lines near 1500 cm^{-1} (Table 2.6). For the case of the stage 1 alkali metal compounds, we note that the $s-p$ alkali metal orbitals on one intercalate layer can overlap with those in adjacent intercalate layers, unlike the behavior for the higher stage compounds [2.7].

Referring to the spectra for the stage 1 alkali metal compounds shown in Fig. 2.15, a sharp spectral feature appears in the vicinity of 570 cm^{-1} for C_8K, C_8Rb, and C_8Cs. This spectral feature has been studied in more detail using a polarization analysis to provide information on the two constituent components which exhibit different basic symmetries [2.62]. These components are separated by a frequency $\Delta\omega$, which is greatest for C_8Cs and smallest for C_8K. The two components are well resolved for C_8Cs with the lower frequency component only observed for the $(\|, \perp)$ geometry, while the upper frequency component is observed for both $(\|, \|)$ and $(\|, \perp)$ scattering geometries. These two components also exhibit a Breit-Wigner-Fano lineshape, with values for Γ/q that are significant at room temperature [2.57, 62], but much smaller than for the line peaking near ~ 1500 cm^{-1}. (Compare Table 2.7, which lists Breit-Wigner-Fano parameters for the ~ 570 cm^{-1} line, with Table 2.6 which gives values for the ~ 1500 cm^{-1} line.)

A weak component at a higher frequency has also been reported by *Caswell* and *Solin* [2.72] for C_8Cs (at 620 cm^{-1}) and for C_8Rb. These authors found that for C_8Cs, the strong component, only observed with $(\|, \perp)$ polarization, exhibits a small discontinuous change by ~ 1.5 cm^{-1} in the frequency ω_0 at the

Table 2.7. Raman lineshape parameters for the intermediate frequency mode for stage 1 alkali metal intercalation compounds[a]

Material	ω_0	Γ	Γ/q	q	Polarization
	[cm^{-1}]				
C$_8$Rb	569 ± 1	3.5 ± 0.5	–		$(\parallel, \parallel)$
	578.5 ± 0.05	3.0 ± 0.5	–		$(\parallel, \parallel), (\parallel, \perp)$
C$_8$Cs	579 ±0.5	2.5 ± 0.5	–		$(\parallel, \parallel)$
	596 ±0.5	2.0 ± 0.5	–		$(\parallel, \parallel), (\parallel, \perp)$
	620^b	0.6 ± 0.1	-0.6 ± 0.2	-1	$(\parallel, \parallel), (\parallel, \perp)$
C$_8$K	565.4 ± 0.5	$5\ \pm0.5$	$-1\ \pm0.2$	-5	$(\parallel, \parallel)$
	566.0 ± 0.5	$5\ \pm0.5$	$-1\ \pm0.2$	-5	$(\parallel, \parallel), (\parallel, \perp)$

[a] The parameters for the Breit-Wigner-Fano lines are ω_0, Γ, and q^{-1} [see (2.6)] where the background continuum is assumed to be frequency-independent. All results, except where otherwise indicated, are for 77 K and from the work of *Eklund* et al. [2.62].

[b] This third component was only reported at room temperature, from the work by *Caswell* and *Solin* [2.72]. These authors also gave room temperature values for ω_0 for the two lower frequency components at 574 and 594 cm^{-1}, with the middle component exhibiting a downshift to 598 cm^{-1} at 600 K and $\Gamma=0.6\pm0.1$ cm^{-1} and $q=-1\pm0.1$ for all lines, independent of temperature.

melting temperature ($T_m = 608$ K), where x ray measurements show the intercalate layer undergoing a phase transition from a (2×2) superlattice to a disordered liquid state. The other Breit-Wigner-Fano parameters Γ and q^{-1}, on the other hand, show no temperature variation near T_m. Based on a comparison of the measured temperature with that determined from the Stokes/anti-Stokes ratios for this line, *Caswell* and *Solin* [2.72] have argued that the continuum background below 600 cm^{-1} is due to a phonon scattering mechanism and not to an electronic mechanism. The observation of a discontinuous change in ω_0 at T_m indicates a significant coupling between the intercalant and graphite bounding layers for this donor compound, in contrast with the behavior reported by *Eklund* et al. [2.70] for the stage 1 acceptor C$_8$AsF$_5$ where no influence of the intercalant on the graphite bounding layer mode was found in the temperature range spanning the order-disorder transition.

Using a polarization analysis of the sharp spectral feature in the vicinity of 570 cm^{-1}, *Eklund* et al. [2.62] identified this feature with a zone folding of the M-point modes of the larger graphite Brillouin zone into the Γ point of the smaller zone for the intercalation compound, in accordance with the (2×2) inplane superlattice structure of C$_8$Cs, C$_8$Rb, and C$_8$K. The zone-folded modes are upshifted by $\delta\omega$ from the M-point graphite mode frequency and exhibit a splitting $\Delta\omega$ of the two components with different symmetries. Both the frequency upshift $\delta\omega$ and the splitting $\Delta\omega$ increase with the increasing size of the intercalate ion and the resulting increase in the intercalant-graphite bounding layer interaction.

The zone-folding argument provides a mechanism for turning on additional lines in the Raman spectra, but does not explain why the modes near 570 cm^{-1} are the only modes that are clearly resolved. The absence of additional Raman structure for these stage 1 compounds has led *Caswell* and *Solin* [2.72] to argue that the modes near 570 cm^{-1} are disorder-induced and reflect maxima in the phonon density of states. One argument against this interpretation is that calculations for the phonon density of states indicate a very high density of states near 570 cm^{-1} for both stage 1 donor as well as acceptor compounds [2.40]. Nor does a density of states argument explain why the stage 1 donor compounds C_6Li [2.66, 77] and C_6Eu [2.78] shows no sharp structure near 570 cm^{-1}. The zone-folding argument, on the other hand, accounts for the absence of the 570 cm^{-1} structure for the stage 1 acceptor compounds which generally have no well-developed intercalate ordering commensurate with the graphite bounding layer, and hence no superlattice and no in-plane zone folding. Since C_6Li and C_6Eu exhibit the ($\sqrt{3} \times \sqrt{3}$) superlattice, zone folding requires the K point rather than the M point to fold into $q = 0$ so that no sharp structure near 570 cm^{-1} would be expected in this case, in agreement with experimental observations [2.66, 77, 78].

Pressure dependent changes of the 570 cm^{-1} modes in the stage 1 compounds were reported by *Wada* [2.74], who found a gradual change in lineshape with increasing pressure and associated these changes with a modification of the broad continuum with pressure. Though pressure is expected to produce more order in the intercalate layer, *Wada* was unable to detect any additional modes in the higher stage compounds associated with in-plane zone folding, and he interpreted this finding as indicating that the 570 cm^{-1} mode in the stage 1 compounds is not associated with in-plane zone folding.

The origin of the 570 cm^{-1} mode thus remains an unanswered question. With the availability of detailed models for the phonon dispersion relations for specific compounds, it should be possible to distinguish between a zone-folding and a density of a states mechanism for the sharp structure observed in the stage 1 compounds C_8Cs, C_8Rb, and C_8K.

The broad asymmetric features of the Raman spectra for C_8K, C_8Rb, and C_8Cs in the vicinity of $\sim 1500 \text{ cm}^{-1}$ have been interpreted in terms of an interaction between a discrete Raman-active graphite phonon mode and a continuum of states. There have, however, been two different proposals for the origin of this continuum. On the one hand, *Eklund* et al. [2.62] have identified this continuum with multi-phonon modes which are coupled because of the reduction in symmetry introduced by the presence of the intercalate layer. This interpretation has been substantiated by a lineshape calculation of the asymmetric Breit-Wigner-Fano line by *Eklund* and *Subbaswamy* [2.79] in which the graphite-intercalate layer interaction is assumed to give rise to a coupling between the zone-center $\hat{E}_{2g_2}$ mode and the continuum of modes arising from multiphonon processes with frequencies extending to $\sim 3700 \text{ cm}^{-1}$. Because of the high two-phonon density of states for $\omega > 1600 \text{ cm}^{-1}$, this coupling gives rise to a downshift of the $\hat{E}_{2g_2}$ mode at the zone center. The good agreement

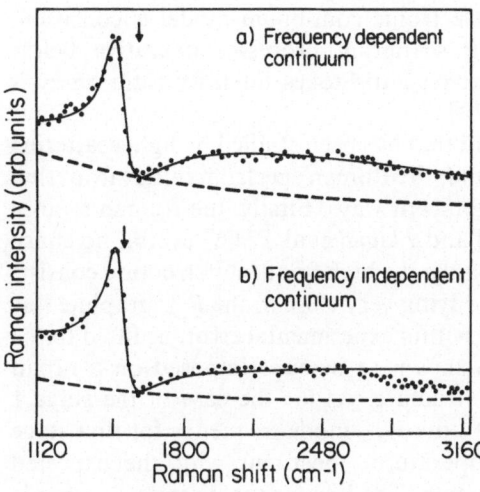

Fig. 2.17a, b. Theoretical fits to the observed room temperature Raman spectrum for first stage C_8Cs in the broad frequency range $1100 \leq \omega \leq 3200$ cm^{-1} using a coupling of the discrete $\hat{E}_{2g_2}$ bounding layer mode to a multiphonon continuum which is (**a**) frequency dependent and (**b**) frequency independent. The arrows in (**a**) and (**b**) indicate the frequency of the discrete (uncoupled) mode. The dashed lines represent the noninterfering continuum in each case. (From *Eklund* and *Subbaswamy* [2.79])

between the calculated and experimental lineshapes over the large frequency range $700 < \omega < 3700$ cm^{-1} (Fig. 2.17) provides strong evidence that the Breit-Wigner-Fano lineshape is due to coupling to a frequency-dependent continuum of phonon states. The connection of this continuum of phonon states to the second-order spectrum is discussed further in Sect. 2.3.5. From the lineshape analysis, *Eklund* and *Subbaswamy* [2.79] were able to deduce the discrete mode frequency $\omega(\hat{E}_{2g_2})$ in the limit of zero coupling to the continuum, yielding a value of 1585 cm^{-1}. This value is in good agreement with that predicted from the $\omega(\hat{E}_{2g_2})$ vs $(1/n)$ plot for donors in Fig. 2.14, so that the same decoupled discrete bounding layer $\hat{E}_{2g_2}$ mode is able to account for the observed Raman spectra for all stages, independent of whether the lineshape is Lorentzian or Breit-Wigner-Fano. It has also been reported from studies on stage 1 and 2 graphite-Li compounds that the second-order spectra are highly temperature dependent [2.66]. Thus, if the continuum were due to multiphonon processes, one would expect a significant temperature dependence of the Breit-Wigner-Fano parameters, in agreement with observation (Table 2.6).

On the other hand, *Nemanich* et al. [2.57] interpreted the Breit-Wigner-Fano lineshape to be due to a continuum of *electronic* states. This interpretation led to a lineshape calculation by *Miyazaki* et al. [2.80] in which the broad Raman line near 1500 cm^{-1} was attributed to the coupling of 3 zone-center modes at frequencies 1580, 1280, and 858 cm^{-1} with an electronic excitation covering the energy range between the filled and empty electronic bands for C_8K. A good fit to the observed lineshape was obtained in the frequency range between $1100 < \omega < 1700$ cm^{-1} using a value of the electron-phonon coupling constant of ~ 0.08 eV. It remains to be demonstrated whether the electronic continuum model can account for the observed lineshape over the broad frequency range over which agreement was achieved with the multiphonon continuum model, whether the electronic continuum model can account for the strong temperature dependence observed for the Breit-Wigner-

Fano parameters, and whether the electronic continuum model is consistent with the dominant phonon-phonon scattering processes operative below $\sim 600\,\text{cm}^{-1}$, as deduced from the Stokes/anti-Stokes intensity ratio analysis for the $\sim 570\,\text{cm}^{-1}$ line in C_8Cs [2.72].

The other stage 1 donor compound that has been studied by light scattering is C_6Li. For this alkali metal compound, the Raman spectrum differs from that for C_8K, C_8Rb, and C_8Cs in two significant ways. Firstly, the Raman spectra for C_6Li taken by $Zanini$ et al. [2.77] and $Eklund$ et al. [2.66] exhibit no sharp lines in the vicinity of 570 cm^{-1}. Secondly, the high frequency structure consists of a single line peaked at $\sim 1600\,\text{cm}^{-1}$ lying very close to the E_{2g_2} graphite line and exhibiting a lineshape that could, within experimental error, be fitted to the Lorentzian form. This line has been shown to be superimposed on a broad temperature-dependent background structure. As for the case of the stage 1 acceptor compound with AsF$_5$ [2.70], the $\hat{E}_{2g_2}$ mode frequency for first stage C_6Li increases with decreasing temperature, consistent with the expected temperature-dependent lattice contraction. The line for the discrete $\hat{E}_{2g_2}$ mode in C_6Li was found to be very broad (HWHM $\sim 45\,\text{cm}^{-1}$), with a temperature-independent linewidth which is one order of magnitude greater than that for the temperature-independent linewidth for the stage 1 acceptor compound with AsF$_5$ (HWHM $\sim 5\,\text{cm}^{-1}$). Although the lineshape of the $\hat{E}_{2g_2}$ mode is Lorentzian within experimental error, the large magnitude of the linewidth suggests that this line too is an incipient Breit-Wigner-Fano line with a coupling parameter Γ/q that is too small to determine quantitatively. $Eklund$ et al. [2.66] noted that the much weaker coupling to the continuum of phonon states in C_6Li was consistent with the very small size of the Li ions which therefore have a smaller effect on lowering the translational symmetry in the graphite layers.

In contrast, stage 1 acceptor compounds (with the intercalants AsF$_5$, AlCl$_3$, and FeCl$_3$) exhibit Lorentzian lineshapes with small linewidths. Since acceptor compounds tend to form intercalate layers that are in most cases incommensurate with the graphite layers, the intercalant is not as effective in lowering the translational symmetry of the graphite layers, as is the case for the alkali metal donor intercalants. Therefore, the coupling between the Raman-active phonon mode and the continuum of phonon modes is not so important in stage 1 acceptor compounds. Also consistent with this interpretation is the observation of well-resolved structure in the second-order Raman spectra (above $\sim 1700\,\text{cm}^{-1}$) for stage 1 acceptor compounds (Sect. 2.3.5), in sharp contrast to the broad continuum observed in the donor compounds and attributed to multiphonon processes.

2.3.4 Raman Scattering by Intercalate Modes

The observation of intercalate-related modes has been reported in several intercalation compounds and they are so identified because of their specificity to intercalate species and their relation to mode frequencies in the parent solid or free molecule. Of these reported Raman observations, the most intensively studied system is the graphite-bromine system for stages $n \geqq 2$. These bromine compounds show a strong resonant enhancement as well as complex spectra, suggesting intercalate layer orderings with a large number of inequivalent molecular sites. Since the fundamental intercalate mode frequencies are expected to occur at relatively low frequencies ($\omega \lesssim 600 \text{ cm}^{-1}$), it is expected that the low frequency graphitic branches of the phonon dispersion relations would be most strongly coupled to the intercalate modes.

The observation of intercalate modes by Raman scattering has proven to be more difficult than for the graphitic modes and this can be understood on the basis of the following simple arguments. The intensity due to the intercalate mode $I(X)$ is expected to be reduced because of the lower intercalate concentration, so that for $C_{8n}X$ stoichiometry, for example, we would expect that $I(X)/I(E_{2g_2}) \propto 1/(8n)$. This low intrinsic intensity of intercalate modes suggests the use of a resonant enhancement mechanism involving resonance between the laser photon energy and an electronic transition to an intercalate level. For most acceptor compounds, the resonant process should occur in the uv, at photon energies much higher than are accessible with an argon ion laser. Thus, most of the work to date on the intercalate modes has been carried out under off-resonance conditions where their observation is difficult.

Another argument for the low peak intensity for intercalate modes is based on structural considerations. It is believed that for many acceptors, the intercalant forms island structures [2.73] with incommensurate ordering of the intercalant relative to the graphite honeycomb lattice. On the other hand, the upshift of the graphite bounding layer mode indicates that the coupling between the intercalant and graphite is significant, resulting in a shift of the intercalate mode frequency. A shift in intercalate mode frequency relative to that in the parent solid is thus expected and is indeed found by direct calculation [2.38]. Thus, an incommensurate intercalant arrangement implies a large variety of intercalate site locations relative to the graphite network, suggesting that wave vector conservation would no longer occur and contributions from a broad intercalate "density-of-states" Raman line would result. A calculation of the Raman spectrum from an ordered graphite network intercalated with a disordered intercalant has not been carried out, though this model might be expected to apply to most acceptor intercalation compounds.

These arguments explain why intercalate modes are difficult to observe in acceptor compounds in general, and why these modes have been successfully observed in the graphite-bromine system where resonant enhancement could be achieved with an argon ion laser and where the bromine molecules are ordered commensurate to the graphite network.

The Raman spectrum associated with the bromine intercalant [2.81] is shown in Fig. 2.18 at 77 K for a graphite-bromine sample (approximately stage 5). The most pronounced low frequency feature in this spectrum is the strong, broad peak at $\omega_0 = 242 \text{ cm}^{-1}$ which, together with its harmonics at $2\omega_0$ and $3\omega_0$, account for the strongest of the spectral features observed by *Eklund* et al.

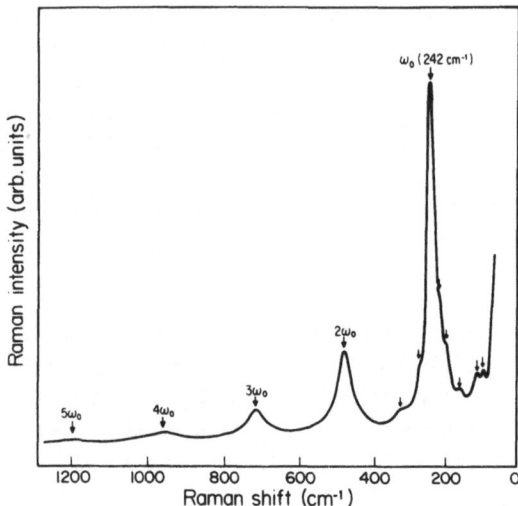

Fig. 2.18. Raman spectrum for a graphite-Br_2 sample at 77 K in the frequency regions where the intercalate modes are prominent. The structure identified with the Br_2 stretch mode in the interaction compound is denoted by ω_0 and its harmonics by $n\omega_0$. Fine structure in the vicinity of ω_0 is indicated by arrows. (From *Eklund* et al. [2.81])

[2.81]. In addition, a number of smaller structures is observed, especially at low phonon frequencies. Consistent with the identification of the dominant structural features in Fig. 2.18 with the intercalant is the correspondence between the frequency of the main spectral feature at 242 cm^{-1} with the molecular stretch mode occurring in solid Br_2 at 300 cm^{-1} [2.82] and in the free Br_2 molecule at 323 cm^{-1} [2.83]. Consistent with this interpretation is the increase in intensity of all the spectral features of Fig. 2.18 with increasing intercalate concentration. In fact, a similar dependence on intercalate concentration is found for the intercalate mode intensities $I(\omega_0)$ relative to $I(\hat{E}_{2g_2})$ for the E_{2g_2} graphitic mode on the bounding graphitic layer [2.81], while the mode frequencies for all the spectral features in Fig. 2.18 are essentially independent of intercalate concentration. This observation is interpreted to indicate that the intercalate force constants are principally determined by intralayer interactions with some perturbations introduced by the bounding graphite layers, but essentially unperturbed by interior graphite layers.

The resonant enhancement effects displayed in Fig. 2.19 facilitate observation of this intercalate mode when the laser excitation energy is close to an electronic transition. Since the ratios of the number of graphite bounding layers and of the number of intercalate layers to the total number of graphite layers are proportional, the plot of $I(\omega_0)/I(\hat{E}_{2g_2})$ vs laser energy in Fig. 2.19 demonstrates a resonant enhancement effect for the intercalant, which is further confirmed by the absence of resonant enhancement effects for the graphite bounding layer $\hat{E}_{2g_2}$ mode. This is shown by the insensitivity to laser excitation energy of the intensity ratio of graphite bounding to interior layers $I(\hat{E}_{2g_2})/I(E^0_{2g_2})$. These observations indicate that the electronic transition responsible for the resonant enhancement phenomenon is an electronic transition involving a bromine level. Similar resonant enhancement effects are indicated by observations made on

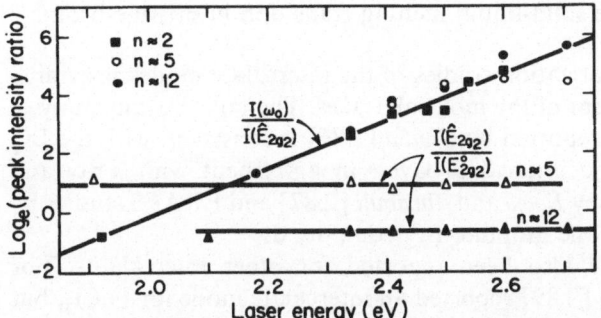

Fig. 2.19. Log_e dependence of Raman peak intensities on laser excitation energy for graphite-bromine samples of several intercalate concentrations (n denotes the stage of the compound). The intensity ratio for the $\hat{E}_{2g_2}$ mode on the graphite bounding layer compared with that for the $E_{2g_2}^0$ mode in the interior layers, denoted by $I(\hat{E}_{2g_2})/I(E_{2g_2}^0)$, is dependent on intercalate concentration but independent of laser excitation energy. In contrast, the intensity of the molecular stretch mode shows large resonant enhancement effects but has the same dependence on intercalate concentration as the graphite bounding layer mode at $\omega(\hat{E}_{2g_2})$. The intensity scale is logarithmic (base e) to cover the two orders of magnitude change in Raman intensity. (From *Eklund* et al. [2.81])

the harmonics at $2\omega_0$ and $3\omega_0$. The insensitivity of the resonant enhancement effect to intercalate concentration has been interpreted by *Eklund* et al. [2.81] as indicating that the electronic transition in $C_n Br_2$ is not between a bromine intercalate level and the Fermi level. The dissociation of the Br_2 molecule has been suggested by *Heflinger* [2.84] as an intermediate state for the resonant enhancement process for the Br_2 intercalate stretch mode because of the broadness in laser excitation energy of the resonant enhancement process.

Recent temperature-dependent studies of the Raman spectra in the graphite-bromine system have provided important information on phase transitions in the intercalate layer [2.85]. Though evidence for phase transitions is found in both the Raman lines associated with the graphite and intercalant layers, the intercalant-bromine stretch mode spectra, as expected, are more sensitive to these phase transitions. A change in lineshape from Lorentzian to Gaussian and a sudden decrease in the Raman scattering intensity in the bromine ω_0 stretch mode is identified with a commensurate-incommensurate structural transition at a temperature T_0 ($T_0 = 335\,\text{K}$ for stage 3 graphite-bromine). This identification is based on the Lorentzian lineshape below T_0, corresponding to a homogeneously broadened line for commensurate bromine-graphite registration. The Gaussian lineshape above T_0 is associated with the incommensurate relation of the bromine molecules to the adjacent graphite layers, giving rise to a variety of bromine site locations and therefore inhomogeneous broadening. The temperature dependence of the mode frequency ω_0 indicates no major change at T_0, but rather a frequency shift over the entire temperature range $345\,\text{K} < T < 380\,\text{K}$. On this basis, a second transition is identified with a continuous melting of the intercalate layer, consistent with the vanishing of $I(\omega_0)$ in this temperature range, and a change in ω_0 of the same magnitude as occurs in the solid-liquid melting transition in pristine bromine [2.82].

Also of interest are polarization studies of the intercalate modes providing information on the alignment of the molecular axes. The polarization study of the graphite-Br_2 spectrum reported by *Chung* [2.86] is consistent with the Br_2 molecular axis lying in the intercalate layer, in agreement with x ray and electron diffraction studies by *Eeles* and *Turnbull* [2.87] and EXAFS studies by *Heald* and *Stern* [2.88] on the graphite-Br_2 compounds.

Intercalate modes have also been reported for other intercalants. For example, *Caswell* and *Solin* [2.89] reported an intercalate mode for $FeCl_3$, but this exhibited essentially no frequency shift from that of solid $FeCl_3$ and no resonant enhancement effects were reported. Moreover, the polarization analysis of lines attributed to the intercalant $FeCl_3$ were related to the polarization dependence of the Raman spectrum of solid $FeCl_3$, indicating that the intercalate layer in graphite-$FeCl_3$ compounds maintains the same layer structure as is found in solid $FeCl_3$ [2.89]. Other workers [2.59] have identified these low frequency modes with a $FeCl_3$ surface deposit. In this connection, it is of interest to note that structure in the low frequency range was found by *Eklund* et al. [2.70] in the case of the intercalants AsF_5, HNO_3, and $SbCl_5$, but by careful study of the difference spectra, these authors showed that the low frequency Raman lines were due to gas in the ampoule in the case of HNO_3 and AsF_5, and to a surface deposit in the case of $SbCl_5$.

Broad ($\sim 50\,cm^{-1}$ linewidth) intercalant modes have recently been reported in stage 1 acceptor compounds with hydrosulfonic acid by *Iskander* et al. [2.90], somewhat downshifted in frequency from those in the liquid. However, this identification awaits confirmation by resonant enhancement studies or observation by other workers.

In addition to the intercalate-related modes discussed above, there have very recently been observations of modes of intercalate layers vibrating against graphite layers, including both compressional and shear mode vibrations. The compressional modes involve vibrations along the c-axis and have been studied by inelastic neutron scattering techniques for the acceptors $FeCl_3$ [2.51, 91] and $SbCl_5$ [2.92] and for the alkali metal donor compounds with K, Rb, and Cs [2.37, 50, 93–96] and with Li [2.97]. The observed dispersion relations are well explained by a k_z-axis zone folding of the acoustic branch of the graphite dispersion relations with splittings observed at the superlattice Brillouin zone boundary and at the zone center. The highest lying mode, exhibiting weak dispersion, is associated with dominantly longitudinal intercalate layer vibrations. Of particular interest is the case of the acceptors $FeCl_3$ and $SbCl_5$ which are shown by x ray diffraction to consist of a three-layer intercalant structure with Cl–Fe–Cl and Cl–Sb–Cl layers, respectively. No sharp intercalant-related normal modes are observed for motion of the iron or antimony layers relative to the surrounding two chlorine layers, or of the two chlorine layers relative to each other, with the iron or antimony layer at rest. These observations are consistent with a rigid motion of the center of mass of the

Cl-metal-Cl intercalate sandwich relative to the graphite layers and a strong damping of the internal intercalate modes.

The mode associated with the transverse vibration of the intercalate layer relative to the graphite bounding layer has been very recently studied by Raman spectroscopy. This mode occurs at very low frequencies and is therefore difficult to study experimentally in opaque materials such as graphite. Such shear modes have recently been observed between 19 cm^{-1} and 41 cm^{-1} in alkali metal graphite intercalation compounds by *Eklund* et al. [2.95, 96] and *Wada* et al. [2.98] using Raman scattering. The corresponding shear mode is found in pristine graphite at 42 cm^{-1} [2.18].

The number of observed shear modes in the spectra for the alkali metal graphite intercalation compounds is generally consistent with the k_z-axis zone folding associated with the staging periodicity, though not all expected modes have been identified. An analysis of the mode frequencies has been carried out [2.95, 96, 99] to yield the interlayer tangential force constants. The results show that the tangential coupling between the graphite bounding layers and the alkali metal layers is much weaker than that between graphite layers in either pristine or intercalated graphite. This result is consistent with the ease of insertion of the intercalant between two graphite layers, and with the change in compressibility [2.100] and elastic constants [2.101] upon intercalation. Also worthy of note is the comparable magnitudes (within 20%) of the interlayer tangential force constants between pristine and intercalated graphite. Unlike the *c*-axis longitudinal collective mode spectra, the tangential shear mode spectra is temperature dependent, showing the attenuation of certain modes and the appearance of new modes as the temperature is lowered below $T_u \sim 130$ K [2.96].

Inelastic neutron scattering experiments to detect phonons with in-plane wave-vector components have been carried out by *Kamitakahara* et al. [2.102] on stage 1 graphite-Rb. Because of the polycrystalline nature of the sample, these experiments are sensitive to an average density of states with an in-plane polarization and wave-vector. Two broad peaks were observed at 40 and 80 cm^{-1} which are consistent with extremely flat dispersion curves for these modes. *Al-Jishi* and *Dresselhaus* [2.99] have shown that when these results are applied to other stages using a k_z-axis zone folding model, a set of three-dimensional dispersion curves can be deduced, consistent with all available Raman and neutron scattering data. From this model, the low temperature phase, consistent with experimental observations [2.96]. These more recent calculations indicate that for the alkali metal intercalates the phonon dispersion curves (see Fig. 2.11a, b) should have intercalate-derived modes below 100 cm^{-1}.

2.3.5 Second-Order Raman Spectra in Graphite Intercalation Compounds

The second-order Raman spectrum from a *c*-face of pristine graphite shown in Fig. 2.20 has several sharp lines, with the dominant structure peaking at 2735 cm^{-1} and peaks of lower intensity (see Table 2.8) at 3248 cm^{-1},

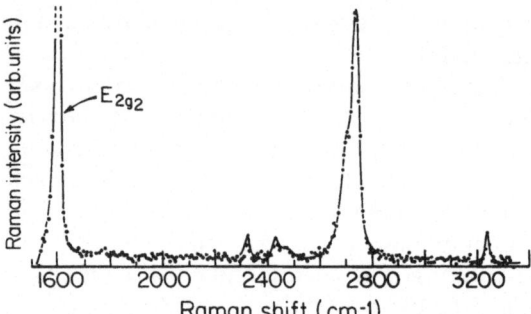

Fig. 2.20. Second-order Raman spectrum for highly oriented pyrolytic graphite (HOPG). The first-order Raman-active E_{2g_2} mode peaks at 1582 cm^{-1} and the second-order peaks are found in the frequency range $1700 < \omega < 3300$ cm^{-1}. See text and Table 2.8 for the second-order peak frequencies. Note that the structure near 2700 cm^{-1} consists of an unresolved doublet structure. (From *Elman* et al. [2.103])

Table 2.8. Second-order Raman frequencies at room temperature for various intercalants[a]

Intercalant	Stage	Peak frequencies	Linewidths
		[cm^{-1}]	
Pristine	∞[b]	3248	10
graphite		2735	20
		2468	30
		2435	10
SbCl$_5$	2[c]	3228	20
		2735	55
		2430	30
	3[c]	3236	20
		2732	50
		2430	30
	4[c]	3246	15
		2730	50
		2430	30
	5[c]	3248	10
		2734	50
		2430	30
Rb	3[d]	3218	20
		2680	50
		2425	30
	4[d]	3228	20
		2690	50
		2425	30

[a] For the low stage alkali metal compounds, no well-resolved second-order peaks can be identified.
[b] From [2.26, 30, 61, 103].
[c] From *Eklund* et al. [2.61].
[d] From the work by *Eklund* et al. [2.95].

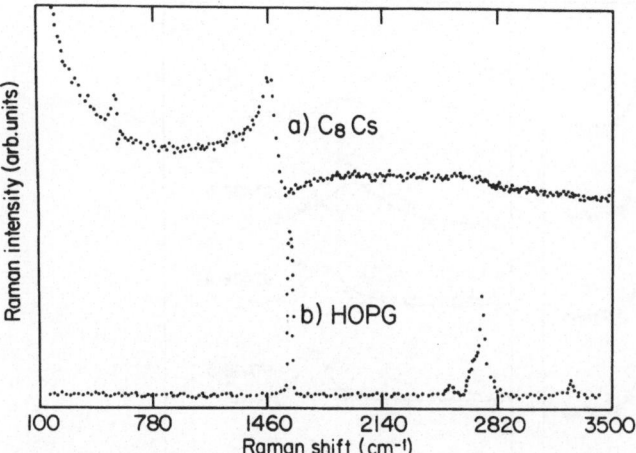

Fig. 21a, b. Room temperature Raman spectra over a long frequency scan $100 < \omega < 3500$ cm^{-1} for (a) first stage C_8Cs and (b) highly oriented pyrolytic graphite (HOPG). Note that the sharp structure in the second-order spectrum of HOPG is not observed in C_8Cs, but rather a broad continuum is found. (From *Eklund* and *Subbaswamy* [2.79])

2435 cm^{-1}, and 2468 cm^{-1} [2.26, 30, 61, 103–105]. The sharpness of the 3248 cm^{-1} line (HWHM of ~ 10 cm^{-1}) relative to the first-order E_{2g_2} mode (HWHM of ~ 14 cm^{-1}) has been emphasized [2.30]. The sharpness of this second-order line reflects the intense sharp structure in the phonon density of states (Fig. 2.6), which in turn is related to the flatness of the modes in the k_z direction. Of particular significance is the observation of second-order frequencies higher than $2\omega(E_{1u})$. This result is consistent with the phonon dispersion relations of the Maeda model [2.31] shown in Fig. 2.5 (and with other proposed models) which show that the maximum phonon frequency does not occur at the Γ point but rather at a point along the Σ axis (Fig. 2.4). Since the second-order spectra are associated with phonons of wave vectors q and $-q$, these measurements are sensitive to the phonon density of states. The main peaks in the phonon density of states occur at 1620 cm^{-1}, 1360 cm^{-1}, 850 cm^{-1}, 830 cm^{-1}, and 570 cm^{-1} (see Fig. 2.6). Thus, $2(1620)$ cm$^{-1} = 3240$ cm^{-1} is identified with two Σ point phonons and $2(1360)$ cm$^{-1} = 2720$ cm^{-1} is identified with two M point phonons, corresponding, respectively, to the peaks at 3248 cm^{-1} and 2730 cm^{-1} in the second-order Raman spectrum. The two peak structure in the density of states at 830 cm^{-1} and 850 cm^{-1} combines with the 1620 cm^{-1} peak to yield 2450 cm^{-1} and 2470 cm^{-1} which are close to the peaks observed in the second-order Raman spectra at 2435 cm^{-1} and 2468 cm^{-1} [2.103].

No sharp second-order spectra have been reported for any of the first and second stage graphite-alkali metal compounds that have been studied to date. Spectra in the region where second-order spectra are expected, $1700 < \omega < 3400$ cm^{-1}, have been taken on the stage 1 compounds with the intercalants Li, K, Rb, and Cs and on stage 2 Li [2.66, 69]. Figure 2.21 shows a comparison of a

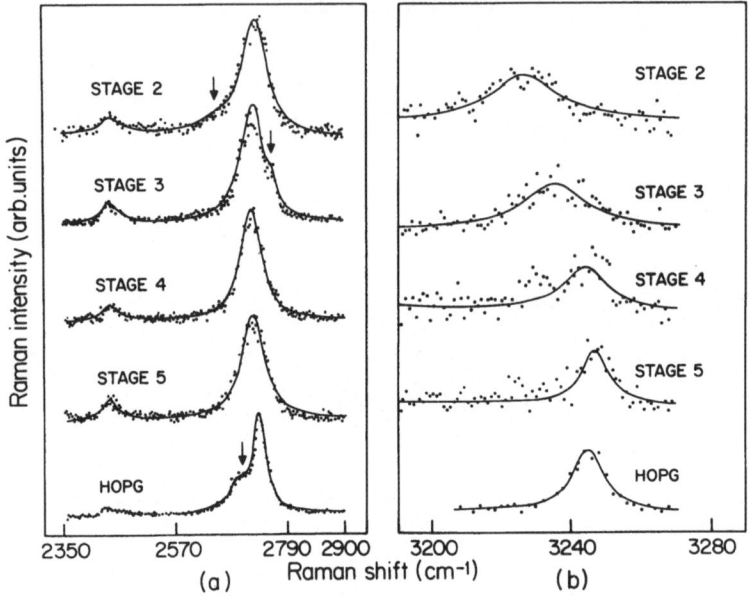

Fig. 2.22. (a) Second-order Raman spectra of stage 2–5 graphite intercalated with SbCl$_5$ and of HOPG in the region 2350–2900 cm^{-1}. The solid lines are the result of a Lorentzian lineshape analysis, on the basis of which the peak positions of unresolved lines are identified and indicated by arrows. (b) Second-order Raman spectra of stage 2–5 graphite intercalated with SbCl$_5$ and of HOPG in the 3200–3280 cm^{-1} region. The solid lines represent a least squares fit of a single Lorentzian line to the data. The intensity of the scattering is $\sim 3\%$ of the first-order peak at 1582 cm^{-1}. (From *Eklund* et al. [2.61])

long frequency scan spectrum for first stage C$_8$Cs with one for HOPG. In this figure the very broad background in the C$_8$Cs trace is evident in the frequency region where the second-order spectrum is observed for HOPG. This very broad, frequency-dependent continuum in the C$_8$Cs trace has been interpreted by *Eklund* and *Subbaswamy* [2.79] as the source of the interference which results in the Breit-Wigner-Fano lineshape for the first-order spectra in the stage 1 alkali metal compounds with K, Rb, and Cs (Sect. 2.3.3). The Breit-Wigner-Fano lineshape that is calculated using the observed frequency-dependent broad continuum in the high frequency regime is shown in Fig. 2.17. Values for the Breit-Wigner-Fano parameters obtained by fitting the experimental data in terms of a frequency-independent continuum are given in Table 2.6. In the case of the spectrum for first stage C$_6$Li, the Breit-Wigner-Fano coupling is very weak, and the second-order spectrum, though still very broad, is also quite intense and temperature dependent. The lines in the second-order spectrum for the stage 2 Li compound are also very broad, and though they differ in detail from the stage 1 spectra, they also exhibit differences in the temperature dependence [2.66].

In contrast to the first and second-order Raman spectra for these alkali metal donor compounds, the stage 1 acceptor compounds *do not* show Breit-

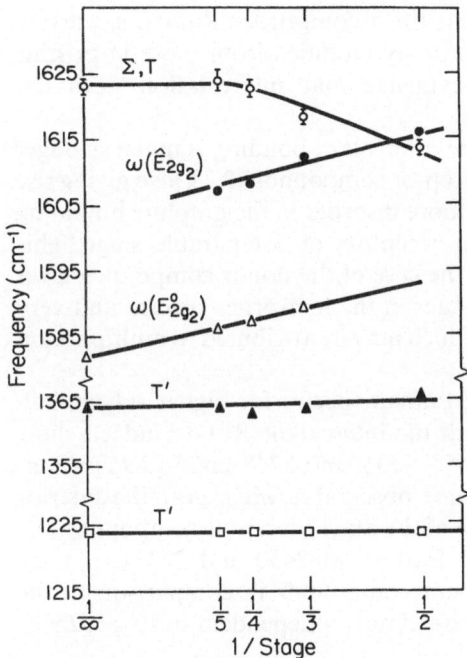

Fig. 2.23. Dependence on reciprocal stage of the high-frequency intralayer phonons studied in the first and second-order Raman spectra of SbCl$_5$ compounds. For each of the second-order features, half of the peak frequency is plotted in the figure. The letters indicate the respective positions in the extended (graphitic) zone (see Fig. 2.4). (From *Eklund* et al. [2.61])

Wigner-Fano lines in the first-order spectra and *do* show discrete lines in the second-order spectra. The most fully documented second-order study for an acceptor compound has been carried out by *Eklund* et al. on the SbCl$_5$ system [2.61]. In this work, the spectral features of the second-order spectra were followed from pristine graphite down to a stage 2 compound, as shown in Fig. 2.22a for the frequency range $2350 < \omega < 2900$ cm^{-1}. The sharp line at 3248 cm^{-1} in pristine graphite was observed to broaden and downshift with increasing intercalate concentration (Fig. 2.22b). A plot of the dependence of *half* the second-order mode frequency vs reciprocal stage ($1/n$) is given in Fig. 2.23, where the corresponding dependence of the first-order graphite bounding ($\hat{E}_{2g_2}$) and interior ($E_{2g_2}^0$) layer mode frequencies are also included. Also given in Fig. 2.23 is the identification by *Eklund* et al. [2.61] of the second-order peaks with specific high symmetry points in the Brillouin zone. The figure shows that some features of the phonon density of states exhibit a significant stage dependence while other features are nearly stage independent. The crossover of half the frequency of the second-order line with the first-order frequency $\omega(\hat{E}_{2g_2})$ in stage 2 SbCl$_5$ compounds indicates (Fig. 2.23) that for such a stage 2 compound, the maximum phonon frequency is associated with the zone center, in contrast to the situation in pristine graphite. We also note that the second-order modes occurring at 2735 cm^{-1} and 2430 cm^{-1} in pristine graphite (Fig. 2.23) are almost independent of intercalate concentration. The lattice dynamical calculations of Sect. 2.2.3 are in general agreement with the trends shown in Fig. 2.23. However, a detailed fit to the second-order spectrum remains to be made. Nevertheless, the observation that the second-order Raman spectra in the acceptor compounds are not substantially modified from those in pristine graphite provides strong additional evidence that intercalation does not significantly modify the graphite layers.

It is well established that the graphite-intercalant bonding is much stronger in the alkali metal compounds than in acceptor compounds [2.7], also giving rise to much stronger carrier scattering and more disorder in the graphite bounding layers of donor compounds relative to acceptors of comparable stage. This strong graphite-intercalant coupling in the case of the donor compounds gives rise to broad Breit-Wigner-Fano lineshapes in the first-order spectra and very broad lines in the second-order spectra which may be attributed to multiphonon processes.

Recent studies on the second order Raman spectra for higher stage alkali metal compounds (stage 3, 4, and 5 with the intercalant Rb) do indeed show well defined second-order features at 2450, 2735, and 3248 cm^{-1} [2.95]. Thus, well-defined second-order features become observable when graphite interior layers are present. In contrast to the case of the SbCl$_5$ acceptor compounds, the mode frequencies for the second-order features at 2450 and 2735 cm^{-1} are strongly stage dependent in the Rb donor compounds in comparison to the SbCl$_5$ acceptor compounds where they are almost independent of stage [2.95].

2.3.6 Raman Scattering from Intercalated Graphite Fibers

Raman scattering has provided an important tool for studying both the lattice and structural properties of intercalated graphite fibers [2.35]. These fibers in their pristine form have important commercial applications because of their high tensile strength (~ 3 GPa), high Young's modulus (~ 400 GPa), light weight and relatively low cost. The high fiber strength (approximately half of the theoretical value for the bulk material) is achieved by aligning the basal planes along the fiber axis, thereby exploiting the strong in-plane carbon-carbon bonding in graphite [2.106–108].

Graphite fibers are prepared by the graphitization of a precursor fiber and subsequent heat treatment. Most commercial fibers are based on polyacrylonitrile (PAN), pitch or rayon precursor materials, and the microstructure of the fiber is strongly dependent on the choice of precursor. For example, the graphitic layers are predominantly aligned circumferentially for the PAN-based fibers and radially for the pitch-based fibers, as shown explicitly in Fig. 2.24 for specific PAN and pitch-based fibers. Though research fibers prepared by graphitization of benzene with hydrogen at 1100 °C and subsequent heat treatment to 2800–3000 °C show a very high degree of circumferential ordering of the graphite layers [2.109], the highest modulus *commercial* fibers presently available show substantial disordered or amorphous regions.

Raman spectroscopy can be used to characterize the extent of the disorder within the optical skin depth (< 1 μm) of pristine graphite fibers (diameter:

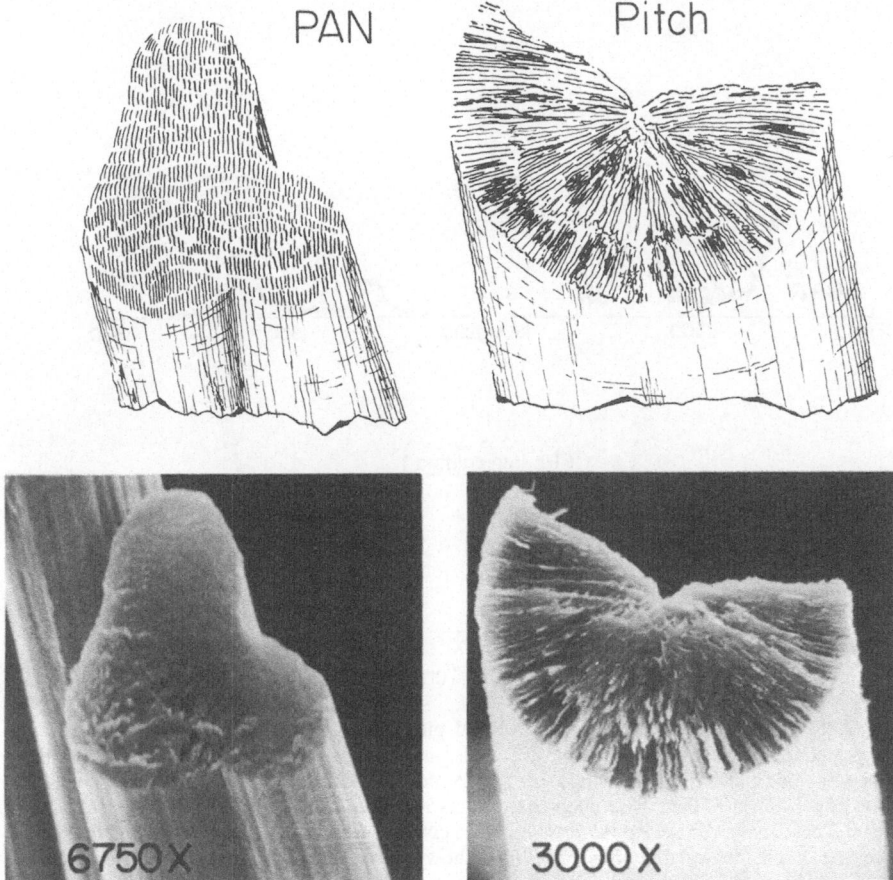

Fig. 2.24. Schematic drawing of graphite fibers prepared from polyacrylonitrile (PAN) and from pitch precursors and actual photographs made by a scanning electron microscope. The PAN-based fiber consists of graphitized layers arranged circumferentially while the pitch-based fiber has the graphite layers arranged radially. (From *Kwizera* et al. [2.113])

~ 10–20 µm). The Raman spectra from pristine commercial graphite fibers (e.g., GY 70) exhibit two peaks, one near $1580 \, \text{cm}^{-1}$ and associated with the Raman-active E_{2g_2} mode and the other, a disorder-induced mode, peaking near the $1355 \, \text{cm}^{-1}$ maximum in the density of states (Fig. 2.25). *Tuinstra* and *Koenig* [2.16] have shown that the intensity ratio $I(1355)/I(1580)$ decreases with increasing microcrystalline ordering and increasing bulk modulus of the fibers, so that Raman scattering can be used to characterize graphite fibers for microcrystalline order. The most highly ordered fibers that have been prepared from the benzene precursor [2.109] show a first-order spectrum essentially indistinguishable from that for HOPG [2.110].

Graphite fibers can be intercalated with both donors and acceptor intercalants, and these intercalated fibers have been discussed for possible

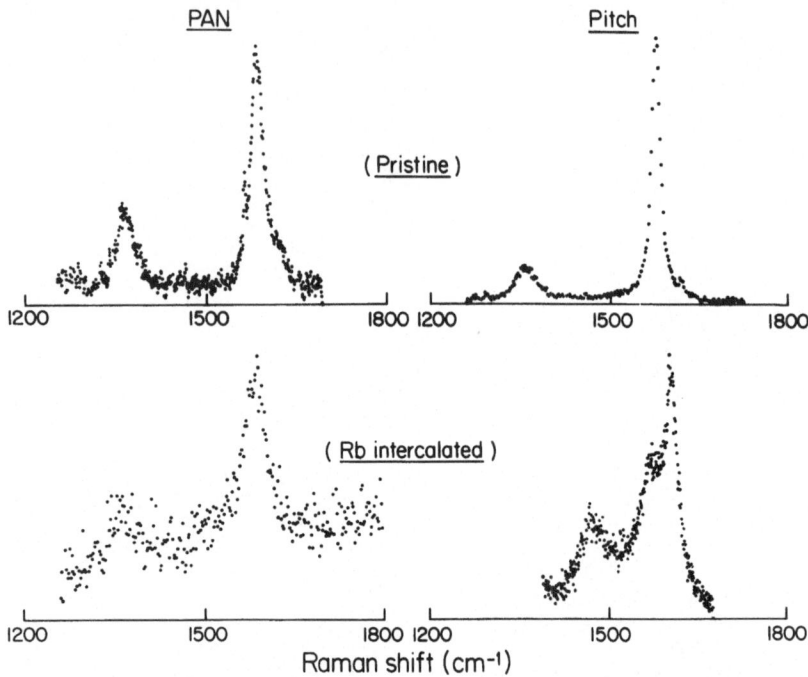

Fig. 2.25. Raman spectra of PAN (GY70) and pitch (UC4104B) based fibers before and after intercalation with Rb. For the pristine fibers, the line at $\sim 1580 \, \text{cm}^{-1}$ is Raman-allowed while the line at $\sim 1355 \, \text{cm}^{-1}$ is disorder-induced. The lower intensity of the $1355 \, \text{cm}^{-1}$ line in the pitch-based fiber indicates that these pitch-based fibers are more highly ordered than the PAN-based fibers. These results for the Rb-intercalated fibers give evidence for mixed staging within the optical skin depth. The intercalated pitch-based fiber shows evidence for a stage 1 region (from the peak near $\sim 1470 \, \text{cm}^{-1}$), and the two-peak structure near $\sim 1600 \, \text{cm}^{-1}$ indicates the presence of higher stages (such as $n = 2, 3$). (From *Kwizera* [2.113])

application as electrical conductors. *Kwizera* et al. [2.35] have shown that Raman scattering, being sensitive to an optical skin depth, can be used to provide complementary information on staging relative to (00*l*) x ray reflections which are sensitive to the bulk fibers. Staging phenomena have been identified for graphite-alkali metal (donor) intercalated fibers [2.111, 112], though Debye-Scherrer x ray patterns indicate that mixed staging is prevalent in the bulk commercial fibers. From the relative intensity of the various components of the structure observed near $1580 \, \text{cm}^{-1}$ in the Raman spectra, *Kwizera* et al. have concluded that for the PAN-based GY 70 and the pitch based UC 4104 B fibers, there is more staging fidelity within the optical skin depth than in the bulk of these alkali metal intercalated fibers, consistent with results obtained from electron microscopy studies [2.35]. The features observed clearly in stage 1 bulk (HOPG) alkali metal intercalated samples (Fig. 2.15) are also seen in the most heavily intercalated alkali metal graphite fibers, though with a poorer signal-to-noise ratio. The more dilute intercalated alkali metal graphite fibers exhibit structure in the Raman spectra which correlate well with those for stage 2 and stage 3 bulk alkali metal intercalation compounds, though the spectra from PAN and pitch-based intercalated fibers (Fig. 2.25) show evidence for mixed stages.

The behavior of the same GY 70 and UC 4104 B fibers intercalated with acceptors such as $FeCl_3$ is markedly different from that for alkali metal intercalated fibers with regard to both x ray and Raman characterization [2.113]. In the case of the $FeCl_3$ intercalated fibers, the Debye-Scherrer patterns exhibit no discernible superlattice lines associated with staging. On the other hand, the Raman spectra yield a broadened spectral feature near $1580 cm^{-1}$ and increased intensity in the disorder-induced $1355 cm^{-1}$ feature (Fig. 2.25). Not only is it more difficult to intercalate these fibers with large acceptor molecules [2.114, 115] than with donor metal ions, but the Raman spectra show that the resulting intercalated fibers are not well staged and exhibit increased lattice disorder arising from the intercalation process itself. It has recently been shown using Raman spectroscopy that the benzene-derived fibers [2.109] heat treated to 2900 °C or higher can be intercalated with acceptors to yield a single-staged intercalated fiber [2.110].

Raman spectroscopy can thus be used to characterize pristine graphite fibers with regard to lattice disorder within the optical skin depth of the fibers. With regard to the intercalated fibers, Raman spectroscopy provides information complementary to x ray diffraction, contrasting the staging fidelity within an optical skin depth to that of the bulk fiber for alkali metal donor intercalants into PAN and pitch-based fibers. For the acceptor-intercalated fibers that do not stage, modification of the Raman spectra by the intercalation process can be used to monitor the intercalation of such fibers.

The new benzene-derived research fibers [2.109, 110] have recently been characterized by Raman spectroscopy to determine the structural order of the pristine fibers as a function of heat treatment temperature, to examine the effect of donor and acceptor intercalation on this structural order, and to characterize the intercalated fibers for staging.

2.3.7 Raman Scattering of Molecules on Graphite Surfaces

Although graphite represents an important substrate for two-dimensional surface phases, surface-enhanced Raman scattering has not yet been extensively exploited on graphite surfaces. Surface-related Raman spectra have been observed from molecular Br_2 on graphite at room temperature and exhibiting resonant enhancement effects as is also observed in the gas phase. At low temperatures ($\sim 100 K$), surface-specific Raman modes have been reported by *Heflinger* [2.84] with frequencies intermediate between $242 cm^{-1}$ for the intercalate Br_2 mode and $300 cm^{-1}$ for solid Br_2. Some workers have identified certain low frequency Raman structure with surface phases of $FeCl_3$ [2.59], ICl

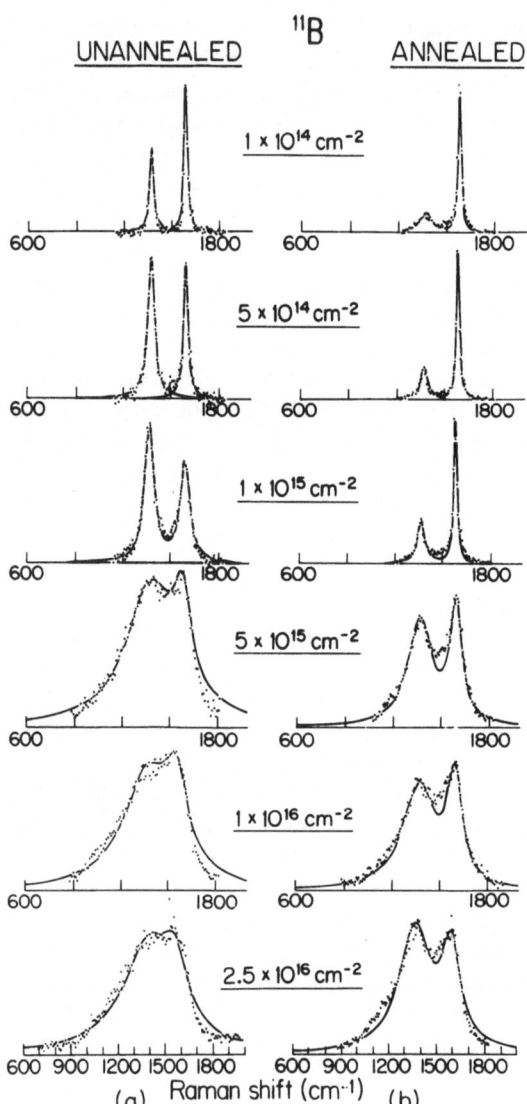

UNANNEALED ^{11}B ANNEALED

Fig. 2.26a, b. Raman spectra for various fluences $(1 \times 10^{14}$ to 2.5×10^{16} ions/cm²) of ^{11}B ions implanted into highly oriented pyrolytic graphite (HOPG) samples. The spectra on the left (**a**) are for the unannealed samples, and on the right (**b**) are for the same samples annealed at 950 °C for 0.5 h. The abscissa is linear in wave number, the points are experimental and the curves are a Lorentzian fit to the data points. (From *Elman* et al. [2.103])

[2.116], and SbCl$_5$ [2.70], though for the case of FeCl$_3$, other workers have associated this structure with an intercalate mode [2.89].

The ability to make ordered microstructures from intercalation compounds suggests that intercalated graphite would be an ideal system for testing the various mechanisms for surface-enhanced Raman scattering. For example, *a*-faces of graphite-FeCl$_3$ can simulate ruled gratings with periodicities up to ~40 Å [2.59] and might provide a useful system for studying surface-enhanced Raman scattering [2.117]. The *a*-faces of pristine graphite exhibit surface roughness which might provide preferred molecular adsorbate sites where high surface electric fields might occur, favoring the surface-enhanced scattering process. Also, intercalation allows preparation of c-faces with a surface conductivity that may be varied by the choice of intercalate species and concentration, and it may be possible to prepare an intercalate superlattice structure that behaves like a two-dimensional grating which couples to a surface plasmon.

2.3.8 Raman Scattering from Ion-Implanted Graphite

Raman scattering has proven to be an important technique for the characterization of ion-implanted materials [2.118]. The technique can be used to study lattice damage and annealing characteristics, to determine conductivity changes in semiconductors where the plasma frequency can be varied by ion implantation to correspond to lattice mode frequencies [2.119], and to study local modes associated with the implanted species so that site symmetries and bonding information can be deduced [2.120].

The application of Raman scattering techniques to ion-implanted and neutron irradiated graphite has been made by several groups [2.103–105, 118, 120, 121]. Raman spectra from the ion-implanted graphite samples offer insight into the properties of intercalated graphite. Some significant results of these ion-implantation studies are shown in Fig. 2.26, where the Raman spectra are shown for HOPG implanted with ^{11}B ions at 100 keV for various fluences [2.103]. In this figure, the magnitude of the ion-induced lattice disorder is determined by the intensity of the 1355 cm^{-1} mode, which corresponds to a peak in the graphite density of phonon states. Annealing the samples at 950 °C for 0.5 h is shown to reduce the lattice damage significantly (Fig. 2.26).

As the ^{11}B ion fluence increases above above 10^{15} ions/cm^2, the implanted graphite samples lose their sharp *second*-order Raman spectra and simultaneously develop a Breit-Wigner-Fano lineshape in their first-order spectra [2.104]. The striking similarity of the first and second-order Raman spectra in heavily ion-implanted graphite to the spectra found in stage 1 alkali metal donor compounds strongly supports a disorder-induced *phonon* mechanism for the origin of the broad continuum which gives rise to the Breit-Wigner-Fano lineshape. In the case of the Raman spectra from ion-implanted graphite, it is unlikely that an electronic process could give rise to a broad continuum in the energy range $1700 < \hbar\omega < 3400$ cm^{-1}.

2.4 Summary

This chapter has reviewed the recent application of Raman scattering to systems of graphitic materials with particular emphasis given to the intercalation compounds which display a c-axis superlattice periodicity due to the staging phenomenon. Although the number of such graphitic materials is vast, Raman scattering represents one experimental technique which is universally applicable to all of these materials. The great stability of the graphitic structure (hexagonal rings of planar bonded carbon atoms) makes it possible to model this large class of materials in terms of the basic graphite structure. The observed Raman spectra confirm the validity of this basic physical approach, often also providing critical input on the magnitude of the perturbation to the basic graphitic structure.

The current state of our understanding of the Raman spectra for the intercalation compounds has established certain systematic trends. For example, the in-plane Raman modes for the donor compounds soften with increasing intercalate concentration while the acceptors stiffen. However, the effect of some of the known, structural phase transitions [2.7] on the Raman spectra has only recently been studied in detail [2.85, 96]. Much work remains to be done in the area of scattering from surface modes and from intercalate molecules. It would appear that an understanding and exploitation of the resonant enhancement process is necessary for an experimental break-through in the study of molecular modes in intercalated graphite. The results of such studies would yield valuable information necessary for the modeling of two-dimensional phases on graphite surfaces or between the graphite layers.

Acknowledgements. The authors wish to thank the members of their research group, both present and past for help with all aspects of this review article, particularly Dr. S. Y. Leung, now at the Bell Telephone Laboratories, and Prof. P. C. Eklund of the University of Kentucky. We also gratefully acknowledge the financial support from ONR (grant # N00014-77-C-0053), which made the writing of this review article possible.

References

2.1   T.J.Wieting, J.L.Verble: In *Electrons and Phonons in Layered Crystal Structures*, Vol. 3, ed. by T.L.Wieting, M.Schlüter (Reidel, Dordrecht, Holland 1979) pp. 321–407
2.2   D.W.Feldman, J.H.Parker, Jr., W.J.Choyke, L.Patrick: Phys. Rev. **173**, 787–793 (1968)
2.3   S.Nakashima: Solid State Commun. **16**, 1059–1062 (1975)
2.4   R.Zallen, M.L.Slade: Solid State Commun. **17**, 1561–1566 (1975)
2.5   C.Colvard, R.Merlin, M.V.Klein, A.C.Gossard: Phys. Rev. Lett. **45**, 298–301 (1980)
2.6   M.S.Dresselhaus, G.Dresselhaus: In *Physics and Chemistry of Materials with Layered Structures*, Vol. 6, ed. by F.Lévy (Reidel, Dordrecht, Holland 1979) pp. 423–480
2.7   M.S.Dresselhaus, G.Dresselhaus: Adv. Phys. **30**, 139–326 (1981)
2.8   M.Zanini, J.E.Fischer: Mat. Sci. Eng. **31**, 169–172 (1977)
2.9   M.Zanini, D.Grubisic, J.E.Fischer: Phys. Stat. Sol. **90**, 151–156 (1978)
2.10   A.Hérold: In *Physics and Chemistry of Materials with Layered Structures*, Vol. 6, ed. by F.Lévy (Reidel, Dordrecht, Holland 1979) pp. 323–422
2.11   J.C.Tsang, M.W.Shafer: Solid State Commun. **25**, 999–1002 (1978)
2.12   J.E.Fischer: In *Physics and Chemistry of Materials with Layered Structures*, Vol. 6, ed. by F.Lévy (Reidel, Dordrecht, Holland 1979) pp. 481–532
2.13   Y.Koike, S.Tanuma, H.Suematsu, I.Higuchi: J. Phys. Chem. Solids **41**, 1111–1118 (1980)
2.14   A.W.Moore: In *Chemistry and Physics of Carbon*, Vol. 11, ed. by P.L.Walker, P.A.Thrower (Dekker, New York 1973) pp. 69–187
2.15   A.N.Berker, N.Kambe, G.Dresselhaus, M.S.Dresselhaus: Phys. Rev. Lett. **45**, 1452–1456 (1980)
2.16   F.Tuinstra, J.F.Koenig: J. Chem. Phys. **53**, 1126–1130 (1970)
2.17   L.J.Brillson, E.Burstein, A.A.Maradudin, T.Stark: In *Proc. Int. Conf. on Semimetals and Narrow Gap Semiconductors*, Dallas, TX (1970), ed. by D.L.Carter, R.T.Bate (Pergamon Press, New York 1971) pp. 187–193

2.18 R.J.Nemanich, G.Lucovsky, S.A.Solin: In *Proc. Int. Conf. on Lattice Dynamics*, ed. by M.Balkanski (Flammarion Press, Paris 1975) pp. 619–621
2.19 R.J.Nemanich, G.Lucovsky, S.A.Solin: Solid State Commun. **23**, 117–120 (1977)
2.20 R.Nicklow, N.Wakabayashi, H.G.Smith: Phys. Rev. B**5**, 4951–4962 (1972)
2.21 N.Wakabayashi, R.M.Nicklow: In *Electrons and Phonons in Layered Structures*, Vol. 3, ed. by T.L.Wieting, M.Schlüter (Reidel, Dordrecht, Holland 1979) pp. 409–464
2.22 A.Yoshimori, Y.Kitano: J. Phys. Soc. Jpn. **11**, 352–361 (1956)
2.23 W.N.Reynolds: *Physical Properties of Graphite* (Elsevier, Amsterdam 1968) p. 34
2.24 W.De Sorbo, W.W.Tyler: J. Chem. Phys. **21**, 1660–1663 (1953)
2.25 J.A.Young, J.U.Koppel: J. Chem. Phys. **42**, 357–364 (1965)
2.26 R.J.Nemanich, J.Lucovsky, S.A.Solin: Mat. Sci. Eng. **31**, 157–160 (1977)
2.27 A.P.P.Nicholson, D.J.Bacon: J. Phys. C **10**, 2295–2306 (1977)
2.28 O.L.Blakslee, D.G.Proctor, E.J.Seldin, G.B.Spence, T.Weng: J. Appl. Phys. **41**, 3373–3382 (1970)
2.29 E.J.Seldin, C.W.Nezbeda: J. Appl. Phys. **41**, 3389–3400 (1970)
2.30 R.J.Nemanich, S.A.Solin: Phys. Rev. B**20**, 392–401 (1979)
2.31 M.Maeda, Y.Kuramoto, C.Horie: J. Phys. Soc. Jpn. **47**, 337–338 (1979)
2.32 S.Y.Leung, M.S.Dresselhaus, G.Dresselhaus: Physica **105**B, 375–380 (1981)
2.33 M.I.Nathan, J.E.Smith, Jr., K.N.Tu: J. Appl. Phys. **45**, 2370 (1974)
2.34 S.A.Solin, R.J.Kobliska: *Proc. 5th Int. Conf. on Amorphous and Liquid Semiconductors* (Taylor and Francis, London 1974) pp. 1251–1258
2.35 P.Kwizera, A.Erbil, M.S.Dresselhaus: Carbon **19**, 144–146 (1981)
2.36 C.Horie, M.Maeda, Y.Kuramoto: Physica **99**B, 430–434 (1980)
2.37 W.D.Ellenson, D.Semmingsen, D.Guérard, D.G.Onn, J.E.Fischer: Mat. Sci. Eng. **31**, 137–140 (1977)
2.38 S.Y.Leung, G.Dresselhaus, M.S.Dresselhaus: Solid State Commun. **38**, 175–178 (1981); Phys. Rev. B**24**, 6083–6103 (1981)
2.39 G.Dresselhaus, S.Y.Leung: Physica **105**B, 495–500 (1981)
2.40 S.Y.Leung: Sc. D. Thesis, MIT (unpublished) (1980)
2.41 S.Y.Leung, G.Dresselhaus, M.S.Dresselhaus: Synthetic Metals **7**, 89–98 (1980)
2.42 S.Flandrois, J.M.Masson, J.C.Rouillon, J.Gaultier, G.Hauw: Synthetic Metals **3**, 1–13 (1981)
2.43 T.Krapchev: M.S.Thesis, MIT (unpublished) (1981)
    T.Krapcher, R.Ogilvie, M.M.Dresselhaus: Carbon (in press)
2.44 D.E.Nixon, G.S.Parry: J. Phys. C**2**, 1732–1741 (1969)
2.45 D.Guérard, C.Zeller, A.Hérold: C. R. Acad. Sci. Paris C**283**, 437–440 (1976)
2.46 L.Pietronero, S.Strässler: J. Phys. Soc. Suppl. A **49**, 895–898 (1980)
    [*Proc. 15th Int. Conf. Physics of Semiconductors. Kyoto, Japan, 1980*]; Phys. Rev. Lett. **41**, 593–596 (1981); In *Physics of Intercalation Compounds*, ed, by L.Pietronero, E.Tosatti, Springer Series in Solid-State Sciences, Vol. 38 (Springer, Berlin, Heidelberg, New York 1981) pp. 23–32
2.47 U.Mizutani, T.Kondow, T.B.Massalski: Phys. Rev. B**17**, 3165–3173 (1978)
2.48 T.Kondow, U.Mizutani, T.B.Massalski: Mat. Sci. Eng. **31**, 267–270 (1977)
2.49 M.G.Alexander, D.P.Goshorn, D.G.Onn: Phys. Rev. B**22**, 4535–4542 (1980)
2.50 A.Magerl, H.Zabel: Phys. Rev. Lett. **46**, 444–446 (1981)
2.51 J.D.Axe, C.F.Majkrzak, L.Passell, S.K.Satija, G.Dresselhaus, H.Mazurek: Extended Abstracts of 15th Biennial Conf. on Carbon, Philadelphia, 1981, pp. 52–53
2.52 G.Dolling, B.N.Brockhouse: Phys. Rev. **128**, 1120–1123 (1962)
2.53 J.Blinowski, H.H.Nguyen, C.Rigaux, J.P.Vieren, R.Le Toullec, G.Furdin, A.Hérold, J.Mélin: J. Physique **41**, 47–58 (1980)
2.54 S.Safran: Synthetic Metals **2**, 1–15 (1980)
2.55 J.J.Song, D.D.L.Chung, P.C.Eklund, M.S.Dresselhaus: Solid State Commun. **20**, 1111–1115 (1976)
2.56 M.S.Dresselhaus, G.Dresselhaus, P.C.Eklund, D.D.L.Chung: Mat. Sci. Eng. **31**, 141–152 (1977)
2.57 R.J.Nemanich, S.A.Solin, D.Guérard: Phys. Rev. B**16**, 2965–2972 (1977)

2.58   S. A. Solin: Mat. Sci. Eng. **31**, 153–156 (1977)
2.59   C. Underhill, S. Y. Leung, G. Dresselhaus, M. S. Dresselhaus: Solid State Commun. **29**, 769–774 (1979)
2.60   G. M. Gualberto, C. Underhill, S. Y. Leung, G. Dresselhaus: Phys. Rev. B**21**, 862–868 (1980)
2.61   P. C. Eklund, D. S. Smith, V. R. K. Murthy, S. Y. Leung: Synthetic Metals **2**, 99–108 (1980)
2.62   P. C. Eklund, G. Dresselhaus, M. S. Dresselhaus, J. E. Fischer: Phys. Rev. B**16**, 3330–3333 (1977)
2.63   S. A. Solin: Physica **99**B, 443–452 (1980)
2.64   L. Pietronero, S. Strässler, H. R. Zeller, M. J. Rice: Phys. Rev. Lett. **41**, 763–767 (1978)
2.65   L. Pietronero, S. Strässler, H. R. Zeller: Solid State Commun. **30**, 399–401 (1979)
2.66   P. C. Eklund, G. Dresselhaus, M. S. Dresselhaus, J. E. Fischer: Phys. Rev. B**21**, 4705–4709 (1980)
2.67   N. Wada, S. A. Solin: Physica **105**B, 353–356 (1980)
2.68   M. R. Corson, S. Millman, G. R. Hoy, H. Mazurek: Solid State Commun. (in press); M. Elaky, C. Nicolini, G. Dresselhaus, G. O. Zimmerman, Solid State Commun. **41**, 289–292 (1982)
2.69   Yu S. Karimov: Zh. Eksp. Teor. Fiz. **66**, 1121–1128 (1974); Soviet Physics JETP **39**, 547–550 (1974)
2.70   P. C. Eklund, E. R. Falardeau, J. E. Fischer: Solid State Commun. **32**, 631–634 (1979)
2.71   D. G. Onn, G. M. T. Foley, J. E. Fischer: Phys. Rev. B**19**, 6474–6483 (1979)
2.72   N. Caswell, S. A. Solin: Phys. Rev. B**20**, 2551–2554 (1979)
2.73   M. S. Dresselhaus, N. Kambe, A. N. Berker, G. Dresselhaus: Synthetic Metals **2**, 121–131 (1980)
2.74   N. Wada: Phys. Rev. B**24**, 1065–1078 (1981)
2.75   M. V. Klein: In *Light Scattering in Solids*, ed. by M. Cardona, Topics Appl. Phys., Vol. 8 (Springer, Berlin, Heidelberg, New York 1975) Chap. 4
        U. Fano: Phys. Rev. **124**, 1866 (1961)
2.76   J. R. Scott: Rev. Mod. Phys. **46**, 83–128 (1974)
2.77   M. Zanini, L.-Y. Ching, J. E. Fischer: Phys. Rev. B**18**, 2020–2022 (1978)
2.78   D. M. Hwang, S. A. Solin, D. Guerard: In *Physics Intercalation Compounds*, ed. by L. Pietronero, E. Tosatti, Springer Series Solid-State Science, Vol. 38 (Springer, Berlin, Heidelberg, New York, 1981) pp. 187–192
2.79   P. C. Eklund, K. R. Subbaswamy: Phys. Rev. B**20**, 5157–5161 (1979)
2.80   H. Miyazaki, T. Hatano, G. Kusunoki, T. Watanabe, C. Horie: Physica **105**B, 381–385 (1980)
2.81   P. C. Eklund, N. Kambe, G. Dresselhaus, M. S. Dresselhaus: Phys. Rev. B**18**, 7069–7079 (1978)
2.82   J. E. Cahill, G. E. Leroi: J. Chem. Phys. **51**, 4514–4519 (1966)
2.83   G. Herzberg: *Spectra of Diatomic Molecules* (Van Nostrand, Princeton, NJ 1945)
2.84   B. L. Heflinger: Ph. D. Thesis, MIT (unpublished) (1979)
2.85   A. Erbil, G. Dresselhaus, M. S. Dresselhaus: Phys. Rev. B**25**, in press (1982)
2.86   D. D. L. Chung: Ph. D. Thesis MIT (unpublished) (1977)
2.87   W. T. Eeles, J. A. Turnbull: Proc. Roy. Soc. London A**283**, 179–193 (1965)
2.88   S. M. Heald, E. A. Stern: Phys. Rev. B**17**, 4069–4081 (1978)
2.89   N. Caswell, S. A. Solin: Solid State Commun. **27**, 961–967 (1978)
2.90   B. Iskander, P. Vast, A. Lorriaux-Rubbens, M. L. Dele-Dubois, Ph. Touzain: Mat. Sci. Eng. **43**, 59–63 (1980)
2.91   G. Dresselhaus, R. Al-Jishi, J. D. Axe, C. F. Majkrzak, L. Passell, S. K. Satija: Solid State Commun. **40**, 229–232 (1981)
2.92   P. C. Eklund, V. Yeh, R. Al-Jishi, G. Dresselhaus, H. G. Smith, R. Nicklow: To be published
2.93   H. Zabel, A. Magerl: In *Physics of Intercalation Compounds*, ed. by L. Pietronero, E. Tosatti, Springer Series in Solid-State Sciences, Vol. 38 (Springer, Berlin, Heidelberg, New York 1981) pp. 180–186
2.94   H. Zabel, A. Magerl: Proc. of the Int. Conf. on Phonon Physics, Bloomington, IN, USA, 1981
2.95   P. C. Eklund, J. Giergiel, P. Boolchand: In *Physics of Intercalation Compounds*, ed. by L. Pietronero, E. Tosatti, Springer Series in Solid-State Sciences, Vol. 38 (Springer, Berlin, Heidelberg, New York 1981) pp. 168–179
2.96   P. C. Eklund, J. Giergiel, R. Al-Jishi, G. Dresselhaus: To be published

2.97 J. Rossat-Mignod, D. Fruchart, M. J. Moran, J. W. Milliken, J. E. Fischer: Synth. Met. **2**, 143–148 (1980)
2.98 N. Wada, M. V. Klein, H. Zabel: In *Physics of Intercalation Compounds*, ed. by L. Pietronero, E. Tosatti, Springer Series in Solid-State Sciences, Vol. 38 (Springer, Berlin, Heidelberg, New York 1981) pp. 199–204
2.99 R. Al-Jishi, G. Dresselhaus: Phys. Rev. (to be published)
2.100 N. Wada, R. Clarke, S. A. Solin: Solid State Commun. **35**, 675–679 (1980)
2.101 D. M. Hwang, B. F. O'Donnell, A. Y. Wu: In *Physics of Intercalation Compounds*, ed. by L. Pietronero, E. Tosatti, Springer Series in Solid-State Sciences, Vol. 38 (Springer, Berlin, Heidelberg, New York 1981) pp. 193–198
2.102 W. A. Kamitakahara, N. Wada, S. A. Solin: Proc. of the Int. Conf. on Phonon Physics, Bloomington, IN, USA, 1981
2.103 B. S. Elman, M. S. Dresselhaus, G. Dresselhaus, E. W. Maby, H. Mazurek: Phys. Rev. B **24**, 1027–1034 (1981)
2.104 B. S. Elman, M. Shayegan, M. S. Dresselhaus, H. Mazurek, G. Dresselhaus: Phys. Rev. B **25**, 4142 (1982)
2.105 B. S. Elman, H. Mazurek, M. S. Dresselhaus, G. Dresselhaus: In *Metastable Materials Formation by Ion Implantation*, Materials Research Society Symposia Proceedings, Vol. 7, ed. by S. T. Picraux, W. J. Choyke (North-Holland, New York 1982) pp. 425–431
2.106 D. Robson, F. Y. I. Assabghy, D. J. E. Ingram: J. Phys. D, Appl. Phys. **5**, 169–179 (1972)
2.107 D. Robson, F. Y. I. Assabghy, E. G. Cooper, D. J. E. Ingram: J. Phys. D, Appl. Phys. **6**, 1822–1834 (1973)
2.108 A. A. Bright, L. S. Singer: Carbon **17**, 59–69 (1979)
2.109 M. Endo, T. Koyama, Y. Hishiyama: Jpn. J. of Appl. Phys. **15**, 2073–2076 (1976)
2.110 T. C. Chien, M. S. Dresselhaus, M. Endo: To be published
2.111 C. Herinckx, R. Perret, W. Ruland: Carbon **10**, 711–722 (1972)
2.112 C. Herinckx: Carbon **11**, 199–206 (1973)
2.113 P. Kwizera, M. S. Dresselhaus, G. Dresselhaus: Carbon (submitted)
2.114 J. G. Hooley: Carbon **13**, 469–471 (1975)
2.115 J. G. Hooley, V. R. Deitz: Carbon **16**, 251–257 (1978)
2.116 A. Erbil, C. Lowe, G. Dresselhaus, M. S. Dresselhaus: Bull. Am. Phys. Soc. **25**, 336 (1980)
2.117 J. C. Tsang, J. R. Kirtley, J. A. Bradley: Phys. Rev. Lett. **43**, 772–775 (1979)
2.118 B. L. Crowder, J. E. Smith, Jr., M. H. Brodsky, M. I. Nathan: Proc. 2nd Int. Conf. on Ion Implantation in Semiconductors (Garmisch-Partenkirchen, Germany 1970) pp. 255–261
2.119 A. Mooradian, G. B. Wright: Phys. Rev. Lett. **16**, 999–1001 (1966)
2.120 R. B. Wright, R. Varma, D. M. Gruen: J. Nucl. Mater. **63**, 415–421 (1976)
2.121 H. Maeta, Y. Sato: Solid State Commun. **23**, 23–25 (1977)

3. Light Scattering from Electronic and Magnetic Excitations in Transition-Metal Halides

D. J. Lockwood

With 15 Figures

The field of light scattering from iron-group transition metal ions in solids has a brief and recent history subsequent to the application of laser sources to Raman spectroscopy. The earliest studies were of magnons in ordered anti-ferromagnets with the first measurements on FeF_2 and MnF_2 reported in 1966 [3.1]. Studies of higher electronic excitations followed on from pioneering work on rare-earth ions [3.2] and the first observations of electronic Raman scattering were made in 1969 on Co^{2+} in MX_2 compounds, where M is a divalent metal ion and X is a halogen ion [3.3]. Since then, numerous papers have been published on theoretical and experimental studies of one and two-magnon scattering in pure antiferromagnets and, more recently, for impure systems. Progress in this field has been reviewed at frequent intervals (e.g., [3.4–8]). Electronic Raman scattering has proved to be more difficult, experimentally and theoretically, but, nevertheless, has now been widely applied to systems containing iron, cobalt and nickel. Progress in this latter subject has never been formally reviewed to the author's knowledge.

In this chapter, emphasis is placed on those research areas that are either new or are in need of review, but other material is included for the sake of completeness and with regard to understanding the recent advances. To accomplish these goals, and yet to remain within reasonable space limitations, only three types of halide systems will be discussed. These are the cubic fluoride perovskites, the tetragonal fluoride rutiles and the trigonal layered compounds. All possess a simple magnetic structure that is theoretically most attractive and for this reason, sufficient work has been performed on these halide systems to demonstrate the physics of interest. Section 3.1 reviews the results on these systems and provides a base for subsequent sections. In Sect. 3.2, recent results from mixed antiferromagnets are reviewed, while Sect. 3.3 summarises the present knowledge of electronic Raman scattering from transition metal ions. Electron-phonon coupling is inherent in all transition metal compounds and the direct evidence for this as determined from Raman spectroscopy is reviewed in Sect. 3.4. Finally, conclusions are drawn in Sect. 3.5 along with a survey of future prospects.

3.1 One- and Two-Magnon Scattering in Pure Antiferromagnets

The halide salts of the iron-group series of transition metal ions are insulators and most become antiferromagnetic at low temperatures. The unpaired $3d$ electrons of the metal ion largely determine the magnetic properties and this leads to an effective spin model in which the "spin" is localised to the magnetic ion lattice site. At sufficiently low temperatures, the interactions between the 3-D array of spins result in an ordered magnetic structure whose unit cell is at least as large as the chemical unit cell. For the antiferromagnets of interest here, the principal spin interaction is the Heisenberg exchange interaction

$$\mathscr{H}_{ex} = \tfrac{1}{2} \sum_{i \neq j} J_{ij} \mathbf{S}_i \cdot \mathbf{S}_j \tag{3.1}$$

between spins S on sites i and j. J_{ij} are the exchange integrals, and the dominant integral is positive for antiferromagnetic ordering. This results in a near antiparallel arrangement of neighboring spins and these can be subdivided into two sublattices according to whether their spin is "up" or "down". The Heisenberg antiferromagnetic ground state is not known exactly and is usually approximated by the Néel ground state where the spins have z component values equal to the maximum value S on one sublattice and the negative of this on the other sublattice. Spin waves, or magnons, are the low-lying excitations associated with z-component deviations in the spins from this static dual-sublattice ordering. The elementary excitations normally involve both sublattices and in the simplest case give rise to a doubly-degenerate spin-wave branch. The degeneracy can be removed by the presence of anisotropy or an external field. At the Brillouin zone boundary, the spin wave propagates entirely on one sublattice or the other with an energy dominated by the Ising terms in the Hamiltonian.

With this brief introduction, we now examine the spin waves in various systems and their observation by light scattering. Readers requiring more detailed information on spin waves are referred to the excellent review article by *Balucani* and *Tognetti* [3.6].

3.1.1 Cubic AMF₃ Compounds

Transition metal fluorides with the perovskite structure shown in Fig. 3.1 are undoubtedly amongst the simplest of antiferromagnets comprising two oppositely-directed sublattices with the magnetic ions located on a face-centred-cubic lattice. Their spin-wave spectra may be characterized by one, or at most two, antiferromagnetic exchange constants together with a uniform field describing the anisotropy. The effective spin Hamiltonian describing the doubly-degenerate spin wave branch may be written as

$$\mathscr{H} = \tfrac{1}{2} \sum_{ij} J_{ij} \mathbf{S}_i \cdot \mathbf{S}_j - H_A \sum_i S_i^z + H_A \sum_j S_j^z, \tag{3.2}$$

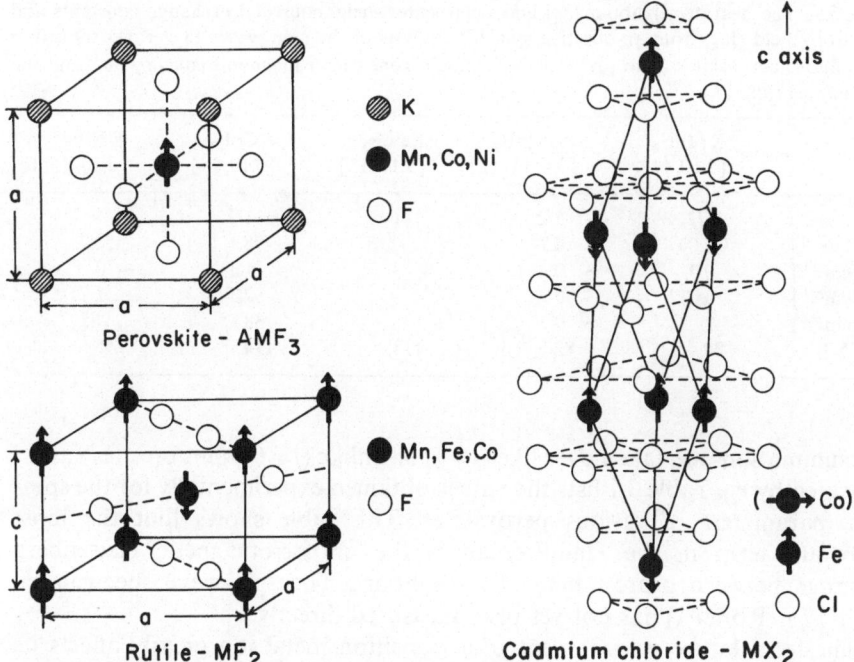

Fig. 3.1. The perovskite, rutile and $CdCl_2$-type crystal structures. The major spin directions are indicated for each structure. Some of the transition metal compounds exhibit small deviations from these ideal structures and spin directions. In other compounds (e.g., $KFeF_3$, NiF_2), the spins have different directions and, in the case of the layered-structure compounds, can have very complex magnetic unit cells

where H_A is an effective anisotropy field in the direction of spin ordering. Single ion anisotropy and weak magnetic dipole-dipole interactions are thus included phenomenologically. With this Hamiltonian, the magnon frequency at wave vector q is [3.9, 10]

$$\omega(q) = [A(q)^2 - B(q)^2]^{1/2}, \tag{3.3}$$

where

$$A(q) = H_A + S \sum_{j=\alpha} J_{ij}[C(q) - 1] + S \sum_{j=\beta} J_{ij}, \tag{3.4}$$

$$B(q) = S \sum_{j=\beta} J_{ij} C(q), \tag{3.5}$$

and

$$C(q) = \exp(iq \cdot r_{ij}). \tag{3.6}$$

Table 3.1. The nearest-neighbour (J_1) and next-nearest-neighbour (J_2) exchange constants and anisotropy field (H_A) from an effective spin (S) analysis of the spin waves in various transition metal fluoro-perovskites. Also given is the Brillouin zone center magnon frequency (ω_0) and the Néel temperature (T_N)

	KMnF$_3$ [3.10, 11]	RbMnF$_3$ [3.6, 11, 12]	KFeF$_3$ [3.10, 13]	KCoF$_3$ [3.10, 14]	KNiF$_3$ [3.10, 15, 16]
S	5/2	5/2	1	1/2	1
J_1 [cm^{-1}]	5.04	4.7	–	78.1	70.9
J_2 [cm^{-1}]	0.0	~ 0	–	1.3	~ 0
H_A [cm^{-1}]	0.032	~ 0	–	2.3	1.7
ω_0 [cm^{-1}]	~ 1	~ 0	–	38	4.5
T_N [K]	88	82	113	114	245

The summations are taken over like ($j = \alpha$) or unlike ($j = \beta$) spins on sites i and j separated by r_{ij}. Table 3.1 lists the values obtained experimentally for the spin-wave parameters of various perovskites. This table shows that the most important term in the Hamiltonian is the antiferromagnetic Heisenberg exchange between nearest magnetic neighbors. The spin-wave spectrum of KFeF$_3$ (or RbFeF$_3$) has not yet been measured directly.

The local lattice environment of a transition metal ion greatly affects its magnetic and electronic properties since the crystal field may or may not quench the orbital angular momentum of the ion. For the case of Mn^{2+}, where the free-ion ground state has zero orbital angular momentum and spin angular momentum of 5/2, neither the spin-orbit interaction nor crystal field effects are important in the spin-wave Hamiltonian. Hence, for Mn^{2+}, the effective spin is also the true spin. For Ni^{2+} ions there is nearly complete quenching of the orbital angular momentum in the $^3A_2$ ground state and the effective and true spins may be equated. However, in Fe^{2+} and Co^{2+} compounds, there is usually an orbital degeneracy in the ground state of the ion. The combined spin-orbit interaction and crystal field produce a manifold of low-lying (0–1200 cm^{-1}) energy levels. Here, the effective spin of the ground state multiplet is no longer simply related to the true spin and interactions between the lowest and higher states must be considered. To a first approximation these interactions are often ignored but their significance is shown in Sect. 3.3.

The discovery in 1966 of one and two-magnon Raman scattering in FeF$_2$ of near equal intensity [3.1] posed new and interesting theoretical problems. The possibility of observing magnetic Raman scattering had been suggested earlier by *Bass* and *Kaganov* [3.17], *Elliott* and *Loudon* [3.18], and by *Shen* and *Bloembergen* [3.19]. There then followed considerably more theoretical work to explain these first results. Only the briefest outline of the theory can be given here: more detailed treatments are given in the reviews [3.6–8] and by *Fleury* and *Loudon* [3.20].

In transition metal fluorides, one-magnon Raman scattering ensues from a third-order electric-dipole interaction mechanism involving indirect coupling to the spins via *spin-orbit coupling* in the excited intermediate states. The electric field of the incident light excites a virtual electronic transition within the metal ion to some excited state which comprises a linear combination of spin and orbital states through the spin-orbit coupling. As the excited state contains S^z components other than that giving rise to the $\Delta S^z = 0$ transition from the ground state of spin S, the electric-dipole interaction permits another $\Delta S^z = 0$ transition back to a level with $S^z = S - 1$ in the ground state. Thus, in the Stokes scattering process, an incident photon of frequency ω_i is absorbed and another emitted of frequency

$$\omega_s = \omega_i - \omega_q^m. \tag{3.7}$$

The energy difference $\hbar\omega_q^m$ corresponds to one elementary spin-wave excitation or magnon of wave vector $q \approx 0$. This mechanism correctly predicts the observed one-magnon Raman cross-section and polarization effects in FeF_2. The more direct magnetic-dipole interaction mechanism [3.17] predicts a scattering cross section that is far too small and cannot reproduce its polarization dependence [3.20]. It is worth noting that the spin-orbit mechanism, as considered here, was applied to those magnetic ions having S-type ($L = 0$) ground states such as Mn^{2+} or orbitally quenched Fe^{2+} in FeF_2.

The first-order spin-orbit scattering mechanism when extended to higher order can give rise to second-order or two-magnon scattering but the cross section is usually several orders of magnitude smaller than the first-order result. The striking result that two-magnon scattering is stronger than the one magnon scattering in the rutile compounds necessitated a new unrelated process called the exchange scattering mechanism [3.20]. In this mechanism, the usual virtual electric-dipole transitions occur between the ground and excited states with an additional exchange interaction between two electrons (spins) in excited states. The exchange is necessarily limited to spins on neighboring magnetic ions on opposite sublattices and results in no change in the z component of the total spin. The mechanism, being of the form $S_i \cdot S_j$, produces pairs of magnons and therefore cannot produce one-magnon scattering. Furthermore, it is restricted to antiferromagnets. In two-magnon Stokes Raman scattering, pairs of magnons are created such that

$$\omega_s = \omega_i - (\omega_q^m + \omega_{-q}^m) \tag{3.8}$$

since the total transferred momentum is small compared with wave vectors in the Brillouin zone. For the degenerate magnon branches in simple antiferromagnets, $\omega_q^m = \omega_{-q}^m$ and (3.8) becomes

$$\omega_s = \omega_i - 2\omega_q^m. \tag{3.9}$$

In principle, all pairs of magnons with equal frequencies and opposite wave vectors can contribute to the scattering to give a broad band of frequencies. However, the scattering is weighted by the magnon density of states which peaks sharply at the Brillouin zone boundary.

At the Brillouin zone centre and for nearest-neighbor exchange only, (3.3) reduces to

$$\omega_0 = [(H_A + SzJ_1)^2 - (SzJ_1)^2]^{1/2}, \tag{3.10}$$

where z is the number of magnetic nearest neighbors. Thus, one-magnon light scattering can give quantitative information on H_A when J_1 is known. Unfortunately, the scattering frequency and intensity both reduce as $H_A^{1/2}$ [3.20] which can lead to experimental difficulties when $H_A \ll H_E$ (H_E is the exchange field). This is usually the case in pure-spin antiferromagnets and, in fact, no first-order magnon scattering has been observed in either $KMnF_3$ or $RbMnF_3$. Such scattering has been observed in the Co and Ni perovskites (see for example [3.14, 15, 21, 22]) where the anisotropy field is larger (see Table 3.1). For both $RbCoF_3$ and $KNiF_3$, the one-magnon frequency decreases to zero as the temperature is increased through T_N [3.15, 21], but much more rapidly than is expected from the sublattice magnetisation behavior. One-magnon Raman experiments do not show any critical magnetic scattering near T_N because they probe the transverse susceptibility, which exhibits no critical behavior in these simple antiferromagnets [3.6].

Most light scattering work on the transition metal ion fluoro-perovskites has centred on the two-magnon excitations. Following experimental work on $RbMnF_3$ [3.23], it was found that the exchange scattering theory failed to predict the observed lineshape, which peaked at a lower frequency. It was soon realised that magnon-magnon interactions were important and the revised theory in a Green's function formalism was found to give good agreement with the low temperature experimental results [3.24]. The strong attractive interaction between magnons occurs because the exchange scattering mechanism requires the spins to be created on neighboring sites on opposite sublattices. Since then, most theoretical work has concentrated on the temperature dependence of the two-magnon scattering observed in $KNiF_3$ [3.25] or $KMnF_3$ [3.26], for example. The two-magnon peak frequency usually decreases much less rapidly with increasing temperature than the one-magnon frequency and the pair excitation can persist *far into the paramagnetic regime*. At the same time, the two-magnon band linewidth increases. This behavior is shown in Fig. 3.2 for $KMnF_3$ where it proved impossible to follow the magnon scattering above T_N because of a structural phase transition at $T_C = 81$ K. The Green's function method, when extended to higher order to allow for interactions between thermally excited magnons, satisfactorily accounts for the observed temperature dependence of the two-magnon spectrum for temperatures up to $\sim 0.8\, T_N$ [3.6, 27]. This is shown, for example, in Fig. 3.2. The computed two-magnon lineshape is in very good agreement with experiment for $T < 0.8\, T_N$ as can be seen in Fig. 3.3, where the theoretical bandshape is least-squares fitted to $KMnF_3$ data at 31.5 K by varying the strength parameter only. Other theoretical approaches using diagrammatic techniques [3.28], the equation-of-motion method [3.29], or where the magnon frequency was

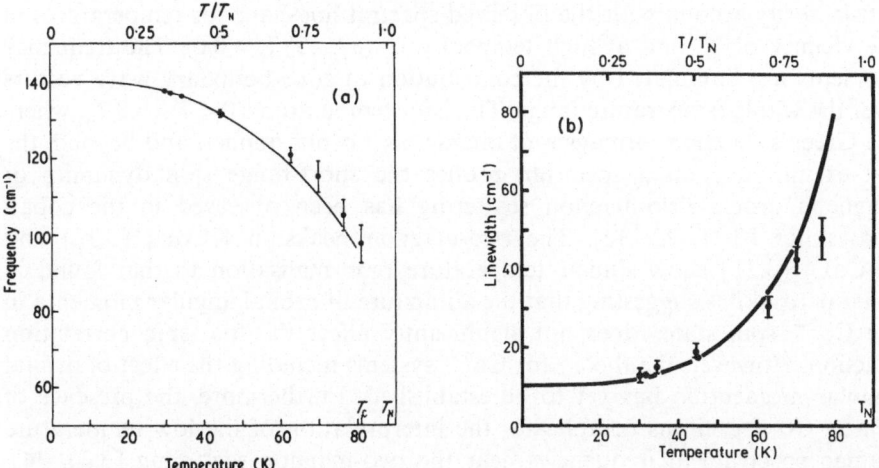

Fig. 3.2a, b. Theoretical curve and experimental points for (a) the two-magnon peak frequency (T_N is the Néel temperature while T_C represents a structural phase transition) and (b) the linewidth versus temperature for $KMnF_3$ [3.26]

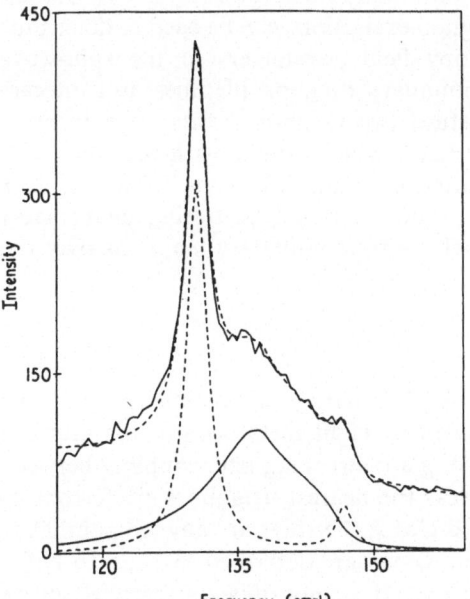

Fig. 3.3. The Raman spectrum of $KMnF_3$ at 31.5 K. The broken line is a fit of theory to experiment, with the two-magnon (——) and other phonon (———) components shown underneath [3.26]

calculated in the mean-field approximation [3.30], have produced similar results, but with damping included phenomenologically. More recently, *Balucani* and *Tognetti* [3.6, 31] have investigated two-spin scattering in the disordered (paramagnetic) phase of $RbMnF_3$ and $KNiF_3$ by means of a modified Markoffian approximation for the Mori continued-fraction representation of the shape function. Their four-spin correlation function approach satisfactorily accounts for the observed spectral lineshape for temperatures in the vicinity of T_N and at high temperatures (e.g., $T/T_N = 3.6$). The frequency moments are dominated by the contribution of zone-boundary wave vectors over the whole temperature range. Thus, for temperatures $0 < T < 0.8\, T_N$, where the Green's function formalism of interacting bosons applies, and beyond, the two-excitation Raman spectrum probes the short-range spin dynamics or magnetic order. Two-magnon scattering has been observed in the cobalt perovskites [3.21, 22, 32]. The two-magnon peaks in $KCoF_3$ [3.32] and $RbCoF_3$ [3.21] show similar temperature renormalisation to that found in other perovskites suggesting that the admixture of orbital angular momenta in the Co^{2+} spin states does not significantly affect the four-spin correlation function. However, the theory for Co^{2+} systems including the effect of orbital angular momentum has yet to be established. Furthermore, the presence of nearby Co^{2+} excitons complicates the interpretation of the low temperature Raman spectrum at frequencies near the two-magnon scattering [3.33, 34]. Finally, the possibility has been raised of observing dynamical critical phenomena through a central peak in the second-order spectrum of $RbMnF_3$ [3.35].

This brief review has established that light scattering from spin waves in the transition metal fluoro-perovskites is now well understood for $T < T_N$. Hence, measurements of the one and two-magnon excitations can be used to determine the principal exchange and anisotropy field parameters in the spin-wave Hamiltonian and to evaluate zone-boundary magnon lifetimes and interactions. For example, the low-temperature two-magnon Raman spectrum of $KMnF_3$ gives $J_1 = 5.00 \pm 0.07\, cm^{-1}$ [3.26], which compares favourably with the value obtained from neutron scattering studies given in Table 3.1. This concept has been applied to $RbMnF_3$ to show that light scattering gives correct values for the frequency and lifetime of the zone boundary magnons over the entire ordered phase [3.36].

3.1.2 Tetragonal MF$_2$ Compounds

The rutile structure depicted in Fig. 3.1 has tetragonal symmetry, with one metal ion at the body centre surrounded by eight metal ions at the unit cell corners. In these fluorides, there is strong antiferromagnetic coupling between next-nearest magnetic neighbors whereas the nearest neighbors are ferromagnetically aligned. For MnF_2, FeF_2, and CoF_2, the spins are aligned parallel or anti-parallel to the c-axis and their spin waves are described by (3.3). In NiF_2, the spins align perpendicular to the c-axis and are not quite antiparallel, giving rise to a weak ferromagnetic moment. Consequently, the two-fold degeneracy for long-wavelength magnons is raised, although they do become degenerate again at the Brillouin zone boundary [3.39]. Measured spin-wave parameters are summarized in Table 3.2. For NiF_2, the lower zone-centre magnon frequency is proportional to the anisotropy in the ab plane, while the higher frequency corresponds to the normal antiferromagnetic resonance observed in the other compounds.

Table 3.2. Magnetic parameters for various transition metal fluorides with the rutile structure. The notation is that given in Table 3.1

	MnF_2 [3.10, 37]	FeF_2 [3.10, 20]	CoF_2 [3.10, 38]	NiF_2 [3.39, 40]
S	5/2	2	1/2	1
J_1 [cm^{-1}]	− 0.44	− 0.05	− 2.00	− 0.22
J_2 [cm^{-1}]	2.48	3.64	12.3	13.9
J_3 [cm^{-1}]	0.06	0.19	–	0.79
H_A [cm^{-1}]	0.74	20.1	12.5	–
ω_0 [cm^{-1}]	8.7	52	37	3.3, 31.1
T_N [K]	68	78	38	73

One and two-magnon Raman scattering has been observed in Fe, Co, and Ni fluorides but only two-magnon scattering has been detected from MnF_2 [3.20, 38, 41, 42]. The one-magnon scattering is expected to be weaker in MnF_2 because H_A is small compared to other fluorides (see Table 3.2) [3.20]. However, a study of the mixed antiferromagnet $Mn_{1-x}Co_xF_2$ showed that the lack of an orbital contribution to the magnon may also diminish the intensity [3.43]. Despite the more complicated magnetic structure of the MF_2 compounds, the theory for light scattering from their magnons parallels that for the fluoro-perovskites. The temperature dependence of the one-magnon frequency in NiF_2 [3.6, 39, 44] and FeF_2 [3.45] approximately follows the sublattice magnetisation and vanishes at T_N. At the same time, the magnon damping increases while the scattering cross section reduces to zero. In NiF_2, the integrated intensity falls as the square of the magnon frequency [3.44]. Very little theoretical work has been undertaken to characterise the temperature dependence of the one-magnon Raman scattering [3.6]. *Cottam* [3.46], by incorporating terms quadratic in the spin operators into the photon-magnon interaction Hamiltonian, has obtained reasonable agreement with experiment for the temperature dependence of the one-magnon light scattering in FeF_2. An earlier theory involving linear terms alone failed completely [3.45]. Thus, at least for FeF_2, quadratic magneto-optical effects are important. Further evidence for this has come recently from a study of the anti-Stokes/Stokes intensity ratio of the magnon scattering in FeF_2 [3.47]. Here, the intensity ratio is very different from the usual Boltzmann thermal population factor. Such a discrepancy is anticipated in magnetic systems [3.48] and is explainable in terms of quadratic magneto-optical coupling [3.47].

The two-magnon Raman spectrum of rutile antiferromagnets can be described at lower temperatures in terms of the theory including magnon-magnon interactions outlined in Sect. 3.1.1, despite the additional complications of lower symmetry and an increased number of exchange constants [3.24, 29, 30, 39]. The two-magnon peak frequency renormalises slowly with increasing temperature and the scattering persists well into the paramagnetic phase. In fact, for MnF_2 and NiF_2, the scattering intensity *increases* with

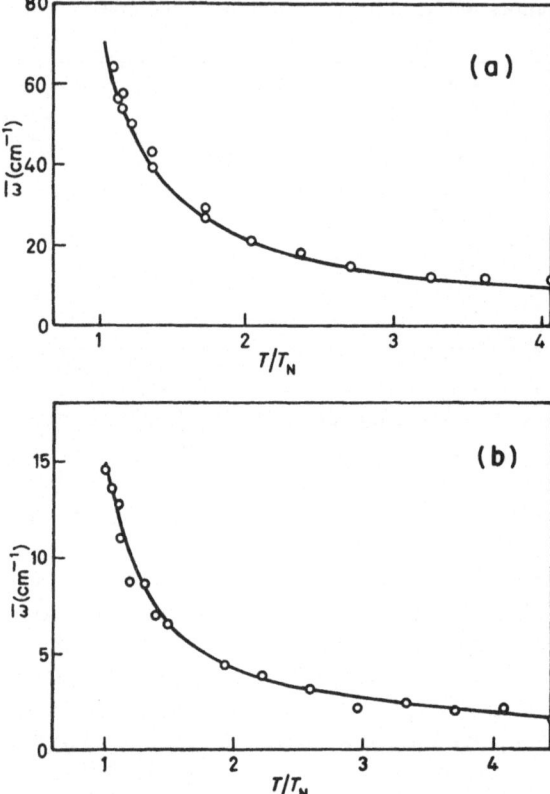

Fig. 3.4a, b. Theory (—) and experiment (○) for the first moment of the two-spin light scattering in (a) paramagnetic NiF_2 and (b) paramagnetic MnF_2 [3.53]

increasing temperature above $\sim T_N$ [3.50, 51]. An improved understanding of these experiments extending into the paramagnetic phase was obtained from a calculation of low-order frequency moments in terms of static multispin correlations [3.52]. The temperature dependence of the first moment is well reproduced by the theory but it is less successful in predicting the integrated intensity behavior [3.52]. The method does not permit a direct comparison between theory and experiment for the spectral lineshape. More recently, this spectral lineshape has been calculated using the continued fraction approach mentioned in Sect. 3.1.1 with some success [3.53]. The first frequency moment of the overall spectrum is found to be in good agreement with experiment for both NiF_2 and MnF_2 for temperatures $T_N \lesssim T \lesssim 4\,T_N$, as can be seen in Fig. 3.4a, b. Resonance enhancement of two-magnon scattering has been observed in MnF_2 [3.54], where the resonant intermediate state is a magnon sideband of a magnetic-dipole transition. This and other resonant light scattering work on magnetic systems has been succinctly reviewed elsewhere [3.55].

Interest in light scattering from magnetic excitations in rutile antiferromagnets has now turned towards the possibility of observing surface modes

Table 3.3. Magnetic parameters for various transition metal halides. The spin ordering direction is either parallel ($\parallel$) or ($\perp$) to the crystal c axis

	$MnCl_2$ [3.59]	$FeCl_2$ [3.62, 63][a]	$FeBr_2$ [3.62, 64][a]	$CoCl_2$ [3.65, 66]	$CoBr_2$ [3.67, 68]	$NiCl_2$ [3.69, 70]	$NiBr_2$ [3.60]
S	5/2	1	1	1/2	1/2	1	1
J_1 [cm^{-1}]	–	~ -2	~ -1	-18	–	-15	–
J_2 [cm^{-1}]	–	~ 0.2	~ 0.4	9	–	0.5	–
ω_0 [cm^{-1}]	–	16.5	17.7	2.3, 19.0	2.0, 22.2	0.4, 2.6	–
T_N [K]	2	23.5	14.2	24.7	19	52.3	52
Spin direction	$\perp$	$\parallel$	$\parallel$	$\perp$	$\perp$	$\perp$	$\perp$

[a] The exchange constants, J, in [3.62] are equivalent to $12 J$ here.

[3.56, 57]. Detailed calculations have predicted the one-magnon spectral lineshape for surface scattering from $RbMnF_3$ and MnF_2 [3.58], but as yet there have been no experimental observations of such modes.

3.1.3 Trigonal MX$_2$ Compounds

The transition metal bromides, chlorides and iodides usually adopt one of two trigonal layered structures. These are the $CdCl_2$ structure shown in Fig. 3.1 or the CdI_2 structure, which differs from $CdCl_2$ in that its anions are in hexagonal rather than cubic closepacking [3.59]. In both structures, layers of metal ions lying perpendicular to the c-axis are sandwiched between two layers of halide ions. The simplest magnetic structure comprises hexagonal sheets of ferromagnetically aligned spins with adjacent sheets having oppositely directed moments, as is shown in Fig. 3.1. Thus, the overall structure is antiferromagnetic, with the spins aligning parallel (easy axis) or perpendicular (easy plane) to the c-axis. The intralayer coupling is much larger than the interlayer coupling, so that the dispersion of magnons along the c-axis direction is much weaker than the in-plane dispersion. In this respect, these compounds approximate $2D$ ferromagnets. Other, more complicated magnetic structures are formed at T_N (e.g., $MnCl_2$ [3.59]) or at lower temperatures (e.g., $NiBr_2$ [3.60]) and these will not be considered here. Models for the spin-wave dispersion in MX_2 compounds have been reviewed elsewhere [3.61] and the effective spin Hamiltonian is complex, containing isotropic (and possibly anisotropic) exchange parameters for ferromagnetic (J_1) and antiferromagnetic (J_2) nearest neighbors plus an (effective) anisotropy parameter. These exchange constants and other parameters characterizing the magnetism in some MX_2 compounds are compared in Table 3.3. For the Fe compounds, the uniaxial anisotropy field (H_A) dominates the magnetic behavior [3.63, 64] and there is a significant antiferromagnetic exchange between next-nearest neighbors in the layers. For the Co compounds, the in-plane anisotropy raises the magnon degeneracy at the zone centre giving the two uniform modes listed in Table 3.3. The degeneracy is also raised at $q = 0$ in $NiCl_2$ through magnetic-dipole interactions [3.70].

Light scattering from magnons in the MX_2 compounds is a relatively new field, with the first observations of zone-centre magnons in $FeCl_2$ and $CoCl_2$ being reported in 1978 [3.65]. Later papers reported the temperature dependence of the one-magnon Raman spectrum in $FeCl_2$ and in $CoBr_2$ [3.67, 71]. For $CoBr_2$, the one-magnon energy renormalises relatively slowly with increasing temperature, whereas the integrated intensity approaches zero at T_N like the magnetisation and the damping increases. This behavior is shown in Fig. 3.5. The one-magnon scattering from $CoCl_2$ exhibits a similar temperature dependence [3.72]. It is noteworthy that the magnon peak frequency does not go to zero at T_N, as is usually found. Instead, the band intensity appears to reflect the magnetic ordering. As yet there is no theoretical explanation for this behavior, although it has been suggested that the weak frequency renormalization may be attributed to the artificially low Néel temperature [3.73]. The peak frequency of the magnon in $FeCl_2$ also renormalises relatively slowly with increasing temperature [$\omega(T)/\omega(0) = 0.7$ for $T = 0.83\, T_N$] in disagreement with the predictions of linear spin-wave theory [3.71, 74]. This discrepancy illustrates the need to use renormalised spin-wave theory for materials like $FeCl_2$ with low dimensionality and low effective spin. The magnon frequency renormalises somewhat faster than the sublattice magnetisation up to $0.9\, T_N$, whereas the exchange energy is essentially constant due to the fact that even at $0.9\, T_N$, only the long-wavelength magnons are appreciably populated [3.63]. Between $0.9\, T_N$ and T_N, the whole magnon branch collapses and there are no propagating modes above the phase transition [3.63]. Again, the one-magnon linewidth increases and the band intensity decreases as T_N is approached. The results of a linear spin-wave calculation of the intensity behavior are shown in Fig. 3.6 and they indicate the importance of quadratic magneto-optical coupling in the light scattering mechanism. Light scattering from magnons in $NiCl_2$ has also been searched for without success [3.69], where it is concluded that their intensity must be less than 1/30th the intensity of the A_{1g} phonon. For comparison, one-magnon scattering in $CoBr_2$ has nearly the same intensity as the $CoBr_2$ A_{1g} phonon [3.75]. This suggests that the one-magnon scattering intensity in the iron and cobalt salts is enhanced by the nonquenching of the orbital angular momentum of the Fe^{2+} and Co^{2+} ions.

Contrary to the behavior found for the perovskite and rutile compounds, the two-magnon scattering is much weaker than one-magnon scattering in these MX_2 compounds. No two-magnon scattering has been observed from the in-plane spin waves of any of the compounds discussed here [3.65, 67, 69] although very weak combination scattering from spin waves near the c-axis zone boundary has been found in $CoBr_2$ [3.67]. The low intensity of in-plane two-magnon scattering may result from the quasi-2D-ferromagnetic ordering which would negate the exchange scattering mechanism (Sect. 3.1.1).

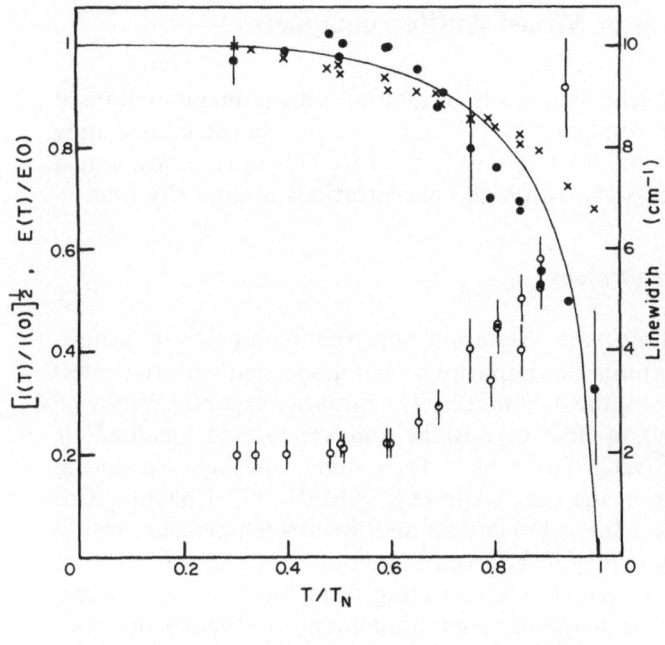

Fig. 3.5. Temperature dependence of the reduced peak frequency $[E(T)/E(0)]$ ($\times$), linewidth ($\circ$), and square root of the reduced integrated intensity $[I(T)/I(0)]^{1/2}$ ($\bullet$) for one-magnon Raman scattering in $CoBr_2$. The solid curve is the reduced sublattice magnetisation $[M(T)/M(0)]$ obtained by neutron diffraction [3.67]

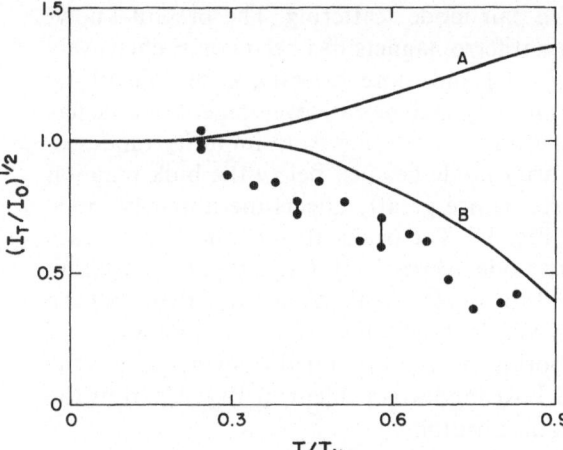

Fig. 3.6. The square root of the reduced integrated intensity $(I_T/I_0)^{1/2}$ of the one-magnon scattering in $FeCl_2$ as a function of the reduced temperature, T/T_N. The theoretical curves, labeled A and B, are calculated for the cases of linear and quadratic magneto-optical coupling, respectively [3.71]

The results obtained so far for Raman scattering from magnons in these trigonal metamagnets are sufficiently different from those obtained from the cubic and tetragonal compounds to warrant further study. In particular, a theoretical investigation of the coupling mechanism between the photons and the magnons is needed to explain the intensities of the one and two-magnon bands and their temperature dependence.

3.2 Magnon Scattering in Mixed Antiferromagnets

This section is concerned with the effects of dilution on the magnon Raman spectrum of a pure antiferromagnet. Such studies began almost immediately after the first experiments on pure materials (e.g., [3.76, 77]) and are now aimed at understanding mixed crystals with high concentrations of impurity ions.

3.2.1 Low Impurity Concentrations

In the dilute impurity limit, the excitation spectrum comprises a group-theoretically determined number of impurity modes associated with the defect and its nearest neighbors plus the host modes. The impurity excitations may lie above or within the band of host excitations and are termed localised or resonance modes, respectively. There have been numerous light scattering studies from point defects in the pervoskite (e.g., $KMnF_3$:Ni, $RbMnF_3$:Co) and rutile (e.g., MnF_2: Ni, MnF_2: Fe) lattices and the low temperature results are well accounted for by theory in a Green's function formalism. Nearly all Raman results appertain to pair-mode scattering from localised excitations, and again, evidence is seen of magnon-magnon interactions affecting the peak frequency and band shape [3.78]. As yet there is no theory to explain in detail the temperature dependence of the pair-mode scattering. The present knowledge of scattering from defects in antiferromagnets has been comprehensively reviewed by *Cowley* and *Buyers* [3.10] and more recently, in less detail, by *Thorpe* [3.7], and *Hayes* and *Loudon* [3.2]. One recent development of note has been the observation of enhanced Raman scattering from impurity modes in FeF_2:Mn [3.79]. The Mn^{2+} impurity mode lies just below the bulk magnon frequency and the impurity-host resonance greatly affects the impurity mode scattering intensity as is shown in Fig. 3.7. The peaks at ~ 49 and $53\,cm^{-1}$ are the Mn^{2+} impurity mode and host mode, respectively. One-magnon scattering from Mn^{2+} compounds is expected to be very weak, as noted earlier, and the high intensity here arises indirectly via the oscillating transverse component of the exchange fields of the neighboring host spins. Further evidence for the interaction between impurity and host modes is evident in Fig. 3.7 from the changes in peak frequency with concentration.

3.2.2 High Impurity Concentrations

For higher concentrations of impurities, the magnetic defects are no longer isolated and disorder ensues. A new theory for the mixed crystal is required, as the disorder obviates the group-theoretical approach used in Sect. 3.2.1. Mixed antiferromagnets are ideal systems for testing the theories of disordered materials. This is partly because the magnetic interactions are of short range and of a well-known form and strength, and partly because the excitations can

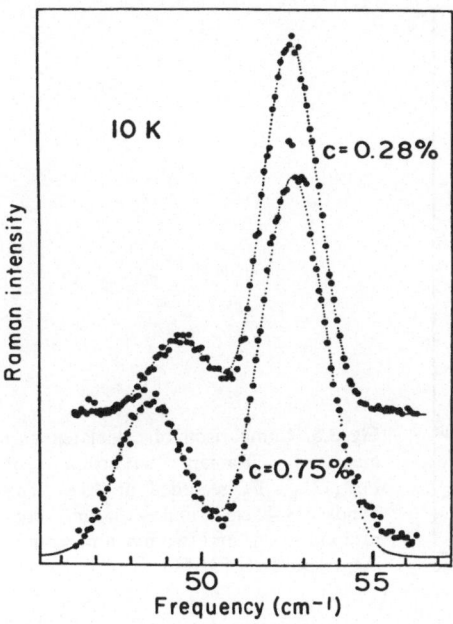

Fig. 3.7. One-magnon Raman scattering from impurity and host modes in $Fe_{1-c}Mn_cF_2$ for two concentrations c. Reprinted with permission from [3.79]

be studied in detail using light or neutron scattering techniques. Furthermore, in the simple magnetic structures considered here, very few parameters are needed to describe the magnetic interactions in the mixed crystals. Simple theories such as the Ising cluster model, or sophisticated ones such as the coherent potential approximation (CPA) model or numerical simulation methods, have been developed to take into account that part due to the disorder in the particular behavior of excitations in these systems [3.80].

In light scattering experiments, the results obtained for high impurity concentrations are not as well understood as for the dilute case and until recently, there has been a paucity of experimental data (see, for example, [3.8]). In the mixed perovskites, there have been Raman studies of $KNi_{1-x}Mg_xF_3$ [3.81], $RbMn_{1-x}Co_xF_3$ and $KMn_{1-x}Co_xF_3$ [3.43, 82], and $KMn_{1-x}Ni_xF_3$ [3.16]. There have also been light scattering studies of mixed crystals possessing the rutile structure ($Mn_{1-x}Zn_xF_2$ [3.83], $Mn_{1-x}Co_xF_2$ [3.43, 82], $Ni_{1-x}Co_xF_2$ [3.82], and $Fe_{1-x}Zn_xF_2$ [3.84]) and the layered MX_2 structure ($Fe_{1-x}Mn_xCl_2$ [3.71], $Fe_{1-x}Co_xCl_2$ and $Fe_{1-x}Cd_xCl_2$ [3.72]). Addition of nonmagnetic ions (Mg or Zn) results in a reduction of T_N until at some critical concentration, x_c, T_N is reduced to zero and the long-range order vanishes. This percolation limit is at $x_c \simeq 0.69$ for the perovskite structure and at $x_c \simeq 0.76$ for the rutile structure. In $Fe_{1-x}Cd_xCl_2$, the ordering temperature falls to zero near $x = 0.45$, close to the $2D$ triangular lattice percolation threshold [3.85]. Zn or Mg form magnetic "vacancies" in the lattice and thus no defect modes can appear at higher frequencies than the pure antiferromagnet magnons. The low-temperature Raman spectrum comprises a

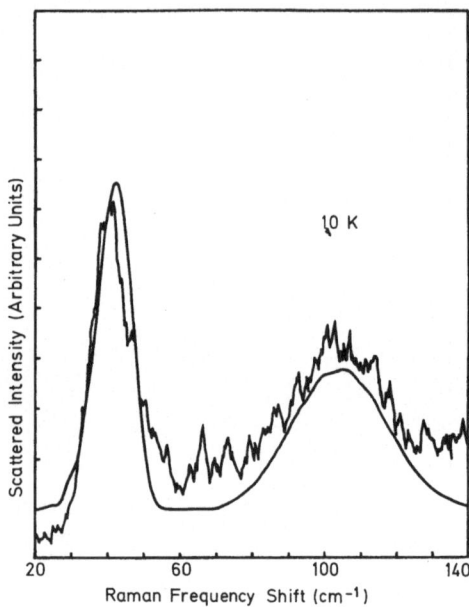

Fig. 3.8. Comparison of calculated and measured Raman spectrum of $Fe_{0.48}Zn_{0.52}F_2$ recorded at 10 K. The bands at $\sim 40\ cm^{-1}$ and $\sim 105\ cm^{-1}$ correspond to one and two-magnon scattering, respectively [3.84]

one-magnon peak and a two-magnon peak (as shown, for example, in Fig. 3.8) whose frequency decreases and damping increases as x increases. The general behavior with increasing x is somewhat like the effect of increasing the temperature of the pure antiferromagnet. The two-magnon results obtained to date have been explained in terms of the Ising cluster model. The reason for using this extremely simple model is that the spin waves at the Brillouin zone boundary dominate the two-magnon spectrum and the energy of these excitations corresponds to the energy of a spin reversal in the Ising model. This excitation energy is just JSz, where J is the dominant antiferromagnetic exchange constant and z is the number of magnetic neighbors. The energy required to create a given pair of spin deviations on neighboring magnetic ions will depend upon the configuration of the surrounding ions on nearest-neighbor magnetic ion sites. Averaging over all these possible clusters gives the mean two-magnon frequency [3.83].

$$\bar{\omega} = 2JSz[1 - x(z-1)/z], \tag{3.11}$$

where magnon-magnon interactions have been ignored. The Ising model can also be used to calculate the spectral lineshape for a given x value by constructing a smoothed histogram where each individual cluster frequency is weighted with the probability of obtaining that cluster. Fits to the data show that the Ising model is in reasonable agreement with experiment, as can be seen in Fig. 3.8. *Buyers* et al. [3.86] have calculated the spin-wave excitations in $Mn_{1-x}Zn_xF_2$ using a coherent potential approximation (CPA) model. The

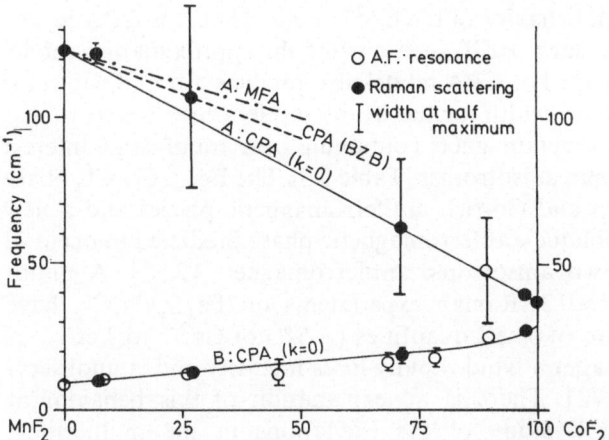

Fig. 3.9. One-magnon Raman scattering in $Mn_{1-x}Co_xF_2$ [3.82]. A denotes the "Co^{2+}" curve and B the "Mn^{2+}" one. The solid lines connecting experimental points are the theoretical predictions of the CPA model [3.87]. MFA-mean field approximation, BZB-Brillouin zone boundary

CPA is used to treat the randomly varying Ising interactions between nearest antiferromagnetic neighbors but the transverse off-diagonal parts of the Heisenberg interaction are included only in an approximate manner. Their calculation of the two-magnon scattering gives better agreement with experiment than the Ising model, particularly in regard to the width of the band [3.86]. There is no critical behavior in the two-magnon Raman cross section as $x \to x_c$ [3.81, 83], since light scattering probes the short-wavelength zone boundary modes which are nearest-neighbor sensitive only. The concentration dependence of the one-magnon scattering in $Fe_{1-x}Zn_xF_2$ is explained reasonably well by a simple cluster model in which the exchange and H_A are varied according to the number of Zn ions on nearest magnetic neighbor sites [3.84]. The computed one-magnon lineshape for one concentration is shown in Fig. 3.8.

Antiferromagnetic alloys possess two well-defined magnon dispersion curves, one (A) corresponding to excitations propagating largely on one kind of magnetic ion, M, and the other (B) to excitations propagating largely on the second kind of magnetic ion, M' [3.80]. Thus, the one-magnon Raman spectrum may comprise two bands corresponding to excitations on branches A and B, respectively. The concentration dependence of such scattering has been studied in $Mn_{1-x}Co_xF_2$, $Ni_{1-x}Co_xF_2$ and $RbMn_{1-x}Co_xF_3$ [3.43, 82], and in $Fe_{1-x}Mn_xCl_2$ and $Fe_{1-x}Co_xCl_2$ [3.71, 72]. The most detailed study of both bands in one alloy is that for $Mn_{1-x}Co_xF_2$ [3.43, 82] and the results obtained are shown in Fig. 3.9. The pure MnF_2 spin wave is not observed because the scattering cross section is too low. However, the Mn^{2+} mode intensity increases rapidly with increasing Co^{2+} concentration being detectable for $x \gtrsim 0.01$. The CPA theory of *Buyers* et al. [3.87] is in good agreement with experiment apart from the low concentration behavior of the Co^{2+} mode. Here, the CPA results at the Brillouin zone boundary (BZB) or a mean field approximation model (MFA) give better agreement. The CPA model also predicts a broad width for the Co^{2+} mode and a narrow width for the Mn^{2+} mode, which is experimentally confirmed. The mixed metamagnets containing iron are of some interest because of the competing spin anisotropies (Table 3.3). The $Fe_{1-x}Co_xCl_2$ phase diagram comprises Fe-rich and Co-rich antiferromagnetic phases and a new phase, which may be the oblique-antiferromagnetic phase predicted to occur in the random mixture of two anisotropic antiferromagnets [3.88]. A multicritical point occurs at $x \simeq 0.3$. Raman experiments on $Fe_{1-x}Co_xCl_2$ have shown that on the addition of small quantities ($\sim 5\%$) of Co^{2+} to $FeCl_2$, or Fe^{2+} to $CoCl_2$, the one-magnon band rapidly loses intensity and is unobservable for $0.05 \lesssim x < 0.9$ [3.72]. There is no explanation of this behavior at present and it prevented a study of the excitations in the multicritical concentration region. More detailed results have been obtained on the Fe^{2+} mode in $Fe_{1-x}Mn_xCl_2$ [3.71]. Here, the one-magnon frequency decreases linearly with increasing Mn^{2+} concentration and then incurves for $x > 0.15$. A simple Ising cluster model reproduces the initial linear concentration dependence, but it fails to predict the absolute value and it cannot explain the change in slope for $x > 0.15$. A more sophisticated theoretical interpretation of the data is required to explain these other aspects. A similar initial linear decrease in the Fe^{2+} mode frequency is seen in $Fe_{1-x}Cd_xCl_2$ [3.72].

The two-magnon Raman spectrum should comprise three bands corresponding to the creation of pair excitations on neighboring M ions, neighboring M and M' ions, and neighboring M' ions, respectively. Of the alloys studied to date, $KNi_{1-x}Mn_xF_3$ has provided the most complete information on the concentration dependence of all three bands [3.16]. The concentration dependence of the two-magnon peak frequencies and widths is shown in Fig. 3.10a, b. The frequencies of these bands are in excellent agreement with the predictions of an average Ising cluster model, provided that allowance is made for the change in exchange constants with lattice parameter. In this model, the exchange is taken between nearest neighbors only, consistent with the known properties listed in Table 3.1. There are three exchange constants J_1^1, J_1^2, and J_1^3 corresponding to nearest neighbors being both Ni^{2+}, both Mn^{2+}, and Ni^{2+} plus Mn^{2+} ions, respectively. By allowing each constant to vary as

$$J_1(x) = J_1(0)(1 + \lambda x),\tag{3.12}$$

the theoretical fit to the data shown in Fig. 3.10a was obtained with $J_1^1(0) = 63.3 \pm 1.0\,cm^{-1}$, $J_1^2(0) = 4.83 \pm 0.10\,cm^{-1}$, $J_1^3(0) = 16.8 \pm 0.3\,cm^{-1}$, and $\lambda = 0.07 \pm 0.005\,cm^{-1}$. $J_1^1(0)$ and $J_1^1(1)$ are in reasonable agreement with the values given for the pure materials in Table 3.1. The cluster model fails to predict the observed linewidths given in Fig. 3.10b, principally because it ignores the intrinsic width evident in the pure materials. A more refined theory is needed to determine the widths in this and other alloys, where similar

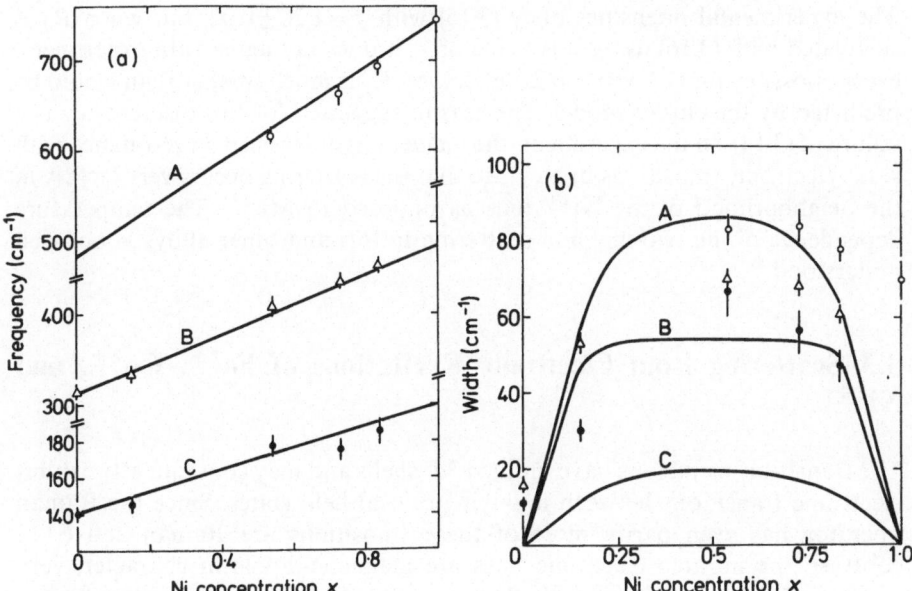

Fig. 3.10a, b. The **(a)** frequency and **(b)** width of the two-magnon bands in $KNi_xMn_{1-x}F_3$ vs concentration x. The full lines were calculated using an Ising cluster model. A and $\bigcirc$: Ni–Ni band; B and $\triangle$: Mn–Ni; C and $\bullet$: Mn–Mn [3.16]

limitations of the cluster model have been noted. The intensities of the three magnetic bands are given in the cluster model by the simple statistical weights of the pairs involved:

$$I_{Ni-Ni} = x^2 F_{Ni-Ni}$$
$$I_{Ni-Mn} = 2x(1-x)F_{Ni-Mn} \tag{3.13}$$
$$I_{Mn-Mn} = (1-x)^2 F_{Mn-Mn},$$

where F denotes the scattering cross section for the appropriate pair of atoms. If the F's are proportional to the superexchange of the virtually excited electronic state (Sect. 3.1.1) and if that superexchange is proportional only to the overlap of the excited states with the fluorine ligand, then

$$F_{Ni-Ni}/F_{Ni-Mn} = F_{Ni-Mn}/F_{Mn-Mn} = f. \tag{3.14}$$

The ratios of the band intensities are thus

$$R_1 = I_{Ni-Ni}/I_{Ni-Mn} = xf/[2(1-x)] \tag{3.15}$$

and

$$R_2 = I_{Mn-Mn}/I_{Ni-Mn} = (1-x)/2xf. \tag{3.16}$$

The experimental intensities obey (3.15) with $f = 1.20 \pm 0.02$, but when R_2 is calculated with (3.16) using this value of f, it does not agree with experiment, except possibly for $(1-x)/x < 0.2$. Otherwise, R_2 is much smaller than would be predicted by the cluster model. The reason suggested for this discrepancy is a failure of (3.14). In the experiment, the incident light is nearly in resonance with some electronic transitions of Ni^{2+}, so that the scattering occurs very largely in the neighborhood of the Ni^{2+} ions as opposed to Mn^{2+}. The temperature dependence of the two-magnon scattering in this and other alloys is not well understood.

3.3 Scattering from Electronic Excitations of Fe^{2+}, Co^{2+}, and Ni^{2+}

The transition-metal ions have unfilled $3d^n$ shells and may consequently exhibit electronic transitions between low-lying crystal field states. Since the Raman operator has even parity most of these transitions are Raman active. In contrast, the infrared electronic lines are magnetic-dipole in character, very weak, and easily confused with their associated vibronics or with impurity activated lines. In recent years following the first Raman observations on Co^{2+} systems [3.3, 38, 89–91], there have been numerous investigations of such transitions in solids. In this section, the salient features of electronic Raman scattering from Fe^{2+}, Co^{2+}, and Ni^{2+} will be reviewed using the perovskite, rutile and $CdCl_2$-type materials as examples.

3.3.1 Diamagnetic Hosts

In the dilute limit, where the transition metal ions are dispersed in a nonmagnetic host, the electronic excitations are localised to the vicinity of the metal ion. The single-ion energy level schemes for Co^{2+} and Fe^{2+} ions in a trigonal crystal field are summarised in Figs. 3.11, 12. Electronic Raman scattering is of low intensity in the fluoride compounds, and although two of the three expected $^4T_{1g}(^4\varGamma_4)$ transitions of Co^{2+} have been observed in cubic $KZnF_3$:Co [3.92], no such scattering from Fe^{2+} in similar hosts has been reported. Electronic Raman scattering has been observed for both Co^{2+} and Fe^{2+} in rutile-structured hosts (e.g., ZnF_2:Co [3.93] and ZnF_2:Fe [3.94]), but the most complete results and understanding have come from studies of MX_2:Co [3.91, 95, 96] and CdI_2:Fe [3.97]. The electronic Raman cross section increases in traversing the halogen series (F < Cl < Br < I) and the Raman linewidth decreases. Thus, the bromides and iodides are more favourable hosts for observing electronic scattering, and this is particularly so for iron. The halogen dependence of the cross section signifies the relative importance of the metal-ligand charge transfer absorption bands as lower intermediate states in the scattering process when using visible excitation frequencies [3.98]. For

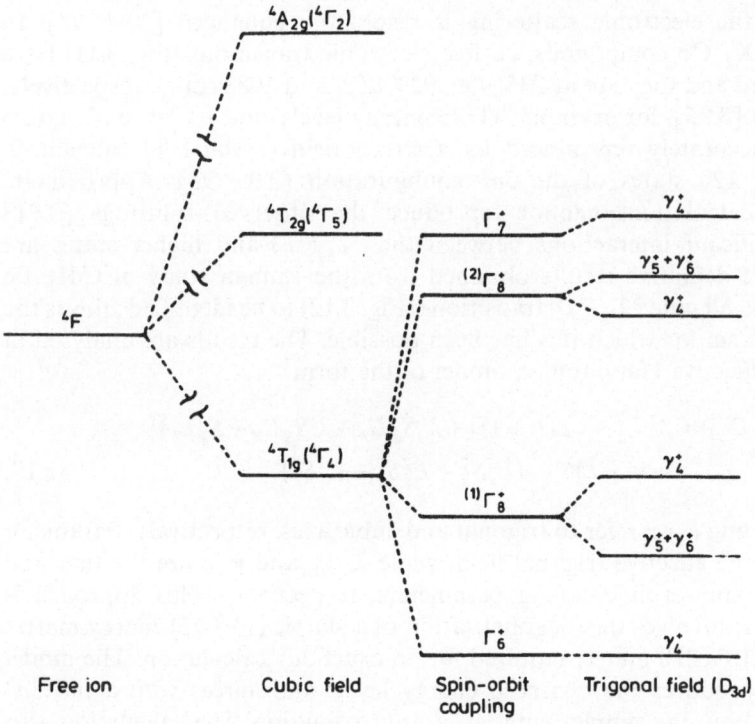

Fig. 3.11. Energy level scheme for the $^4T_{1g}(^4F)$ term of a Co^{2+} ion in a trigonal field. The overall splitting is $\sim 1300\ cm^{-1}$ [3.75]

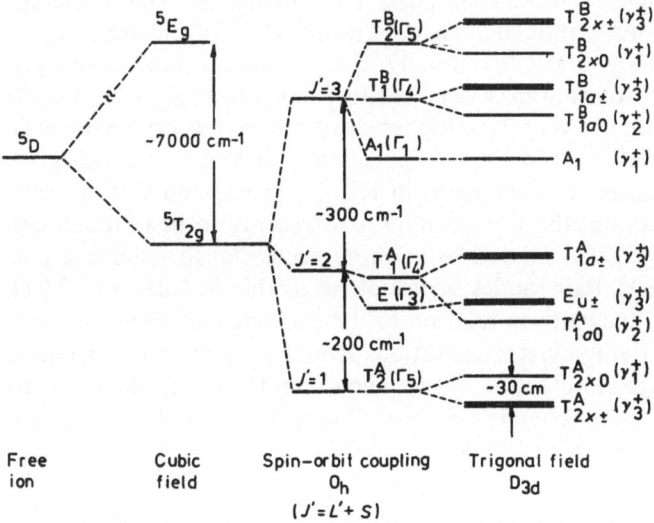

Fig. 3.12. Energy level scheme for the $^5T_{2g}(^5D)$ term of a Fe^{2+} ion in a trigonal field. The overall splitting is $\sim 700\ cm^{-1}$ [3.108]

the iodides, the electronic scattering is resonantly enhanced [3.97, 98]. In studies of MX_2:Co compounds, all five electronic transitions (Fig. 3.11) have been identified and they are at 235, 499, 923, 953, and 1088 cm^{-1}, respectively, in $CdCl_2$:Co [3.95], for example. These energy levels, and the ground state g values, are accurately reproduced by a strong-field crystal field calculation involving all 120 states of the $3d^7$ configuration [3.95, 96]. Approximate crystal-field calculations cannot reproduce the observed splittings [3.91] because significant interactions between the $^4T_{1g}(^4F)$ and higher states are excluded. The definitive results obtained from the Raman study of CdI_2:Fe [3.97] enabled all nine $^5T_{2g}(^5D)$ transitions (Fig. 3.12) to be identified; this is the first Fe^{2+} system for which this has been possible. The results are analysed in terms of an effective Hamiltonian model of the form

$$\mathscr{H}_{eff}(^5T_{2g}, D_{3d}) = \Delta[L_Z^2 - \tfrac{1}{3}L(L+1)] + \lambda_0 S_Z L_Z + \lambda(S_X L_X + S_Y L_Y)$$
$$+ \kappa(\boldsymbol{S}\cdot\boldsymbol{L})^2 + \varrho(L_x^2 S_x^2 + L_y^2 S_y^2 + L_z^2 S_z^2), \tag{3.17}$$

where (XYZ) and (xyz) refer to trigonal and cubic axes, respectively. Parameter Δ represents the effective trigonal field, while λ, λ_0, and κ, ϱ are the first and second-order spin-orbit coupling parameters, respectively. This approach is used because it involves the diagonalisation of a simple (15×15) energy matrix rather than 210×210 matrix required for an exact $3d^6$ calculation. The model accurately reproduces the observed energy levels and agrees with a full $3d^6$ calculation within the simpler cubic field approximation. The calculation also shows that admixtures from higher-lying states are important and must be included for a correct description.

As the concentration of transition metal ions increases, the probability of the occurrence of higher clusters of such ions increases. The exchange interactions between ions within such clusters result in shifts and splittings of former "single-ion" energy levels. Eventually, the percolation limit is reached and antiferromagnetic ordering may occur at low temperatures. It is of some interest to study the excitations of these higher clusters and thus bridge the wide gap between our presently good knowledge of the single-ion and collective excitations. Pair excitations of Co^{2+} ions in $KZnF_3$ [3.92] and $CdCl_2$ [3.91, 95] have been observed, but the Raman data is too scanty to be of much use. Infrared studies of $CdCl_2$:Co proved to be more successful for studying pair mode interactions [3.99]. Pair modes were not resolvable in CdI_2:Fe [3.97]. The problem is one of linewidth in relation to the exchange interaction, and a suitable system for studying cluster excitations would be a bromide or iodide with a large exchange interaction. One such candidate appears to be $CsMgBr_3$:Co, as the pure cobalt salt has a very high antiferromagnetic exchange constant ($J = 56$ cm^{-1}) [3.100].

3.3.2 Paramagnetic and Antiferromagnetic Hosts

Electronic Raman scattering from Fe^{2+} or Co^{2+} in paramagnetic hosts can be successfully interpreted in terms of the single-ion model. This is because the effect of short-range exchange interactions between the ion and its magnetic neighbors appears mainly in line broadening, i.e., the crystal field effects and spin-orbit interaction dominate the Hamiltonian. Of the pure perovskites, a very broad intense band exhibiting some structure was found in $RbFeF_3$ and is interpreted as an electronic Raman transition within the $^5T_{2g}$ ground state manifold [3.101]. Other broad structure, also electronic in origin, is observed in paramagnetic $KCoF_3$ and $RbCoF_3$ (see [3.33, 34] and references therein). While some detail is found in the electronic Raman spectrum of paramagnetic FeF_2 [3.102, 103] and CoF_2 (see, for example, [3.104]), more information comes from studies of the other halides $FeCl_2$, $FeBr_2$, $CoCl_2$, and $CoBr_2$ [3.62, 75, 95]. The bromides give the best results, although many lines are still rather broad, as can be seen from the spectra of Fig. 3.13. The $CoBr_2$ data were analysed with various models [3.75] and the complete $3d^7$ calculation or an effective Hamiltonian model in second order [similar to (3.17)] produced the best results giving equally good agreement with experiment. The effective Hamiltonian approach has the distinct advantage of requiring less computational effort, although the admixtures of higher states into the ground term are included only through effective parameters. The full $3d^7$ calculation clearly demonstrates the importance of including all higher-lying states in the calculation. For $FeBr_2$ [3.62], a dynamic Jahn-Teller calculation using a full lattice model gave the best description of the observed electronic transitions within the $^5T_{2g}(^5D)$ manifold of Fig. 3.12. However, the subsequent work on CdI_2:Fe [3.97] showed that a *static* effective Hamiltonian was sufficient to explain the results in this case.

At sufficiently low temperatures, antiferromagnetic ordering occurs and the long-range order is seen in the coupling of single-ion electronic levels to give whole crystal excitations or excitons. These exciton states may exhibit marked wave vector dispersion, especially the lowest one (the magnon). The effect of antiferromagnetic ordering on the Raman spectrum of transition metal ions is demonstrated in Fig. 3.13. Paramagnetic bands may shift to new frequencies and split in the ordered phase. The theoretical interpretation of the low energy excitons in Fe^{2+} and Co^{2+} compounds is more difficult than for electronic excitations in the paramagnetic phase. The problem lies in the use of spin-only operators between spinor states. A few attempts have been made to extend spin-wave theory to include higher excitons (e.g. $KCoF_3$ [3.105] and CoF_2 [3.106]) but their limited success probably results from the use of approximate wave functions in the paramagnetic state. There have been numerous Raman investigations of Co^{2+} excitons in various antiferromagnets, but surprisingly few studies of the Fe^{2+} compounds [3.62, 102]. *Macfarlane* [3.107] has observed inter-term Raman scattering from $^5T_{2g} \rightarrow {}^5E_g$ (Fig. 3.12) excitons in FeF_2 at the huge inelastic shift of ~ 7000 cm^{-1}. The spectrum is in the form of a sharp zero-phonon response accompanied by a broad quasicontinuum of exciton plus multiple-phonon scattering. The results are used to determine excited state Debye-Waller factors, which are not normally obtainable from measurements of one-photon absorption between states of the same parity. The comprehensive results obtained for paramagnetic $FeCl_2$ and $FeBr_2$, mentioned

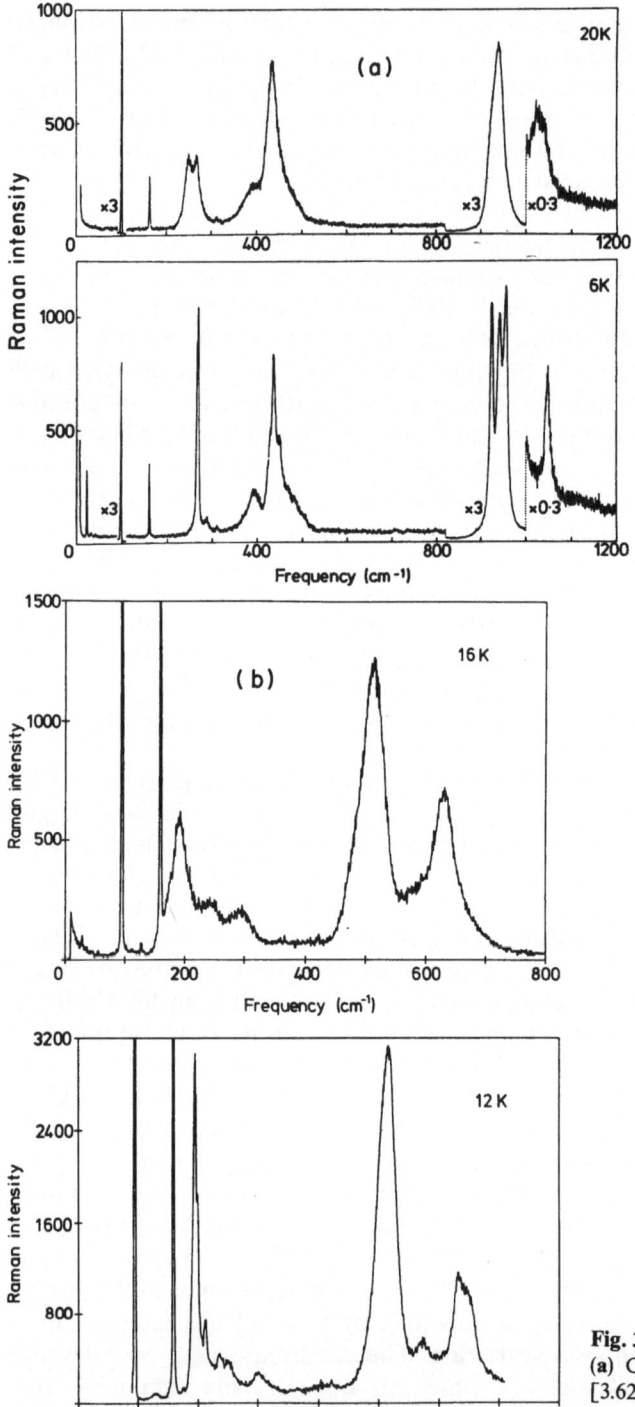

Fig. 3.13a, b. Raman spectra of (a) CoBr$_2$ [3.75] and (b) FeBr$_2$ [3.62] recorded at temperatures above and below T_N

earlier, extend through into the antiferromagnetic phase, where about two-thirds of the expected number of excitons were observed [3.62]. Their energies are reasonably well reproduced by a molecular-field approximation model. The expected discrepancies for the lower states can be attributed to the use of an isotropic exchange Hamiltonian for calculating the energy levels of an orbital triplet. It is the off-diagonal part of the inter-ion coupling that gives rise to the collective nature of the excitations. Electronic excitations of Fe^{2+} ions in the mixed crystal $Fe_{1-x}Mn_xCl_2$ have also been measured for temperatures $T \gtrless T_N$ [3.108]. In the paramagnetic phase, for $0 < x \lesssim 0.75$, the position and linear concentration dependence of observed transitions are successfully predicted using a single-ion effective Hamiltonian model which assumes that the Fe^{2+} environment in the disordered crystal can be represented by an average crystalline potential. For $T < T_N$, the single-ion model is extended within a molecular field approximation to describe the linear concentration dependence ($0 \leqq x \leqq 0.25$) of the exciton frequencies. The magnon, however, exhibits a weaker linear concentration dependence probably as a result of significant $Fe^{2+} - Mn^{2+}$ spin correlations. These effects are negligible at higher energies since $J_{Fe-Fe} \gg J_{Fe-Mn}$.

Despite considerable efforts by several research groups, the exciton Raman spectrum of $KCoF_3$ and of $RbCoF_3$ remains incompletely understood. Many of the discrepancies between the results of earlier workers have been attributed to the use of strained samples [5.33] and the most informative results have come from a study of a single domain sample of $RbCoF_3$ [3.34, 109]. However, except for the magnon, the observed features fit poorly into the exciton energy level scheme of *Buyers* et al. [3.105]. There have been several measurements of the excitons in CoF_2 [3.38, 41, 104, 110], all in general accord. The measured low-temperature cross sections [3.38] are in satisfactory agreement with the calculations of *Ishikawa* and *Moriya* [3.111], whose scattering mechanism involves direct coupling of the electric vector of the incident light to the orbital part of the excitations. This mechanism produces a higher scattering cross section than the indirect spin-orbit coupling mechanism of Sect. 3.1.1 for these low-lying spinor states. The temperature dependence of the exciton frequency and damping arises principally from coupling between exciton and exciton-magnon states constructed from physically renormalised magnons [3.110]. Calculations using a mean-field approximation with a restricted effective Hamiltonian are only in fair agreement with the observed exciton energies in CoF_2 and for Co^{2+} in MnF_2 [3.41]. The low-lying excitons observed in $CoCl_2$ [3.95, 112] and $CoBr_2$ [3.75] have been analysed in terms of a single-ion molecular field Hamiltonian of the form [3.75, 95]

$$\mathscr{H}_{mf} = -2J_{eff}\langle S_X \rangle_0 S_X, \tag{3.18}$$

where $\langle S_X \rangle_0$ is the zero-temperature expectation value of S_X for the ground state of the cobalt ion, and J_{eff} is the effective exchange interaction including both the intralayer and interlayer contributions. This interaction was diagonalised between eigenvectors corresponding to the twelve lowest eigenvalues of the full 120-dimension free-ion and crystal-field matrices appropriate to the cobalt salts in their paramagnetic phase. All of the observed energy levels in $CoBr_2$ and all but the magnon in $CoCl_2$ can be accounted for by this approach using only the one parameter, J_{eff}. The agreement between theory and experiment in $CoBr_2$ is remarkable for such a simple model and is probably fortuitous for the magnon.

High energy excitons have been observed in nickel compounds by using ultraviolet excitation [3.113]. Light scattering spectra of $KNiF_3$ and NiF_2 show sharp excitonic lines accompanied by broad vibronic bands exhibiting pronounced structure. Transitions are observed from the $^3A_{2g}$ ground state to $^3T_{2g}$ (~ 6500 cm^{-1}) and to $^3T_{1g}^a$ ($\sim 11,000$ cm^{-1}) excited states. The g values have been determined for some excitons and such information is useful for further tests of crystal-field models.

3.4 Scattering from Electron-Phonon Coupled Modes

All antiferromagnetic materials have the potential to exhibit magnetoelastic coupling, because the spin-wave and acoustic-phonon branches usually cross at some point in the Brillouin zone. The occurrence of such an interaction is governed by the symmetries of the excitations. This kind of magnon-phonon coupling has been observed in $FeCl_2$ [3.114], for example. First-order Raman scattering cannot probe these interactions directly because of the $q \simeq 0$ selection rule. However, as all Fe^{2+} and Co^{2+} compounds have significant orbital contributions to their magnetic moments and a corresponding complexity of low-lying electronic energy levels, there is the possibility of observing other strong electron-phonon interactions. This is especially so for more complex crystal structures where there are more optic modes at lower frequency. Strictly speaking, all electronic states are coupled to phonons to give vibronic eigenstates via the dynamic Jahn-Teller effect [3.115]. However, here we are concerned only with examples of strong interactions localised in frequency and wave vector space.

Neutron scattering experiments on $KCoF_3$ [3.116] have shown that a strong interaction exists between the magnon and Γ_{25} phonon near the paramagnetic Brillouin zone boundary R point or magnetic Brillouin zone centre. *Suzuki* and *Takaoka* [3.117] use an interaction Hamiltonian incorporating the phonon-modulated crystal field and the spin-orbit interaction as perturbations to explain the neutron scattering results. They find a coupling constant of 5 cm^{-1} gives the observed mode separation of ~ 15 cm^{-1} at the crossover point $q = (2\pi/a)(0.46, 0.46, 0.46)$. The effect of the magnon-phonon coupling is also reflected in the Raman spectrum, where scattering from the normally forbidden phonon at the zone boundary appears at 57 cm^{-1} beside the magnon at 38 cm^{-1} [3.14]. The induced phonon Raman intensity is proportional to the degree of admixture of the magnon mode and *Suzuki* and *Takaoka* [3.117] predict a magnon-to-phonon band intensity ratio of ~ 6 comparable with the experimental value of ~ 4.

Strong coupling between an electronic transition in the $^5T_{2g}(^5D)$ manifold of Fe^{2+} ions in $FeCl_2$ and the E_g optical phonon has been observed by light scattering [3.118]. The low-temperature Raman spectrum near $150\ cm^{-1}$ is strongly temperature dependent, as can be seen in Fig. 3.14, and shows a marked change at T_N. Well above T_N, the spectrum contains a symmetric band which becomes increasingly asymmetric to higher frequency when the temperature is lowered. This asymmetry is apparent in the 24 K spectrum of Fig. 3.14. The major part of this behavior was explained in terms of a resonant electron-phonon interaction between the $T^A_{1a0}(\gamma_2^+)$ Fe^{2+} level (Fig. 3.12) and the E_g phonon [3.118]. The Hamiltonian describing this interaction for $T > T_N$ can be written as

$$\mathcal{H}_{int} = V[E_\theta(a_\theta^\dagger + a_\theta) + E_\varepsilon(a_\varepsilon^\dagger + a_\varepsilon)], \tag{3.19}$$

where E_θ, E_ε are electronic orbital operators connecting the ground and $T^A_{1a0}(\gamma_2^+)$ electronic levels, $a^\dagger$ and a are phonon creation and annihilation operators for representations θ and ε of the doubly-degenerate phonon, and V is the strength of the interaction. The eigenvalues of the perturbed states are

$$E_{1,2} = \tfrac{1}{2}\hbar(\Omega_0 + \omega_0) \pm [2V^2 + \tfrac{1}{4}\hbar^2(\Omega_0 - \omega_0)^2]^{1/2}, \tag{3.20}$$

where $\hbar\Omega_0$ and $\hbar\omega_0$ are the energies of the uncoupled phonon and electronic excitation, respectively. The envelope of the coupled mode spectrum was obtained by mapping a broad Gaussian line profile at $\hbar\omega_0$ onto the solutions E and then adding a sharper component at $\hbar\Omega_0$ from unperturbed phonon transitions. The spectral lineshape at 24 K (Fig. 3.14) was reproduced when $\hbar\Omega_0 = 146\ cm^{-1}$, $\hbar\omega_0 = 149\ cm^{-1}$, and $V = 5\ cm^{-1}$. At higher temperatures, $\omega_0 \simeq \Omega_0$ and a symmetric lineshape is obtained. Thus, the temperature dependence of the lineshape was explained in terms of the temperature dependences of the resonating excitations. The analysis for $T < T_N$ is more complicated because of the magnetic ordering. However, the major features in the spectrum including the two sharp lines observed for temperatures just below T_N (see Fig. 3.14) could be explained by an extension of the above theory to include the molecular field. More recent Raman experiments on the mixed crystal $Fe_{1-x}Mn_xCl_2$ [3.108] have revealed the presence of another electronic level, $E_{u\pm}(\gamma_3^+)$, at slightly higher frequency (Fig. 3.12), which may also couple to the E_g phonon. The structure observed on the broad band near $165\ cm^{-1}$ at 12 K in Fig. 3.14 is part of the evidence for this additional interaction, which has not yet been formulated in detail.

Another kind of electron-phonon coupling has been observed in $CoBr_2$ [3.75]. The phenomenon is evident in the Raman spectrum through an unusual

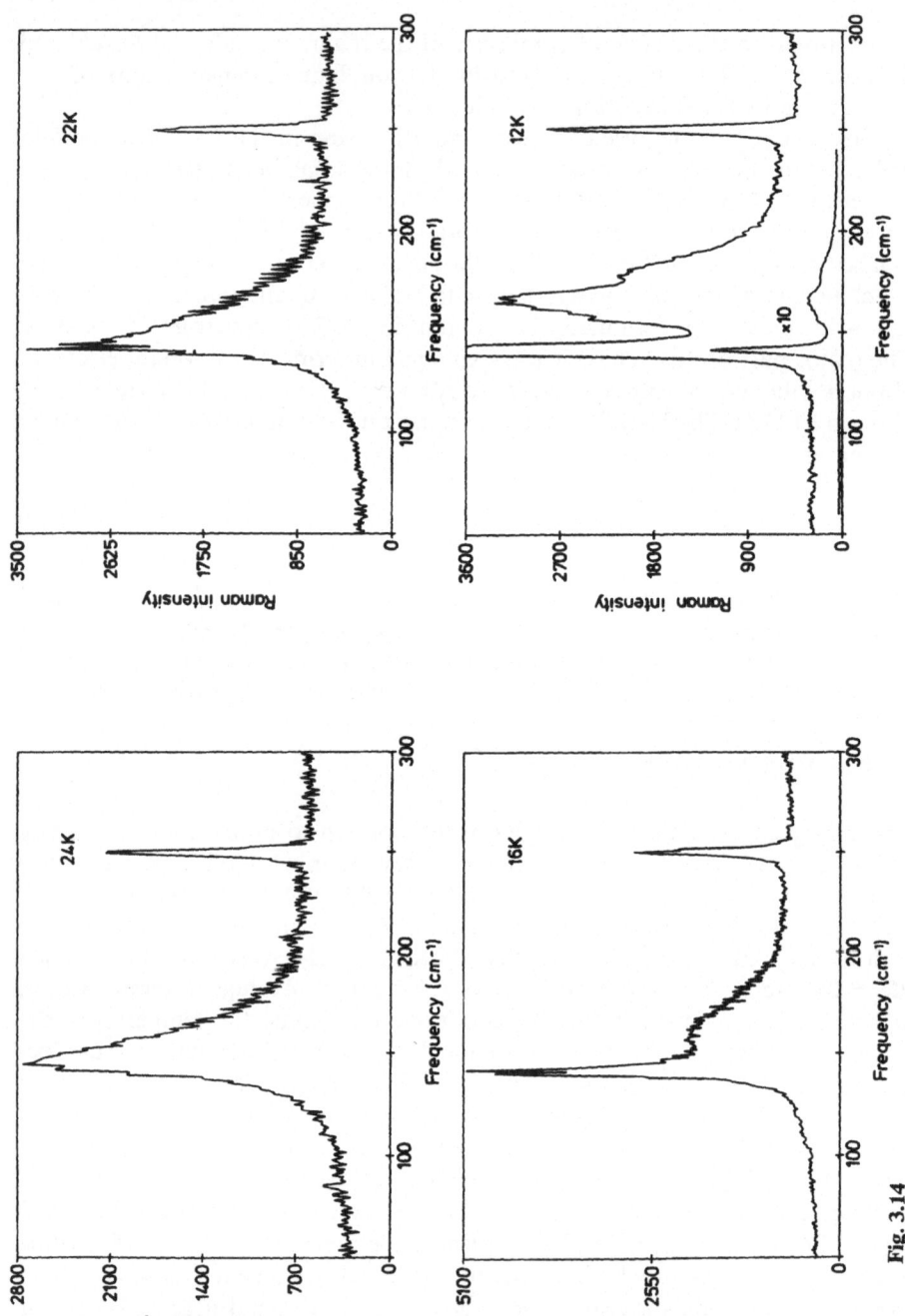

Fig. 3.14. The Raman spectrum of $FeCl_2$ showing the temperature dependent electron-phonon coupling in the 150 cm^{-1} region ($T_N = 23.5$ K) [3.118]

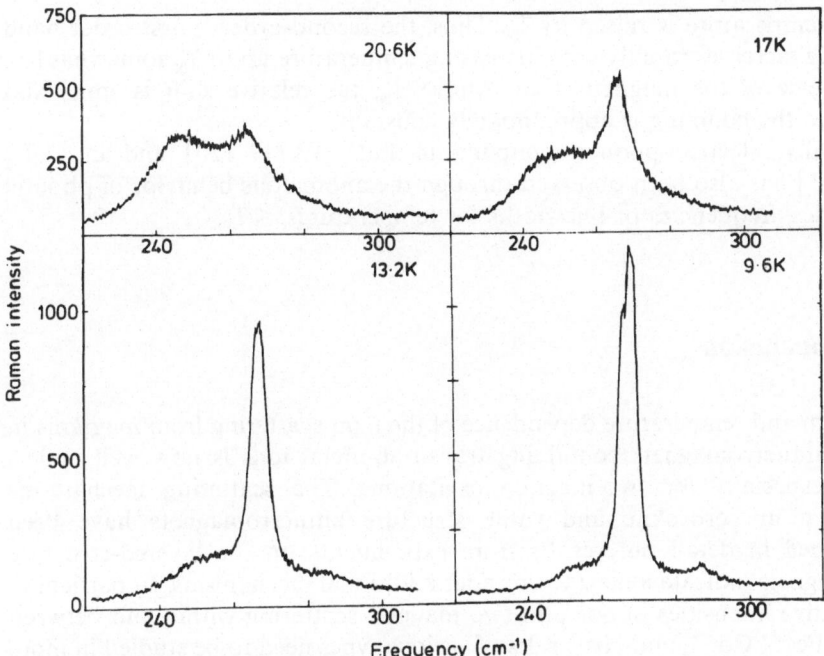

Fig. 3.15. The Raman spectrum of $CoBr_2$ showing the temperature dependent electron-phonon coupling in the 265 cm^{-1} region ($T_N = 19 \text{ K}$) [3.75]

temperature dependence as is shown in Figs. 3.15 and 3.13a. There are three main features at ~ 250, ~ 267, and $\sim 288 \text{ cm}^{-1}$ which exhibit differing temperature dependences. The centre peak is readily assigned to the Co^{2+} electronic transition Γ_6^+, $\gamma_4^+ \rightarrow {}^{(1)}\Gamma_8^+$, $\gamma_5^+ + \gamma_6^+$ (Fig. 3.11). This peak shifts and splits into two in the antiferromagnetic phase, as expected. The weaker peak to higher energy is due to combination scattering from the exciton pair at $\sim 265 \text{ cm}^{-1}$ plus the zone-centre magnon. The 250 cm^{-1} band is more unusual in character. The integrated intensity of this band has near-zero intensity at 6 K and increases rapidly in intensity up to $T_N = 19 \text{ K}$, whereupon there is a relatively slow increase in intensity with increasing temperature. The band moves slightly to lower frequency with increasing temperature. Such behavior was explained by resonant coupling between the 250 cm^{-1} band and the exciton pair/electronic band at $\sim 267 \text{ cm}^{-1}$ [3.75]. The 250 cm^{-1} band was attributed to two-phonon scattering from the A_{1g} plus E_g phonons and has the correct symmetry for Fermi resonance with the electronic excitation. Thus, the combination band gains intensity at the expense of the resonating first-order band, while simultaneously maintaining constant total integrated intensity. The temperature dependence of the intensity is explained by the relative band positions. The 250 cm^{-1} band is relatively insensitive to change in temperature, whereas the exciton pair shifts to lower frequency and the splitting disappears as the temperature is raised to T_N. Thus, the second-order – first-order band coupling increases rapidly with increasing temperature up to T_N somewhat like the inverse of the magnetisation. Above T_N, the relative shift is small and therefore the coupling is approximately constant.

Finally, electron-phonon coupling in FeF_2 [3.119, 120] and in CoF_2 [3.38, 41] has also been observed through the anomalous behavior of phonon intensities, frequencies or linewidths for temperatures $\lesssim T_N$.

3.5 Conclusion

The form and temperature dependence of the light scattering from magnons in *simple* antiferromagnets containing transition-metal ions is now well understood, especially for two-magnon excitations. The scattering mechanisms operating in perovskite and rutile structure antiferromagnets have been established in detail but results from experiments on the layered-structure metamagnets indicate a need to consider additional mechanisms. In particular, the relative intensities of one and two-magnon scattering within and between Mn^{2+}, Fe^{2+}, Co^{2+}, and Ni^{2+} salts of various types need to be studied in more detail, especially with regard to resonance effects. The role of spin-orbit coupling in the scattering mechanism also deserves further investigation. Mixed magnetic systems are not so well understood, except for the isolated impurity limit. The theoretical understanding of low temperature results is patchy and there is a need for further, more detailed calculations of the spectral lineshapes. Discrepancies between observed and expected relative intensities of two-magnon excitations in $KNi_{1-x}Mn_xF_3$ [3.16] form another example of the unknown role of resonance effects on the scattering intensity. The detailed form of the magnetic scattering at higher temperatures has yet to be investigated theoretically.

Electronic Raman scattering from transition metal ions in paramagnets and at low concentrations in diamagnetic hosts is now well established as a technique for studying low-lying excitations. Observed ground-term energy levels of Co^{2+} and Fe^{2+} ions are readily understood in terms of single-ion crystal field theory, provided that due allowance is made for their interactions with higher-lying terms. Questions that remain are the precise role of the dynamic Jahn-Teller effect within the ground multiplet of Fe^{2+} ions and the microscopics of the electron-photon scattering mechanism. Recent work on CdI_2:Fe suggests that third, fourth or even fifth-order mechanisms need to be considered [3.98]. Results of light scattering studies of excitons in antiferromagnetically-ordered compounds have been explained in terms of single-ion theory within a molecular field approximation. More exact calculations are needed by extending spin-wave theory to include these higher-lying spin-orbit coupled states. Interest is now centred on the excitations of isolated clusters of magnetic ions in a diamagnetic host for impurity concentrations up to the percolation limit. The limited results obtained to date show that a large inter-ion exchange constant is needed for Raman experiments to be able to resolve excitations arising within the various clusters.

Surface physics has become a topical subject with prospects for sustained interest [3.121]. The study of surfaces involves a determination of their atomic and electronic structure and energy levels. Light scattering has already made considerable contributions to this field [3.122] and in view of the recent observations of light scattering from magnons on the surfaces of ferromagnets [3.123], it would be informative to study surface excitations of antiferromagnets.

Acknowledgement. I am grateful to my colleague I. W. Johnstone for helpful comments and criticism.

References

3.1   P. A. Fleury, S. P. S. Porto, L. E. Cheeseman, H. J. Guggenheim: Phys. Rev. Lett. **17**, 84–87 (1966)

3.2   See, for example, the review by W. Hayes, R. Loudon: *Scattering of Light by Crystals* (Wiley-Interscience, New York 1978) Chap. 7, pp. 288–292. The first observation of electronic Raman scattering from a rare-earth ion was made in this laboratory in 1963 [J. T. Hougen, S. Singh: Phys. Rev. Lett. **10**, 406–407 (1963)]

3.3   D. J. Lockwood: "Raman Spectra of Cadmium Chloride and Cadmium Bromide", in *Light Scattering Spectra of Solids*, ed. by G. B. Wright (Springer, Berlin, Heidelberg, New York 1969) pp. 75–84; D. J. Lockwood, J. H. Christie: Paper presented at the 2nd Int. Conf. on Raman Spectroscopy, Oxford, Sept. 13–17 (1970)

3.4   P. A. Fleury: Int. J. Magnetism **1**, 75–83 (1970)

3.5   S. R. Chinn, R. W. Davies, H. J. Zeiger: "Recent Topics in Two-Magnon Optical Raman Scattering", in Am. Inst. Phys. Conf. Proc., Vol. 5 (1972) pp. 317–336

3.6   U. Balucani, V. Tognetti: Riv. Nuovo Cim. **6**, 39–104 (1976)

3.7   M. F. Thorpe: "Magnetic Excitations", in *Correlation Functions and Quasiparticle Interactions in Condensed Matter*, ed. by J. Woods Halley (Plenum Press, New York 1978) pp. 261–305

3.8   W. Hayes, R. Loudon: *Scattering of Light by Crystals* (Wiley-Interscience, New York 1978) Chap. 6, pp. 239–285

3.9   C. Kittel: *Quantum Theory of Solids* (Wiley, New York 1963) p. 60

3.10 R. A. Cowley, W. J. L. Buyers: Rev. Mod. Phys. **44**, 406–450 (1972)

3.11 D. E. Cox: IEEE Trans. MAG-**8**, 161–182 (1972)

3.12 D. T. Teaney, M. J. Freiser, R. W. H. Stevenson: Phys. Rev. Lett. **9**, 212–214 (1962)

3.13 G. R. Davidson, M. Eibschütz, H. J. Guggenheim: Phys. Rev. B**8**, 1864–1880 (1973)

3.14 P. Moch, C. Dugautier: Phys. Lett. **43A**, 169–170 (1973)

3.15 P. Moch, C. Dugautier: "One Magnon Light Scattering in KNiF₃", in Proc. 1973 Int. Conf. Magnetism, Vol. 1 (Nauka, Moscow 1974) pp. 185–188

3.16 D. J. Lockwood, G. J. Coombs, R. A. Cowley: J. Phys. C**12**, 4611–4622 (1979)

3.17 F. G. Bass, M. I. Kaganov: Zh. Eksp. Tear. Fiz. **37**, 1390–1393 (1959) [English transl.: Sov. Phys.-JETP **37**, 986–988 (1960)]

3.18 R. J. Elliott, R. Loudon: Phys. Lett. **3**, 189–191 (1963)

3.19 Y. R. Shen, N. Bloembergen: Phys. Rev. **143**, 372–384 (1966)

3.20 P. A. Fleury, R. Loudon: Phys. Rev. **166**, 514–530 (1968)

3.21 J. Nouet, D. J. Toms, J. F. Scott: Phys. Rev. B**7**, 4874–4883 (1973)

3.22 D. J. Toms, J. F. Ryan, J. F. Scott, J. Nouet: Phys. Lett. **44A**, 187–188 (1973)

3.23 P. A. Fleury: Phys. Rev. Lett. **21**, 151–153 (1968)

3.24 R. J. Elliott, M. F. Thorpe: J. Phys. C **2**, 1630–1643 (1969)

3.25 S. R. Chinn, H. J. Zeiger, J. R. O'Connor: Phys. Rev. B **3**, 1709–1735 (1971)

3.26 D.J.Lockwood, G.J.Coombs: J. Phys. C **8**, 4062–4070 (1975)
3.27 U.Balucani, V.Tognetti: Phys. Rev. B **8**, 4247–4257 (1973)
3.28 R.W.Davies, S.R.Chinn, H.J.Zeiger: Phys. Rev. B **4**, 992–1004 (1971)
3.29 C.R.Natoli, J.Ranninger: J. Phys. C **6**, 345–369 (1973)
3.30 M.G.Cottam: J. Phys. C **5**, 1461–1474 (1972)
3.31 U.Balucani, V.Tognetti: Phys. Rev. B **16**, 271–279 (1977)
3.32 S.Racine, J.Cipriani, C.Pontikis: C. R. Acad. Sci. (Paris) B **274**, 16–18 (1972)
3.33 G.H.Johnson: "Raman Scattering in Antiferromagnetic Perovskites: $KCoF_3$ and $RbCoF_3$"; Ph. D. Thesis, Cornell University (1975)
3.34 L.S.Lichtmann, H.Temkin, D.B.Fitchen: Solid State Commun. **25**, 127–131 (1978)
3.35 J.W.Halley: Phys. Rev. Lett. **41**, 1605–1608 (1978)
3.36 F.Barocchi, P.Mazzinghi, V.Tognetti, M.Zoppi: Solid State Commun. **25**, 241–243 (1978)
3.37 F.M.Johnson, A.H.Nethercot: Phys. Rev. **114**, 705–716 (1959)
3.38 R.M.Macfarlane: Phys. Rev. Lett. **25**, 1454–1457 (1970)
3.39 M.T.Hutchings, M.F.Thorpe, R.J.Birgeneau, P.A.Fleury, H.J.Guggenheim: Phys. Rev. B **2**, 1362–1373 (1970)
3.40 P.L.Richards: Phys. Rev. **138**, A1769–A1775 (1965)
3.41 P.Moch, J.P.Gosso, C.Dugautier: "Raman Scattering by Excitons in CoF_2 and Localised Modes in $MnF_2:Co^{2+}$", in *Light Scattering in Solids*, ed. by M.Balkanski (Flammarion, Paris 1971) pp. 138–145
3.42 P.A.Fleury: Phys. Rev. **180**, 591–593 (1969)
3.43 J.P.Gosso, P.Moch: "Raman Scattering by Magnons in Substitutionally Disordered Co–Mn Fluorides", in *Light Scattering in Solids*, ed. by M.Balkanski, R.C.C.Leite, S.P.S.Porto (Flammarion, Paris 1976) pp. 214–218
3.44 D.-M.Hwang, T.T.Chen, H.Chang: Solid State Commun. **18**, 1099–1101 (1976)
3.45 P.A.Fleury: "Antiferromagnetic Magnons at Finite Temperatures and in High Magnetic Fields", in *Light Scattering in Solids*, ed. by M.Balkanski (Flammarion, Paris 1971) pp. 151–156
3.46 M.G.Cottam: J. Phys. C **8**, 1933–1949 (1975)
3.47 P.A.Bates, M.G.Cottam, S.R.P.Smith: Solid State Commun. **33**, 129–132 (1980)
3.48 R.Loudon: J. Raman Spectr. **7**, 10–14 (1978)
3.49 M.F.Thorpe: J. Appl. Phys. **41**, 892–893 (1970)
3.50 W.J.Brya, R.R.Bartkowski, P.M.Richards: "Light Scattering from Spin Fluctuations in MnF_2", in Am. Inst. Phys. Conf. Proc., Vol. 5 (1972) pp. 339–343
3.51 W.J.Brya, P.M.Richards, P.A.Fleury: "Frequency Moments of Magnetic Light Scattering in NiF_2", in Am. Inst. Phys. Conf. Proc., Vol. 10 (1973) pp. 729–733
3.52 W.J.Brya, P.M.Richards: Phys. Rev. B **9**, 2244–2263 (1974)
3.53 U.Balucani, V.Tognetti, M.G.Pini: Nuovo Cimento B **46**, 81–97 (1978)
3.54 N.M.Amer, T.-C.Chiang, Y.R.Shen: Phys. Rev. Lett. **34**, 1454–1457 (1975)
3.55 R.Merlin: J. Physique **41**, C5, 233–239 (1980)
      G.Güntherodt, R.Merlin: In *Light Scattering in Solids IV*, ed. by M.Cardona, G.Güntherodt, Topics Appl. Phys., Vol. 54 (to be published)
3.56 M.G.Cottam: J. Phys. C **11**, 151–164 (1978)
3.57 R.E.Camley: Phys. Rev. Lett. **45**, 283–286 (1980)
3.58 M.G.Cottam: J. Phys. C **11**, 165–181 (1978)
3.59 M.K.Wilkinson, J.W.Cable, E.O.Wollan, W.C.Koehler: Oak Ridge Nat. Lab. Phys. Div. Semiann. Prog. Rep. ORNL-2430, 65–74 (1957)
3.60 A.Adam, D.Billerey, C.Terrier, R.Mainard, L.P.Regnault, J.Rossat-Mignod, P.Mériel: Solid State Commun. **35**, 1–5 (1980)
3.61 D.J.Robbins, P.Day: J. Phys. C **9**, 867–882 (1976)
3.62 I.W.Johnstone, D.J.Lockwood, G.Mischler: J. Phys. C **11**, 2147–2164 (1978)
3.63 R.J.Birgeneau, W.B.Yelon, E.Cohen, J.Makovsky: Phys. Rev. B **5**, 2607–2615 (1972)
3.64 W.B.Yelon, C.Vettier: J. Phys. C **8**, 2760–2768 (1975)
3.65 D.J.Lockwood, I.W.Johnstone, G.Mischler, P.Carrara: Solid State Commun. **25**, 565–568 (1978)
3.66 K.R.A.Ziebeck, C.Escribe, J.P.Redoules, J.Gelard: Solid State Commun. **23**, 867–870 (1977)

3.67   G.Mischler, M.C.Schmidt, D.J.Lockwood, A.Zwick: Solid State Commun. 27, 1141–1146 (1978)
3.68   J.Margarino, J.Tuchendler, A.R.Fert, J.Gelard: Solid State Commun. 23, 175–178 (1977)
3.69   D.J.Lockwood, D.Bertrand, P.Carrara, G.Mischler, D.Billerey, C.Terrier: J. Phys. C 12, 3615–3620 (1979)
3.70   P.A.Lindgard, R.J.Birgeneau, J.Als-Nielsen, H.J.Guggenheim: J. Phys. C 8, 1059–1069 (1975)
3.71   G.Mischler, D.Bertrand, D.J.Lockwood, M.G.Cottam, S.Legrand: J. Phys. C 14, 945–960 (1981)
3.72   G.Mischler, A.Zwick, D.J.Lockwood, S.Legrand: J. Phys. C 15, L187–L191 (1982)
        D.J.Lockwood, G.Mischler, A.Zwick, I.W.Johnstone, G.C.Psaltakis, M.G.Cottam, S.Legrand, J.Leotin: J. Phys. C 15, 2973–2992 (1982)
3.73   M.T.Hutchings: J. Phys. C 6, 3143–3155 (1973)
3.74   A.Latiff Awang, M.G.Cottam: J. Phys. C 12, 137–147 (1979)
3.75   D.J.Lockwood, G.Mischler, I.W.Johnstone, M.C.Schmidt: J. Phys. C 12, 1955–1975 (1979)
3.76   P.Moch, G.Parisot, R.E.Dietz, H.J.Guggenheim: Phys. Rev. Lett. 21, 1596–1599 (1968)
3.77   A.Oseroff, P.S.Pershan: Phys. Rev. Lett. 21, 1593–1596 (1968)
3.78   M.F.Thorpe: Phys. Rev. Lett. 23, 472–474 (1969)
3.79   S.M.Rezende, Cid B. de Araujo, E.Montarroyos, V.Jaccarino: Solid State Commun. 35, 627–630 (1980)
3.80   R.A.Cowley: "Excitations of Substitutionally Disordered Antiferromagnets", in Am. Inst. Phys. Conf. Proc., Vol. 29 (1976) pp. 243–247
3.81   P.A.Fleury, W.Hayes, H.J.Guggenheim: J. Phys. C 8, 2183–2190 (1975)
3.82   J.P.Gosso, P.Moch: Physica 89B, 209–212 (1977)
3.83   M.Buchanan, W.J.L.Buyers, R.J.Elliott, R.T.Harley, W.Hayes, A.M.Perry, I.D.Saville: J. Phys. C 5, 2011–2026 (1972)
3.84   E.Montarroyos, Cid B. de Araujo, S.M.Rezende: J. Appl. Phys. 50, 2033–2035 (1979)
3.85   D.Bertrand, A.R.Fert, S.Legrand, J.P.Redoules, M.C.Schmidt: J. Phys. C 14, 1789–1797 (1981)
3.86   W.J.L.Buyers, D.E.Pepper, R.J.Elliott: J. Phys. C 6, 1933–1952 (1973)
3.87   W.J.L.Buyers, D.E.Pepper, R.J.Elliott: J. Phys. C 5, 2611–2628 (1972)
3.88   T.Tawaraya, K.Katsumata: Solid State Commun. 32, 337–341 (1979)
        P.Wong, P.M.Horn, R.J.Birgeneau, C.R.Safinya, G.Shirane: Phys. Rev. Lett. 45, 1974–1977 (1980)
3.89   A.Azima, P.Grunberg, J.Hoff, J.A.Koningstein, J.Preudhomme: Chem. Phys. Lett. 7, 565–566 (1970)
3.90   J.H.Christie, D.J.Lockwood: Chem. Phys. Lett. 8, 120–122 (1971)
3.91   D.J.Lockwood, J.H.Christie: Chem. Phys. Lett. 9, 559–563 (1971)
3.92   D.J.Lockwood, N.L.Rowell, W.J.L.Buyers: "Electronic Raman Scattering from Co^{2+} Clusters in $KZnF_3$", in Proc. 7th Int. Conf. Raman Spectroscopy, ed. by W.F.Murphy (North-Holland, Amsterdam 1980) pp. 154–155
3.93   J.P.Gosso, J.Quazza, P.Moch, J.Labbe: Physica 86–88B, 722–724 (1977)
3.94   Cid B. de Araujo: Phys. Rev. B 22, 266–272 (1980)
3.95   J.H.Christie, I.W.Johnstone, G.D.Jones, K.Zdansky: Phys. Rev. B 12, 4656–4665 (1975)
3.96   G.D.Jones, C.W.Tomblin: Phys. Rev. B 18, 5990–5994 (1978)
3.97   I.W.Johnstone, L.Dubicki: J. Phys. C 13, 121–130 (1980)
3.98   I.W.Johnstone, L.Dubicki: "Resonance Enhanced Electronic Raman Scattering in $CdI_2:Fe^{2+}$", in Proc. 7th Int. Conf. Raman Spectroscopy, ed. by W.F.Murphy (North-Holland, Amsterdam 1980) pp. 152–153
3.99   I.W.Johnstone, G.D.Jones: Phys. Rev. B 15, 1297–1306 (1977)
3.100   I.W.Johnstone, D.J.Lockwood, M.W.C.Dharma-wardana: Solid State Commun. 36, 593–597 (1980)
3.101   B.F.Gächter, J.A.Koningstein: Chem. Phys. Lett. 23, 28–29 (1973)
3.102   S.R.Chinn, H.J.Zeiger: "Electronic Raman Scattering in FeF_2", in Am. Inst. Phys. Conf. Proc., Vol. 5 (1972) pp. 344–348
3.103   J.T.Hoff, J.A.Koningstein: Chem. Phys. 1, 232–237 (1973)

3.104 J.T.Hoff, P.A.Grunberg, J.A.Koningstein: Appl. Phys. Lett. **20**, 358–359 (1972)
3.105 W.J.L.Buyers, T.M.Holden, E.C.Svensson, R.A.Cowley, M.T.Hutchings: J. Phys. C **4**, 2139–2159 (1971)
3.106 R.A.Cowley, W.J.L.Buyers, P.Martel, R.W.H.Stevenson: J. Phys. C **6**, 2997–3019 (1973)
3.107 R.M.Macfarlane: Solid State Commun. **15**, 535–538 (1974)
3.108 I.W.Johnstone, D.J.Lockwood, D.Bertrand, G.Mischler: J. Phys. C **13**, 2549–2566 (1980)
3.109 R.M.Couto, J.H.Nicola: Solid State Commun. **30**, 713–715 (1979)
3.110 R.M.Macfarlane, H.Morawitz: "Light Scattering and Exciton Relaxation in CoF$_2$", in *Light Scattering in Solids*, ed. by M.Balkanski (Flammarion, Paris 1971) pp. 133–138
3.111 A.Ishikawa, T.Moriya: J. Phys. Soc. Japan **30**, 117–128 (1971)
3.112 J.H.Christie, D.J.Lockwood: "Electronic Raman Scattering in CoCl$_2$", in *Light Scattering in Solids*, ed. by M.Balkanski (Flammarion, Paris 1971) pp. 145–150
3.113 A.T.Abdalian, P.Moch: J. Appl. Phys. **49**, 2189–2191 (1978)
3.114 G.Laurence, D.Petitgrand: Phys. Rev. B **8**, 2130–2138 (1973)
3.115 C.A.Bates: Phys. Reports **35**, 187–304 (1978)
3.116 T.M.Holden, W.J.L.Buyers, E.C.Svensson, R.A.Cowley, M.T.Hutchings, D.Hukin, R.W.H.Stevenson: J. Phys. C **4**, 2127–2138 (1971)
3.117 N.Suzuki, Y.Takaoka: Solid State Commun. **21**, 365–370 (1977)
3.118 I.W.Johnstone, D.J.Lockwood, G.Mischler, J.R.Fletcher, C.A.Bates: J. Phys. C **11**, 4425–4438 (1978)
3.119 J.L.Sauvajol, R.Almairac, C.Benoit, A.M.Bon: "Absolute First Order Raman Polarizabilities in MgF$_2$, MnF$_2$, and FeF$_2$", in *Lattice Dynamics*, ed. by M.Balkanski (Flammarion, Paris 1978) pp. 199–203
3.120 A.Stasch: "Theory of Raman Scattering by Coupled Phonon-Magnon Excitations in FeF$_2$", in *Lattice Dynamics*, ed. by M.Balkanski (Flammarion, Paris 1978) pp. 273–274
3.121 The National Academy of Sciences: *Science and Technology: A Five-Year Outlook* (Freeman, San Francisco 1979) pp. 163–170
3.122 See, for example, *Proc. 7th Int. Conf. Raman Spectroscopy*, ed. by W.F.Murphy (North-Holland, Amsterdam 1980) pp. 345–448
3.123 J.R.Sandercock: "Light Scattering from Thermally Excited Surface Phonons and Magnons", in *Proc. 7th Int. Conf. Raman Spectroscopy*, ed. by W.F.Murphy (North-Holland, Amsterdam 1980) pp. 364–367

Additional References with Titles

Ariai, J., Bates, P.A., Cottam, M.G., Smith, S.R.P.: The effects of linear and quadratic magneto-optical coupling in the one-magnon Raman spectrum of FeF$_2$. J. Phys. C**15** (1982, in press)
Brady Moreira, F.G., Fittipaldi, I.P.: New effective-medium approach for randomly diluted uniaxial antiferromagnets: application to Fe$_{1-x}$Zn$_x$F$_2$. Phys. Rev. B**24**, 6596–6599 (1981)
Izyumov, Yu.A., Gurin, O.B., Petrov, S.B.: The coupled magnon-phonon modes in magnetic crystals with Shubnikov symmetry. J. Magn. Magn. Mater. **24**, 291–298 (1981)
Johnstone, I.W., Jones, G.D., Lockwood, D.J.: Raman scattering from Co^{2+} excitations in the disordered 1D antiferromagnet CsMg$_{1-x}$Co$_x$Cl$_3$. J. Phys. C**15**, 2043–2058 (1982)
Pini, M.G., Rastelli, E., Tassi, A., Tognetti, V.: The influence of the anisotropy on the temperature renormalisation of the magnetic excitations in FeCl$_2$. J. Phys. C**14**, 3041–3055 (1981)
Pini, M.G., Tognetti, V., Giunti, C.: Magnetic excitations in the trigonal compounds CoBr$_2$ and CoCl$_2$. Solid State Commun. (1982, in press)
Pini, M.G., Tognetti, V., Rastelli, E., Tassi, A.: Temperature renormalization of the magnon energies of FeBr$_2$. Solid State Commun. **40**, 421–423 (1981)
Psaltakis, G.C., Cottam, M.G.: Theory of spin-wave interactions in $S=1$ two-sublattice uniaxial magnets. J. Phys. C**15** (1982, in press)
Rastelli, E., Tassi, A.: Ladder approximation for uniaxial metamagnets. J. Appl. Phys. **53**, 1870–1872 (1982)
Rastelli, E., Tassi, A., Pini, M.G., Rettori, A., Tognetti, V.: Kinematic consistency in ferromagnets with single-ion anisotropy: application to FeCl$_2$. J. Appl. Phys. **52**, 2246–2248 (1981)
Rezende, S.M., de Araujo, Cid B., Montarroyos, E.: Raman scattering by magnetic excitations in disordered FeF$_2$. J. Raman Spectrosc. **10**, 173–177 (1981)

4. Light Scattering by Superionic Conductors

W. Hayes

With 19 Figures

In general, ionic solids have values of the electrical conductivity σ immediately below the melting temperature T_m about four orders of magnitude smaller than in the melt. In LiF, for example, σ increases from about $10^{-8}\,\mathrm{Sm}^{-1}$ at room temperature to $10^{-1}\,\mathrm{Sm}^{-1}$ just below $T_m = 1140$ K. On melting, σ increases discontinuously to $10^3\,\mathrm{Sm}^{-1}$. However, a few types of ionic solids have values of σ in the crystalline state comparable with that in the melt and are referred to as superionics or fast-ion conductors. This high value of σ is a consequence of extensive disorder in a component sublattice of the solid and, looked at from a fundamental point of view, the study of superionics may be regarded as a study of disorder [4.1].

Although there are materials which show both high electronic and high ionic conductivity simultaneously, we shall confine our attention here to materials with negligible electronic conductivity. For a single mobile species, the dc conductivity $\sigma(0)$ is given by

$$\sigma(0) = N\mu(Ze), \tag{4.1}$$

where N is the concentration of the mobile species per unit volume, μ is the mobility of the ionic species and Ze is the ionic charge. Superionics are characterised by relatively large N ($\gtrsim 1$ particles per cent).

At finite temperatures, crystals contain defects such as vacancies and interstitials [4.2, 3] and these provide mechanisms for ionic movement. The ionic conductivity quite often follows an Arrhenius relationship of the form

$$\sigma(0) = (\sigma_0/T)\exp(-E_A/k_B T), \tag{4.2}$$

where E_A is composed of two parts, E_F determined by the energy of formation of a defect, and E_M, its activation energy for motion. A plot of $\log[\sigma(0)T]$ against $1/T$ will give a straight line whose slope is determined by E_A. In practice, relatively simple ionic solids such as PbF_2 have regions with different slopes. At low temperatures (region I, Fig. 4.1), conductivity arises from defects existing in the crystal because of the presence of impurities, so that E_A is determined primarily by the activation energy for motion only. Region II

---

1 The MKS unit of conductivity is Siemens per metre (Sm^{-1}); $1\,(\Omega\,\mathrm{cm})^{-1}$ is equivalent to $10^2\,\mathrm{Sm}^{-1}$.

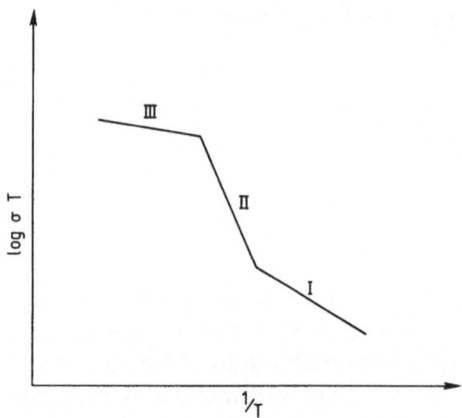

Fig. 4.1. Schematic representation of the dependence of ionic conductivity on temperature (see text)

(Fig. 4.1) is the region of intrinsic conductivity in which $\sigma(0)$ is dominated by defects created thermally in the lattice and there is a higher activation energy involving both the energy of creation and of motion. In most ionic solids, e.g., LiF, region II is followed by melting. However, in some ionic materials such as PbF_2, region II terminates with a specific heat anomaly associated with an order-disorder transition (Sect. 4.4). Region III then represents a highly-disordered crystalline state with a relatively low E_A, determined largely by E_M.

For noninteracting defects, the conductivity $\sigma(0)$ is related to a diffusion coefficient D for the mobile charge-carrying species through the Nernst-Einstein relation[2]

$$\sigma(0) = \frac{DN(Ze)^2}{k_B T} \tag{4.3a}$$

$$\sim \frac{vl^2}{6} \frac{N(Ze)^2}{k_B T}, \tag{4.3b}$$

where N and Ze are defined as in (4.1) and v^{-1} is the time it takes an ion to hop an average distance l. For a detailed understanding of superionics a knowledge of $\sigma(\omega)$ is required, covering a wide range of frequencies extending from microwave frequencies into the infrared [4.1].

For conventional ionic materials, e.g., alkali halides, values of E_A are $\sim 1\,\text{eV}$. By contrast, superionics have a relatively low E_A, $\sim 0.1\,\text{eV}$. With defect concentrations of the order of one mole per cent or greater, one gets $\sigma(0) \gtrsim 1\,\text{Sm}^{-1}$ at room temperature and $D \gtrsim 10^{-10}\,\text{m}^2\,\text{s}^{-1}$.

The superionic materials which have been subjected to the most intense study at a fundamental level in recent years may be classified as follows:

a) Silver and copper-based compounds, e.g., AgI, $RbAg_4I_5$, and CuI, in which the disorder occurs in the silver and copper sublattices. Here the

---

2 This value of D is not necessarily equal to the diffusion coefficient D^* measured by tracer methods, the ratio $D/D^* = H_R$ being known as the Haven ratio. Neutral defects can contribute to D^* but not to D [4.4].

Table 4.1. Approximate melting temperatures T_m of fluorite crystals and temperatures T_c for transition to the superionic state

Crystal	T_m [K]	T_c [K]
CaF_2	1633	1430
SrF_2	1723	1400
BaF_2	1553	1235
$SrCl_2$	1146	1000
PbF_2	1158	705

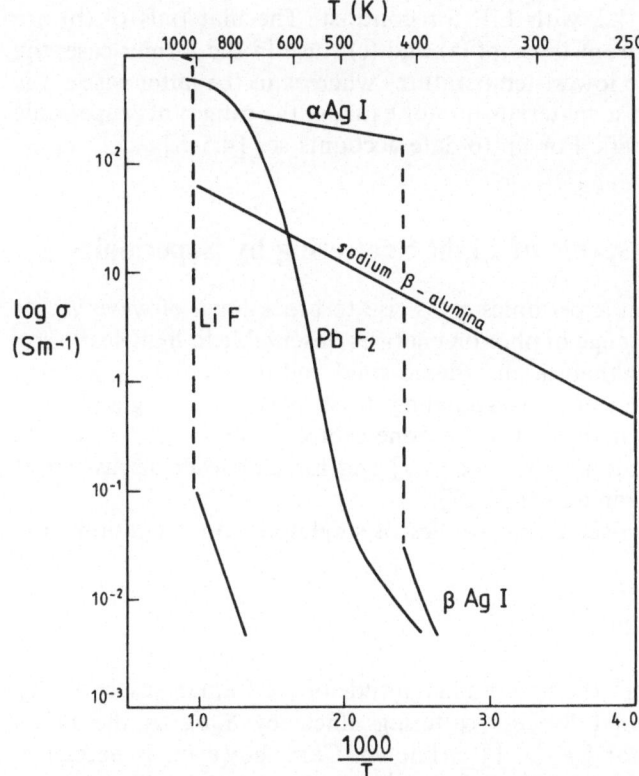

Fig. 4.2. Temperature dependence of ionic conductivity of three superionic materials and of a normal ionic material (LiF)

prototype material is AgI. At room temperature, AgI has the hexagonal wurtzite structure (space group C_{6v}^4). At $T_c = 147\,°C$, a first-order structural phase change occurs to a body-centred-cubic structure, referred to as α-AgI. The value of $\sigma(0)$ just below T_c is $\sim 3 \times 10^{-2}\,Sm^{-1}$. It increases abruptly at T_c to $\sigma(0) \simeq 1.3 \times 10^2\,Sm^{-1}$, with little change thereafter with increasing T [$\sigma(0)$ drops by about 12 % on going into the molten state].

b) Hexagonal compounds with the β-alumina structure. The prototype material here is sodium β-alumina which, if available in stoichiometric form, would have the formula $Na_2O:11\ Al_2O_3$. However, as-grown crystals are not stoichiometric, crystals from the melt containing up to 30 % excess sodium. This is one of the few materials which has appreciable ionic conductivity at room temperature, due largely to nonstoichiometry.

c) Fluorites, such as CaF_2, $SrCl_2$, and PbF_2, which exhibit broad specific heat anomalies at temperatures T_c well below T_M (Table 4.1). This well-defined premelting phenomenon is due to the development of extensive disorder in the anion sublattice, giving rise to a value of $\sigma(0)$ between T_c and T_M comparable to that of the melt. Some oxides with the fluorite structure, doped with aliovalent cations, are also good ionic conductors (Sect. 4.4).

The temperature dependence of $\sigma(0)$ for some of the superionics discussed above is shown in Fig. 4.2, with LiF for contrast. The materials of (b) are fundamentally different from those of (a) and (c) since in the former case, the disorder is present at the lowest temperatures whereas in the latter cases, it is thermally generated. These materials do not exhaust the range of superionic investigated in recent years. For up-to-date accounts see [4.5–8].

4.1 Some General Aspects of Light Scattering by Superionics

The extensive disorder in superionics gives rise to breakdown of wave vector conservation rules. The range of phonon energies observable in light scattering can, therefore, be greater than in an ordered solid and in some cases, we may expect to observe scattering corresponding to a perturbed single-phonon density of states rather than to discrete zone-centre phonons (for a general account of light scattering in solids see [4.9] and for an earlier discussion of light scattering by superionics, see [4.10]).

Frequently in Raman-scattering studies of single-phonon excitations, the relationship [4.9]

$$S(\omega) \propto |\chi'(\omega)|^2 [n(\omega) + 1] \tag{4.4}$$

is used to calculate $|\chi'(\omega)|^2$, the squared magnitude of the Raman susceptibility tensor, from the measured Stokes scattering efficiency $S(\omega)$, as the Bose-Einstein population factor $[n(\omega)+1]$ is known. Care, however, is necessary when comparing the reduced spectrum $|\chi'(\omega)|^2$ with theory for small ω because of the overlap of spurious Rayleight scattering and the contribution to $\chi'(\omega)$ due to hopping motion (see below).

The mobile ion in a superionic may stop at a lattice site sufficiently long to execute a number of cycles of vibration. A parameter of some consequence is the ratio of the time spent at a lattice site, τ_d, to the time spent in flight between sites, τ_f. From simple diffusion theory, describing the motion of uncorrelated particles we get

$$v^{-1} = \tau_d = l^2/6D, \tag{4.5}$$

where the parameters have already been defined for (4.3). Simple considerations suggest

$$\tau_f \sim \left(\frac{ml^2}{k_B T}\right)^{1/2}, \tag{4.6}$$

where m is the mass of the diffusing ion. If we make the crude approximation of applying (4.5, 6) to superionics and take $l \simeq 2\,\text{Å}$ and $D \simeq 10^{-10}\,\text{m}^2\,\text{s}^{-1}$ as typical values, we find $\tau_d \simeq 3\,\text{ps}$ from (4.5) and $\tau_f \simeq 0.5\,\text{ps}$ from (4.6). To the extent that $\tau_d > \tau_f$, a hopping or jump-diffusion model is appropriate to describe the dynamics of the moving ions [4.11].

Light scattering is caused by changes of electronic polarizability [4.9]. In superionics, polarizability changes associated with movement from site to site may cause quasi-elastic scattering. It is convenient to discuss this type of scattering in terms of the so-called relaxation modes of the moving ions [4.12]. These modes have well-defined symmetries and hence predictable light-scattering selection rules. For Raman scattering by phonons, only small displacements about equilibrium positions are involved. However, a variety of experiments suggest that for Ag^+ in α-AgI (Sect. 4.2.1), τ_d is only about 2.5 times greater than τ_f so that time of flight should be of consequence, leading to continuous models for the dynamics of superionics [4.13]. A general feature of the application of jump diffusion models and continuous models to the calculation of light scattering by superionics [4.11–13] is the prediction of a peak in the region of vibrational excitation ($\sim 100\,\text{cm}^{-1}$) corresponding to oscillatory motion at a lattice site, and also a quasi-elastic peak (centred on the laser frequency) arising from diffusive motion. It will be apparent later that reliable measurement of the quasi-elastic peak presents difficulties and that interpretation of low-frequency data is not without problems.

Nuclear magnetic resonance (nmr) techniques are useful for the study of the dynamics of superionics and we now make a slight deviation to discuss them. In a rigid lattice, the magnetic dipole-dipole interaction between neighboring nuclei gives a Gaussian linewidth, to a good approximation, and its width is of the order of the local magnetic field produced at the site of the spin by its neighbors, i.e., in the range 10^{-4} to 10^{-3} Tesla. However, in liquids and gases where fast relative motion of the ions occurs, linewidths are an order of magnitude smaller than in a rigid lattice and the lineshape is closer to Lorentzian than to Gaussian. If the nuclear spins are in rapid relative motion, the local magnetic field seen by a spin will fluctuate rapidly in time. Only the relatively small average value taken over a time which is long compared to the fluctuations will be seen. The rate of fluctuation of the local field is described by a fluctuation time τ_c which, for simple diffusion theory, is the same as τ_d discussed above. By measuring the temperature dependence of the nuclear spin-lattice relaxation time, the temperature dependence of τ_d can be determined. This is sometimes found to obey an Arrhenius relationship of the form

$$\frac{1}{\tau_d} = v_0 \exp(-U/k_B T), \tag{4.7}$$

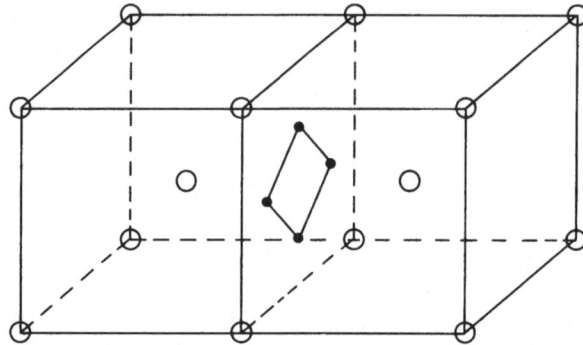

Fig. 4.3. Schematic representation of the structure of α-AgI. The open circles represent I⁻ ions on a *bcc* lattice. For clarity only, four of the sites with approximate tetrahedral symmetry available to Ag⁺ ions, are shown by filled circles. There are twelve such sites (12*d*) in the unit cell

where v_0 is an "attempt frequency" and U is the height of the barrier against diffusion.

Both v_0 and U can be extracted from nmr data. A question often asked is whether or not v_0 is related to the vibrational frequency $\bar{v}$ of a mobile ion residing in a potential well between hops. Simple classical diffusion theory for uncorrelated motions predicts that $v_0 = \bar{v}$ [4.13a]. We shall see later, however (Sect. 4.3), that care is required in making comparisons between v_0 measured by nmr and $\bar{v}$ measured by Raman scattering.

4.2 Light Scattering by Silver and Copper Compounds

4.2.1 Silver Compounds

a) Silver Iodide

Superionic silver iodide has received considerable attention over a period of more than fifty years. The structure of the room-temperature (β)-phase is now generally accepted to be ideal wurtzite. Single crystals do not normally survive the first-order transition from the β-phase to the superionic α-phase. However, techniques were recently developed [4.14] which enable small single crystals to survive the phase change and neutron diffraction studies of these crystals resulted in a reliable structural determination of the α-phase. This work [4.14] showed that in the α-phase the iodine ions form a fairly rigid *bcc* lattice and that the silver ions occupy interstices between them. There are 42 well-defined interstitial sites for Ag⁺ ions in the unit cell composed of 12 equivalent tetrahedrally-coordinated positions (12*d*) (Fig. 4.3), 24 three-fold coordinated sites (24*h*) and 6 octahedrally-coordinated sites. The largest available volume for an Ag⁺ ion is inside a tetrahedron and the smallest inside an octahedron. The neutron studies show that the silver ions are located primarily in the tetrahedral sites $\frac{1}{4}$, 0, $\frac{1}{2}$. The octahedral sites are minima in the silver density. Appreciable silver density occurs at 0.4, 0, 0.4, a three-fold coordinated site

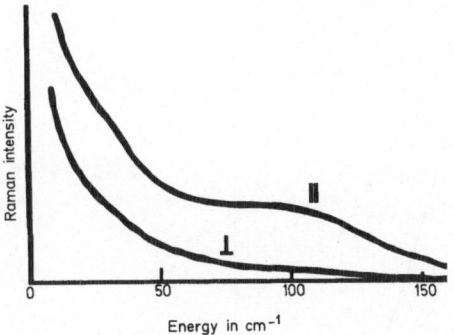

Fig. 4.4. Raman scattering from polycrystalline α-AgI at 340 °C. Polarized and depolarized Raman spectra are measured with the polarization of the incident and scattered light parallel and perpendicular, respectively, and are denoted ∥ and ⊥ [4.18]

bridging two tetrahedral sites. Structural studies based on EXAFS methods reach similar conclusions [4.15]. A variety of experiments suggests that motion of Ag^+ ions between neighboring tetrahedral sites is the basic step in migration. It would seem that nearest-neighbor tetrahedral sites cannot be occupied simultaneously since their separation is 1.79 Å, whereas the diameter of Ag^+ is 2.52 Å.

The wurtzite structure of β-AgI gives rise to polar Raman scattering (for a discussion of this type of scattering, see Chap. 4 of [4.9]). The Raman spectrum has been investigated by a number of authors (see, e.g., [4.16]) showing sharp A_1, E_1, and E_2-type optic modes at low temperatures in the range 106–124 cm^{-1} and, in addition, a sharp mode of E_2 symmetry at 17 cm^{-1}. A neutron study of the vibrations of β-AgI by *Bührer* and *Brüesch* [4.17] showed that the 17 cm^{-1} mode is dispersionless. These authors suggested that this low-frequency mode arises from Ag^+–I^- bond bending and that its eigenvector is such that it is intimately involved in the β→α structural phase transition.

In the α-phase, the Raman scattering at low energy is intense (Fig. 4.4; [4.18]). The spectrum is broad, extending out to ~240 cm^{-1}. There is a weak shoulder in the scattering at 30 cm^{-1} and a readily observable shoulder at ~110 cm^{-1}. There has been some controversy over the interpretation of this spectrum, with *Delaney* and *Ushioda* [4.19] taking the view that the observed intensity is mostly due to multiphonon processes. However, *Alben* and *Burns* [4.20] take a different view. The latter authors calculated the harmonic lattice dynamics of β-AgI using the equation-of-motion method. The crystal structure is approximated by 1000–2000 atoms with periodic boundary conditions. They use a simple potential which includes only nearest-neighbor forces and manage to reproduce the measured phonon dispersion curves of β-AgI. The same potential is used to calculate the lattice dynamics of α-AgI. This phase is modelled with Ag^+ ions in 12d sites only (Fig. 4.3) and it is assumed that the nearest-neighbor sites of an occupied 12d site are empty (see above). The reduced Raman intensity (for a definition see Sect. 6) is calculated assuming static disorder for Ag^+ ions and this is compared in Fig. 4.5 with the reduced intensity obtained from the measured data.

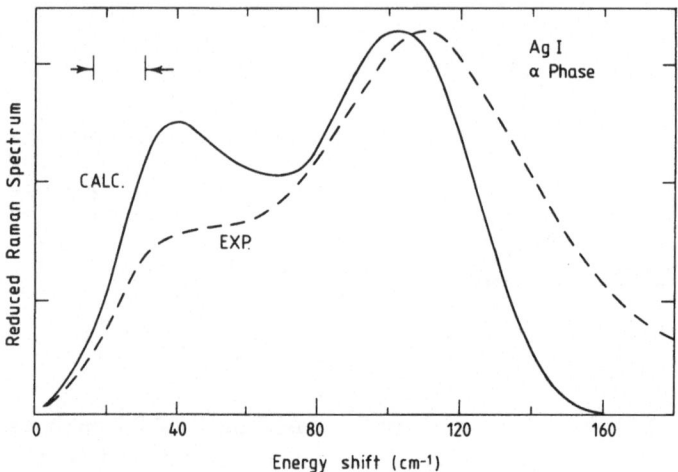

Fig. 4.5. Comparison of calculated and observed reduced Raman spectrum of α-AgI [4.20]

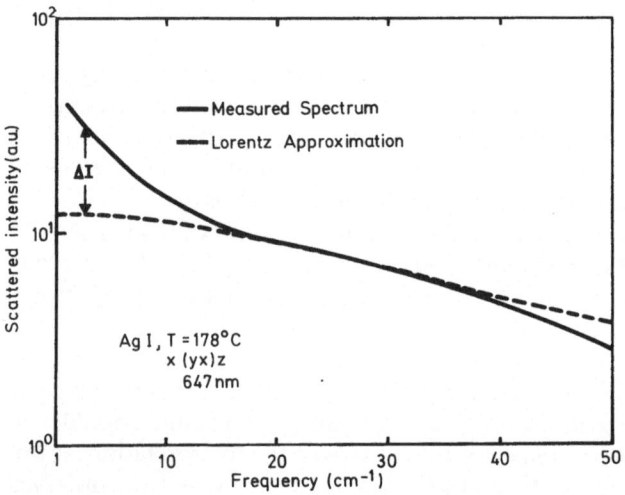

Fig. 4.6. Depolarized Raman scattering of α-AgI at low frequency. The dashed curve is a Lorentzian approximation to the broad component of the scattering [4.22]

It is apparent (Fig. 4.5) that there is reasonable agreement between calculations and observation below about $120\,cm^{-1}$. The two main calculated peaks arise from acoustic modes associated with Ag^+–I^- bond bending (low frequency peak) and from optic modes associated with symmetric breathing of iodine tetrahedra surrounding the silver ions (high frequency peak). The scattering intensity at energies higher than $120\,cm^{-1}$ may be due to multiphonon processes [4.19].

The diffusive motion of the Ag^+ ion in the α-phase is expected to give scattering with energies of the order of a few cm^{-1} [4.21]. The quasi-elastic light-scattering spectrum of α-AgI was measured by *Winterling* et al. [4.22]. They used an unoriented single crystal of α-AgI and 647 nm laser light, with a double monochromator for $\omega > 2\,\text{cm}^{-1}$ and a triple-pass Fabry-Perot interferometer in tandem with a monochromator for $\omega < 2\,\text{cm}^{-1}$. They fitted their experimental data with two Lorentzians centred on $\omega = 0$ (Fig. 4.6). One had a FWHM $= 32\,\text{cm}^{-1}$ at $178\,^\circ\text{C}$. The other, superimposed on the first, had a FWHM $= 3.8\,\text{cm}^{-1}$. No clear-cut explanation of the wider curve was offered. It was suggested that the narrow component was due to hopping of Ag^+ ions between neighboring $12d$ sites. The relationship (6.5) with $D = 2 \times 10^{-10}\,\text{m}^2\,\text{s}^{-1}$ [4.23] and with $\tau_d = 1.4\,\text{ps}$ obtained from the relation FWHM $= (2\pi\tau_d)^{-1}$, gives $l = 1.4\,\text{Å}$. This average jump distance is comparable with the separation of nearest-neighbor $12d$ sites (1.79 Å). It should be said that this estimate is considerably smaller than a value $l = 5\,\text{Å}$ obtained from a somewhat similar analysis of quasi-elastic neutron scattering measurements of polycrystalline α-AgI [4.24].

b) Rubidium Silver Iodide

The silver-based compound RbAg_4I_5 has also been the subject of detailed light-scattering studies. The room-temperature α-phase has cubic (O) symmetry [4.25] and is superionic. In this phase, the crystal contains four formula units per unit cell with the $16\,\text{Ag}^+$ ions distributed over 56 interstitial sites forming three inequivalent sets. As in α-AgI, the Ag^+ ions in α-RbAg_4I_5 are in slightly distorted iodine tetrahedra.

When α-RbAg_4I_5 is cooled to 209 K, a nearly second-order phase transition occurs to a rhombohedral β-phase. The skeleton structure in the β-phase is not greatly different from that in the α-phase and there is no sudden change in σ. At 122 K a first-order transition occurs to a γ-phase and σ falls by about two orders of magnitude. The structure of the γ-phase is trigonal but is not well understood [4.26].

The Raman spectrum of α-RbAg_4I_5 has been studied by *Gallagher* and *Klein* [4.27]. The allowed Raman modes for O symmetry are A_1, E, and F_2 (Fig. 4.7). RbAg_4I_5 is optically active and allowance must be made for this in the investigation of polarization spectra. It is apparent from Fig. 4.7 that there are two main regions of scattering, near 20 and $105\,\text{cm}^{-1}$. As in the case of α-AgI (Sect. 4.2.1a), the lower-energy structure is assigned to vibration of Ag^+ within I^- tetrahedra which is expected to give roughly equal intensity for E and F_2 symmetries, whereas the symmetric A_1-type at $105\,\text{cm}^{-1}$ is assigned to the breathing mode of the tetrahedron.

Quasi-elastic scattering of α-RbAg_4I_5 was measured at $90\,^\circ\text{C}$ by *Field* et al. [4.28]. This was dominated by a very intense Rayleigh peak of unknown origin with FWHM $< 10\,\text{MHz}$. Removal of this peak with an iodine filter revealed a dynamic central peak. This could be broken down into two Lorentzians with FWHM $= 9$ and $40\,\text{GHz}$. As in the case of α-AgI (Sect. 4.2.1a), the in-

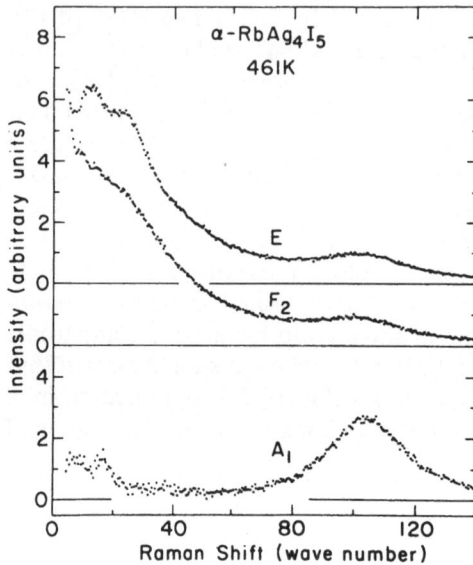

Fig. 4.7. Raman spectra of RbAg$_4$I$_5$ at 495 K [4.27]

Table 4.2. Temperatures $T_c [K]$ at which phase transitions occur in cuprous halides

	$\gamma - \beta$	$\beta - \alpha$
CuI	642	680
CuBr	658	742

terpretation of these peaks is not clear cut, since it is not easy to establish the contributions of defect-induced acoustic-mode scattering and phonon-difference processes. Assuming that the narrow peak is due to diffusive motion of the silver ions, one estimates $\tau_d = 18$ ps from the FWHM. Using $D = 3.7 \times 10^{-12} \text{m}^2\text{s}^{-1}$ [4.29] and (4.5) gives the reasonable value of $l = 1.8$ Å for the average jump distance of Ag$^+$ ions. It seems surprising, however, that τ_d for RbAg$_4$I$_5$ should be an order of magnitude larger than for α-AgI (Sect. 4.2.1a).

Raman studies have also been made of RbAg$_4$I$_5$ in the β- and γ-phases and of the related compounds NH$_4$Ag$_4$I$_5$ and KAg$_4$I$_5$ [4.27].

4.2.2 Cuprous Halides

The cuprous halides have the cubic zinc-blende structure (γ-phase, space group T_d^2) at room temperature. On heating, the γ phase changes to a β-phase with the wurtzite structure (Table 4.2). The γ phase is a good ionic conductor due to the motion of Cu$^+$ ions in a relatively fixed halogen sublattice. On changing from

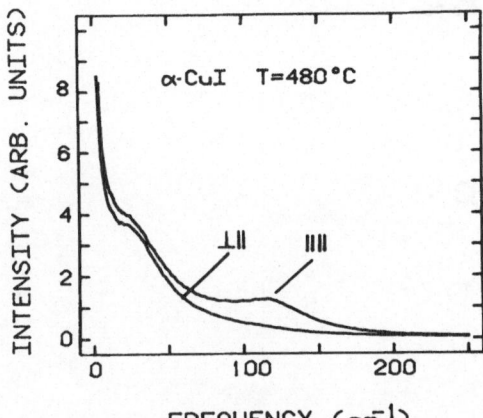

Fig. 4.8. Polarized and depolarized Raman spectra of polycrystalline α-CuI [4.33]

the γ-phase to the β-phase, σ increases by a small factor. On further heating, CuBr changes to an α phase (space group O_h^9) in which the bromines form a bcc sublattice similar to AgI (Sect. 4.2.1a), whereas CuI changes to an α phase (space group O_h^5) in which the iodines form an fcc sublattice (Table 4.2). In the α phase of both compounds, the ion conductivity approaches a value of $\sim 10^2\,\mathrm{Sm}^{-1}$.

We shall confine our remarks on light scattering to α-CuI although studies of CuBr and CuCl have been made [4.30–33]. EXAFS studies [4.15] suggest that in α-CuI the Cu$^+$ ions occupy tetrahedral and octahedral sites with an occupation probability of 70 % and 30 %, respectively (but see [4.31]). *Nemanich* et al. [4.33] investigated the shape of the light-scattering intensity of polycrystalline α-CuI with a triple grating monochromator at energies down to $2\,\mathrm{cm}^{-1}$ (Fig. 4.8). The shape is similar to that of α-AgI (Fig. 6.4), displaying a shoulder in the optic phonon region and a rapidly-rising scattering at low frequency. This quasi-elastic scattering has been discussed at some length by *Nemanich* et al. [4.33] who suggest the possibility that interactions between the mobile Cu$^+$ ions may have an important role to play (see also [4.34, 35]).

In concluding this section, we mention that *Grieg* et al. [4.36] have studied the effect of hydrostatic pressure on the Raman spectra of the superionics Ag$_2$HgI$_4$ and Cu$_2$HgI$_4$ (see also [4.37]).

4.3 β-Alumina Compounds

4.3.1 General Background

As-grown crystals of sodium β-alumina have the formula $(1 + x)\mathrm{Na_2O}:11\,\mathrm{Al_2O_3}$ where x represents departure from stoichiometry, having a value of ~ 0.3 for melt-grown crystals. The crystals have a centrosym-

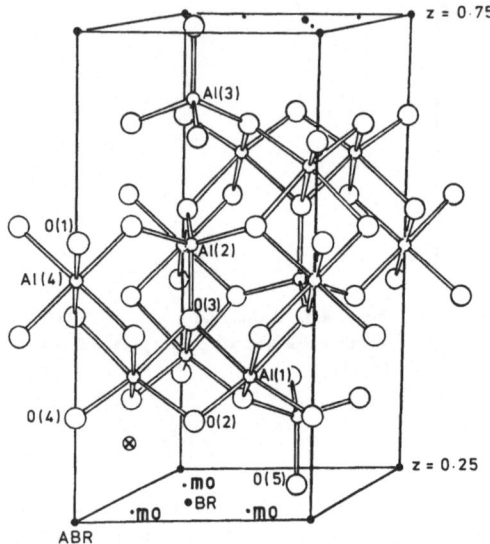

Fig. 4.9. Schematic representation of the β-alumina structure showing BR, aBR, mo, and O(5) sites in the mirror plane

metric structure with hexagonal symmetry. They consist of spinel-like ($Al_{11}O_{16}$) blocks separated by mirror planes containing sodium and oxygen ions (Fig. 4.9). The general features of the structure were solved by *Bragg* et al. [4.38] and by *Beevers* and *Ross* [4.39]. The spinel blocks are 11.25 Å thick along the c-axis. The oxygen ions in the mirror plane [O(5) in Fig. 6.9] are coordinated by Al^{3+} ions in the spinel blocks, joining the blocks together. In effect, each mirror plane is a hexagonal network of O(5) ions interspersed with cation sites referred to as Beevers-Ross (BR) and anti-Beevers-Ross (aBR), which have site symmetry D_{6h}, and mid-oxygen (mo) which has site symmetry C_{2v} (Fig. 4.9).

Neutron diffraction studies by *Roth* et al. [4.40] on sodium β-alumina at room temperature indicate that $\sim 66\%$ of the sodium ions are near BR sites, $\sim 30\%$ near mo sites and a relatively small part ($\sim 4\%$) are near aBR sites. These authors also considered the charge-compensation mechanism for the excess sodium in the mirror plane arising from nonstoichiometry and suggested that this occurs through interstitial oxygen ions (O_i^{2-}) in mo sites, bound by two aluminium ions [Al(1) in Fig. 4.9] displaced from their normal positions toward the O_i^{2-}.

The measured activation energy E_A for ionic conductivity in sodium β-alumina at room temperature is about 0.16 eV. The mechanism giving rise to this low value is of fundamental interest. *Wang* et al. [4.41] calculated an activation energy of similar magnitude using a model of two sodium ions occupying nearest-neighbor mo sites and assuming migration by a mechanism based an interstitials. This approach was extended by *Wolf* [4.42] who explored the consequences of interaction between mo-mo pairs and O_i^{2-}.

It has recently been shown [4.43, 44] that it is possible to prepare β-alumina compounds in a form approximating stoichiometry (in future, such compounds

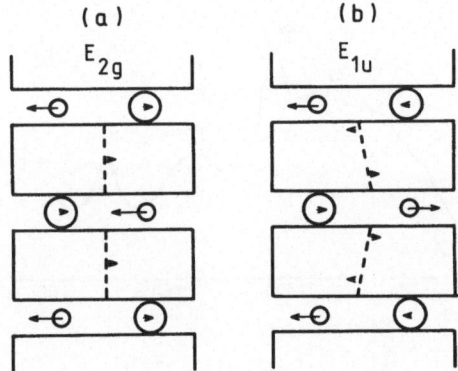

Fig. 4.10a, b. Schematic representation of the eigenvectors of (**a**) the Raman-active E_{2g} mode and (**b**) the infrared active E_{1u} mode in β-alumina. The small circles represent cations and the large circles oxygen ions in the mirror plane. Associated distortions of the spinel blocks are represented

will be given the prefix S). *Hayes* et al. [4.45] found that the ionic conductivity of S-sodium β-alumina was about an order-of-magnitude smaller than that of the normal compound and E_A was a factor of about four higher. We shall compare spectroscopic properties of these materials in Sect. 4.3.2.

The distribution of cations in the mirror plane of other β-alumina isomorphs differs from that of the sodium compound [4.46, 47]. For example, x ray studies suggest [4.46] that in potassium β-alumina at room temperature, the number of K^+ ions near BR and mo sites are comparable but that they are absent from aBR sites. However, x ray studies on silver β-alumina at room temperature have revealed three silver locations with approximately equal occupancy: (i) midway between BR and mo sites (henceforth referred to as BR/mo), (ii) close to aBR sites, and (iii) close to BR sites (see [4.47] for references).

4.3.2 Sodium β-Alumina

The optic-mode vibrations of ions in the spinel blocks of sodium β-Alumina occur at frequencies higher than $\sim 200\,\mathrm{cm}^{-1}$ [4.48] and are not of great significance for understanding properties of the mirror plane. The potentials for the various possible cation sites in the mirror plane are relatively shallow leading to vibrational frequencies below $\sim 200\,\mathrm{cm}^{-1}$. It is therefore possible to study the vibrations of the cations of importance for conductivity without interference from other modes, although shear modes of the spinel blocks do occur at $\sim 100\,\mathrm{cm}^{-1}$. It happens that the coupling between neighboring cations in the mirror plane is weak so that vibrations can be considered to be local modes, with very little dispersion.

To a first approximation, when considering the vibrations of Na^+ ions near BR sites, we can ignore the mirror-plane oxygens, O(5). The in-plane motion of the sodium ions then results in E_{2g} and E_{1u} zone-centre vibrations for the D_{6h} point group. The E_{2g} mode is Raman active in XX and XY polarization and is characterised by displacements in opposite directions in the two nearest-neighbor mirror planes; for the E_{1u} vibrations, the displacements are in phase (Fig. 4.10). Similarly, cation displacements along the $c(z)$-axis lead to a Raman

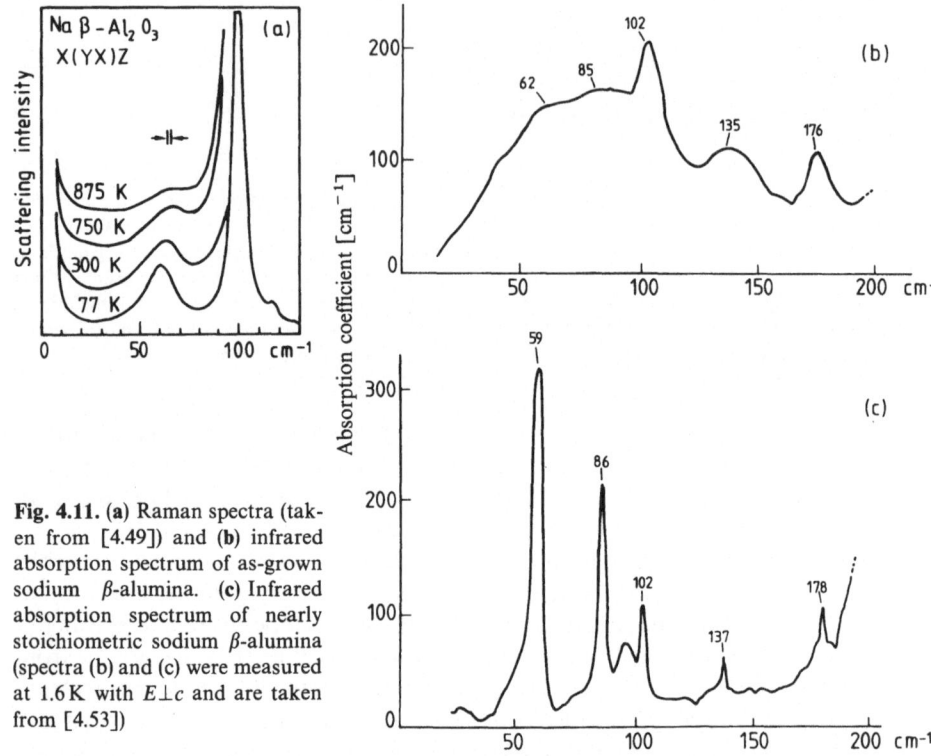

Fig. 4.11. (a) Raman spectra (taken from [4.49]) and (b) infrared absorption spectrum of as-grown sodium β-alumina. (c) Infrared absorption spectrum of nearly stoichiometric sodium β-alumina (spectra (b) and (c) were measured at 1.6 K with $E \perp c$ and are taken from [4.53])

inactive B_{2g} mode and an A_{2u} vibration with, again, antiparallel and parallel cation displacements in neighboring mirror planes. Inclusion of O(5) oxygens in the mirror plane doubles the number of modes of each symmetry. It is expected that the oxygen mode of each symmetry pair should have the higher frequency.

A study of the Raman spectrum of sodium β-alumina [4.49, 50] shows that at 77 K, there are peaks at 61 and 100 cm^{-1} (Fig. 4.11a). The peak at 61 cm^{-1} has E_{2g} symmetry and was assigned by *Chase* et al. [4.49] to vibrations of Na$^+$ ions in BR sites. However, it was subsequently suggested on the basis of infrared absorption studies [4.51] and also of Raman studies of mixed sodium-potassium β-alumina [4.52] that the peak at 61 cm^{-1} was due to centre-of-mass motion of the interstitial pairs postulated by *Wang* et al. [4.41] and that the vibrations of single ions in BR sites occurred at a higher frequency.

For comparison with the Raman spectrum, we show in Fig. 4.11b the infrared absorption spectrum of sodium β-alumina at 1.6 K [4.53] with the electric vector of the infrared radiation in the mirror-plane. Varying the angle of the electric vector relative to the c-axis changes the intensity of this spectrum, but not its shape, indicating that the electric dipole moment of the vibrations lies in the mirror plane. It is possible to convert as-grown sodium β-alumina into an approximately stoichiometric (S) material: the infrared absorption of the S compound [4.53] is shown in Fig. 4.11c. It is apparent that the bands in Fig. 4.11b have sharpened to lines in Fig. 4.11c, indicating that the mirror plane is much more ordered in the S compound. The peak at $59\,\mathrm{cm}^{-1}$ in Fig. 4.11c is dominant and is clearly the E_{1u} infrared-active mode of Na^+ ions near BR sites. Since the E_{1u} and E_{2g} modes (Fig. 4.10) should be close in frequency [4.49, 50], there seems little doubt that the $61\,\mathrm{cm}^{-1}$ Raman active vibration is the E_{2g} mode of single Na^+ ions, as suggested by *Chase* et al. [4.49].

The origins of the other peaks in Fig. 4.11b, c have been discussed by *Hayes* and *Holden* [4.53] in the context of complexes expected in the mirror plane, arising from the presence of charge-compensating O_i^{2-}. These authors have also suggested, on the basis of temperature-dependence studies, that the complexes giving rise to absorption near $100\,\mathrm{cm}^{-1}$ in Fig. 4.11b are effective in contributing to ionic conductivity. It is not clear why the mirror-plane excitations at energies greater than $62\,\mathrm{cm}^{-1}$ in Fig. 4.11b do not show up in Raman scattering (Fig. 4.11b; see below).

The most intense peak in the Raman spectrum (Fig. 4.11a) occurs at $100\,\mathrm{cm}^{-1}$ and has E_{2g} symmetry. It has been assigned by *Chase* et al. [4.49] to shearing of spinel blocks, possibly combined with motion of the mirror-plane ions, and may be associated with a transverse zone-centre mode observed at $12\,\mathrm{meV}$ ($96\,\mathrm{cm}^{-1}$) by inelastic neutron scattering [4.54]. *Hao* et al. [4.50] showed that a Raman mode of E_{1g} symmetry occurs in sodium β-alumina at $114\,\mathrm{cm}^{-1}$ and this is also assigned to a spinel block shear.

The sharp vibrational excitations of sodium β-alumina (Fig. 4.11) have prompted correlation with attempt frequencies obtained from nmr measurements. *Walstedt* et al. [4.55] found that a fit of (4.7) to the measured correlation time $\tau_c (= \tau_d)$ for Na^+ ions in sodium β-alumina gave an attempt frequency ν_0 of $6\,\mathrm{cm}^{-1}$, an order of magnitude smaller than the frequencies of Fig. 4.11. This result has been regarded as curious and attempts to rationalise it have been made by a number of authors (see, e.g., [4.56]). However, it would appear that a direct comparison between the nmr value $\nu_0 \simeq 6\,\mathrm{cm}^{-1}$ and the lowest vibrational frequency so far identified ($\sim 60\,\mathrm{cm}^{-1}$) is not justified since the $\sim 60\,\mathrm{cm}^{-1}$ vibration does not appear to be of immediate relevance to conduction mechanisms. So far, no identification has been made of a vibrational frequency associated with centre-of-mass motion of the interstitial pairs postulated by *Wang* et al. [4.41]. It is to be expected, however, that such a frequency will be considerably smaller than $60\,\mathrm{cm}^{-1}$.

4.3.3 Silver β-Alumina

Silver β-alumina is isomorphic with sodium β-alumina and may be prepared from the sodium compound by ion exchange. The XY Raman spectrum of silver β-alumina at a variety of temperatures is shown in Fig. 4.12a. Again for comparison, we show the infrared absorption spectrum of silver β-alumina at

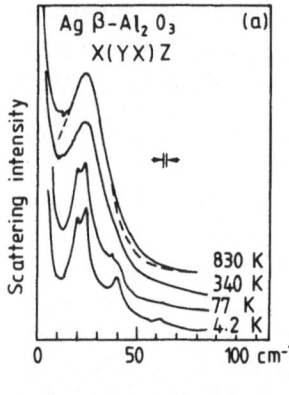

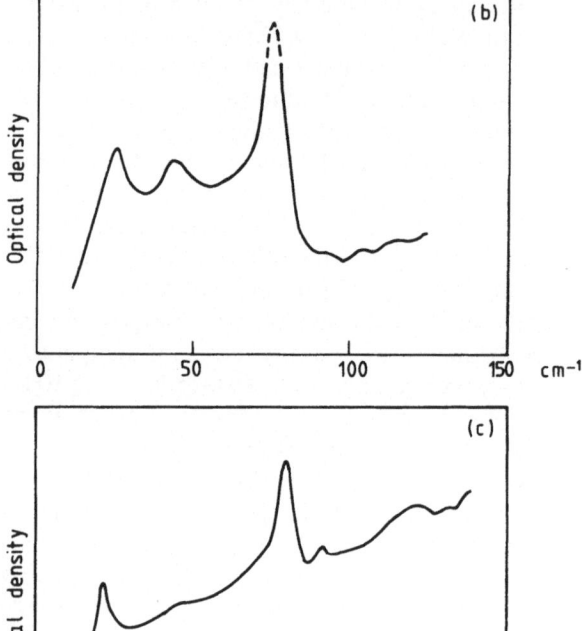

Fig. 4.12. (a) Raman spectra (taken from [4.49]) and **(b)** infrared absorption of nonstoichiometric silver β-alumina. **(c)** Infrared absorption of nearly stoichiometric silver β-alumina (spectra (b) and (c) were measured at 1.6 K with $E \perp c$ and are taken from [4.47])

1.6 K in Fig. 4.12b with the electric vector of the infrared radiation perpendicular to the $c(z)$-axis. It is possible to prepare silver β-alumina in a form approximating stoichiometry and the infrared absorption of this (S) material [4.47] is shown in Fig. 4.12c.

The most pronounced feature of the Raman spectrum at 4.2 K (Fig. 4.12a) is an E_{2g} vibrational mode at $25 \, \text{cm}^{-1}$, split by about $6 \, \text{cm}^{-1}$. In addition, there are weaker Raman peaks at 40 and $60 \, \text{cm}^{-1}$. The infrared spectrum (Fig. 4.12b) shows structure at 25, 43, and $74 \, \text{cm}^{-1}$; there are also shoulders at 21 and $65 \, \text{cm}^{-1}$ which are revealed as actual structure by lineshape analysis using overlapping Gaussians [10.47]. The infrared spectrum of the S-material (Fig. 4.12c) is weaker, showing two main peaks at 21 and $78 \, \text{cm}^{-1}$ and a much weaker absorption at $43 \, \text{cm}^{-1}$.

A recent neutron diffraction study of S-silver β-alumina at 4 K [4.57] indicates that there is a single silver location with unit occupancy in BR/mo positions (see Sect. 4.3), where silver-oxygen distances are minimised. It seems clear that the peak at $21 \, \text{cm}^{-1}$ in Fig. 4.12c is due to in-plane motion of Ag$^+$ ions in these sites. The absence of any structure in the line suggests that effects of departure from axial symmetry associated with the BR/mo site are small.

This excitation is analogous to the excitation at $59 \, \text{cm}^{-1}$ in S-sodium β-alumina (Fig. 4.11c). It has been suggested [4.47] that the excitation at $78 \, \text{cm}^{-1}$ (Fig. 4.12c) is due primarily to in-plane motion of O(5) ions, the relatively low frequency being due to bonding to silver ions. No corresponding excitation has been found so far in sodium β-alumina, presumably because the frequency is sufficiently high to be overlaid by spinel-block vibrations.

The infrared spectrum of Fig. 4.12b shows three principal silver configurations, in agreement with results of x ray structural studies (Sect. 4.3.1). These are associated with:

(a) The 25 and $74 \, \text{cm}^{-1}$ peaks, analogous to the 21 and $78 \, \text{cm}^{-1}$ peaks in Fig. 4.12c. The fact that the two sets of frequencies are different is due to different local environments. In non stoichiometric silver β-alumina, we expect each silver ion in a BR/mo site to have one or two neighboring aBR sites occupied, a situation that does not occur in the S compound. The absence of a Raman peak corresponding to the $74 \, \text{cm}^{-1}$ band is unexpected.

(b) The $43 \, \text{cm}^{-1}$ peak which is associated with in-plane vibrations of Ag^+ ions in aBR sites [4.47]. There is a corresponding Raman peak in Fig. 4.12a.

(c) The poorly resolved shoulders at 21 and $65 \, \text{cm}^{-1}$ which have been assigned [4.47] to vibrations of Ag^+ ions near BR sites, in association with charge-compensating O_i^{2-} ions. There is a weak Raman peak in Fig. 4.12a corresponding to the $65 \, \text{cm}^{-1}$ excitation. The 21 and $65 \, \text{cm}^{-1}$ peaks lose much of their intensity on warming to room temperature due to the break-up of the $Ag^+ - O_i^{2-}$ complexes and there is a corresponding growth of intensity in the $43 \, \text{cm}^{-1}$ peak and to a much smaller extent in the 25 and $74 \, \text{cm}^{-1}$ peaks. The activation energy associated with this decomposition is $\sim 0.03 \, \text{eV}$.

Barker et al. [4.58] measured the activation energies for microwave conductivity of silver β-alumina. They found that $E_A = 0.06 \, \text{eV}$ between 100 and $250 \, \text{K}$ and $E_A = 0.15 \, \text{eV}$ above $250 \, \text{K}$. The lower value of E_A may be connected with dissociation of the $Ag^+ - O_i^{2-}$ complexes, analogous to the dissociation of centres responsible for the absorption at $\sim 100 \, \text{cm}^{-1}$ in sodium β-alumina (Sect. 4.2.2).

Studies of x ray diffuse scattering by silver β-alumina [4.59–61] show the build up of two-dimensional short-range order between the mobile silver ions as the temperature is lowered from 700 to $77 \, \text{K}$. A superlattice forms in the ordered regions with dimensions $\sqrt{3a} \times \sqrt{3a}$, i.e., three times larger than the ordinary lattice cell. Recently, diffuse x ray scattering studies of S-silver β-alumina [4.62] have shown the onset of long-range three-dimensional order in the silver sublattice on cooling through room temperature. The superlattice is the same as that observed in nonstoichiometric material and it has been suggested by *Newsam* and *Tofield* [4.57] that it is a consequence of the location of Ag^+ ions in BR/mo sites with associated displacements of O(5) ions.

It has been proposed [4.49, 50] that the splitting of the $25 \, \text{cm}^{-1}$ Raman peak (Fig. 4.12a) might be a consequence of superlattice formation. However, the infrared studies [4.47] suggest that this splitting may be due to the presence

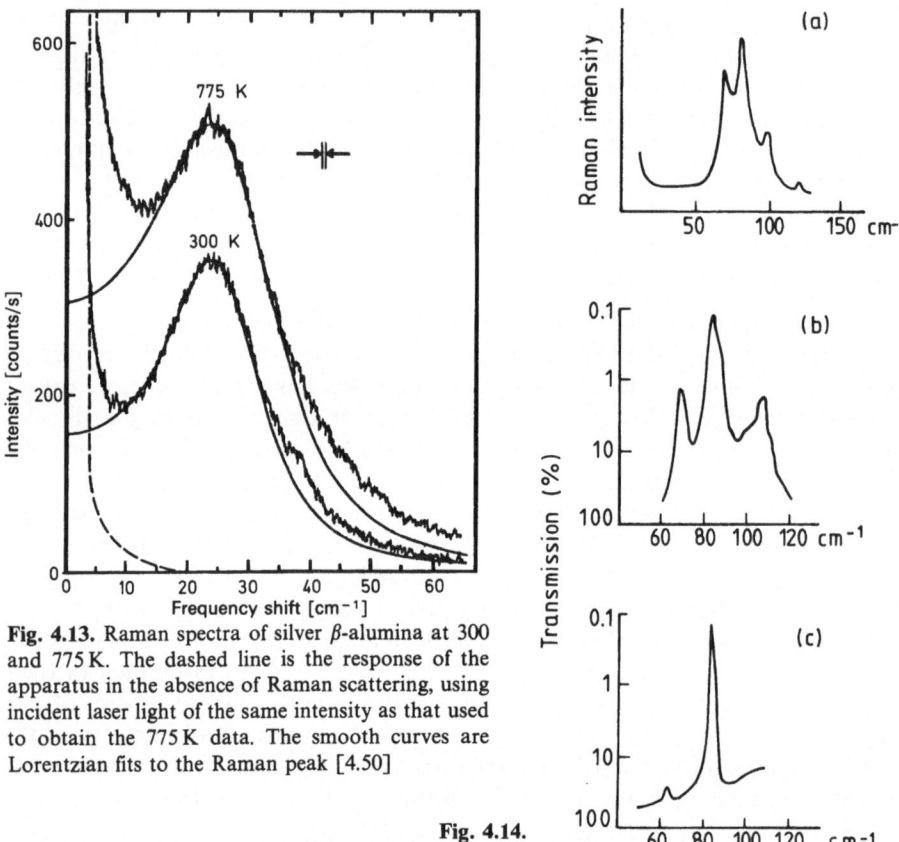

Fig. 4.13. Raman spectra of silver β-alumina at 300 and 775 K. The dashed line is the response of the apparatus in the absence of Raman scattering, using incident laser light of the same intensity as that used to obtain the 775 K data. The smooth curves are Lorentzian fits to the Raman peak [4.50]

Fig. 4.14.

Fig. 4.14. (a) Raman spectra (taken from [4.52]) and (b) infrared absorption of nonstoichiometric potassium β-alumina. (c) Infrared absorption of nearly stoichiometric potassium β-alumina (spectra (b) and (c) were measured at 1.6 K with $E \perp c$ and are taken from [4.44])

of O_i^{2-} ions, giving rise to absorption by silver ions in two different enviroments.

So far in Sect. 4.3 we have confined our discussion to vibrations about equilibrium configurations, ignoring the effects of hopping motion. Raman scattering associated with hopping motion has not yet been studied in detail in β-alumina isomorphs. Preliminary measurements were reported by *Hao* et al. [4.50] for silver β-alumina. This material shows intense quasi-elastic light scattering at ∼ 600 K and above (Fig. 4.13). However, the higher resolution of a Brillouin scattering apparatus, combined with the use of an iodine filter [4.63], would appear to be required for reliable results. The intensity of quasi-elastic scattering by sodium β-alumina at high temperatures is much smaller than that of silver β-alumina, presumably because of the lower polarizability of Na^+ [4.50].

4.3.4 Other Isomorphs of β-Alumina

The low-energy end of the Raman spectrum of potassium β-alumina shows a strong triplet of lines with E_{2g} symmetry whose frequencies were found by *Hao* et al. [4.50] to be at 72, 82, and 98 cm^{-1} at room temperature and by *Klein* et al. [4.52] to be at 69, 80, and 98 cm^{-1}, also at room temperature (Fig. 4.14a). The infrared absorption spectrum at 1.6 K with the electric vector perpendicular to the $c(z)$-axis also shows three peaks at 69, 84, and 107 cm^{-1} (Fig. 4.14b; [4.44]). *Klein* et al. [4.52] report the corresponding peaks at room temperature to be at 69, 80, and 108 cm^{-1}. *Klein* et al. [4.52] also reported infrared and Raman studies of mixed crystals of Na_xK_{1-x} β-alumina with x covering the range 0–1 and concluded that the peaks observed by them at 69 and 80 cm^{-1} for $x=0$ were due to pairs of K$^+$ ions, and that the peak at 108 cm^{-1} was due to singly-occupied BR sites. However, *Kaneda* et al. [4.64] found that replacement of K by Sn or Li in potassium β-alumina results in a reduction of intensity of both the 69 and 104 cm^{-1} Raman bands relative to the 80 cm^{-1} band, leading them to suggest that the 69 cm^{-1} band is due to pairs of K$^+$ ions and that the 80 cm^{-1} band is due to singly-occupied K$^+$ sites.

Hayes et al [4.44] found that the 69 and 107 cm^{-1} bands (Fig. 4.14a) are absent from the infrared spectrum of S potassium β-alumina (Fig. 4.14c) and concluded immediately, in agreement with *Kaneda* et al. [4.64], that the 80 cm^{-1} band was due to vibrations of K$^+$ ions in singly-occupied BR sites. We have already seen (Sect. 4.3.1) that in nonstoichiometric potassium β-alumina, there is a comparable number of K$^+$ ions in BR and mo sites and that they are largely absent from aBR sites. Although well-defined models for the centres giving rise to the 69 and 108 cm^{-1} bands are not available, it seems likely that they may involve pairs of K$^+$ ions in mo sites, in association with O_i^{2-}.

It is a matter of some interest that the energy of the Raman mode of K$^+$ ions in BR sites (~ 80 cm^{-1}) is considerably larger than the corresponding energy for Na$^+$ (61 cm^{-1}; Sect. 4.2.1). This result is presumably due to the larger size of K$^+$ and hence stronger coupling to mirror-plane oxygens. This behavior is not inconsistent with the observation that the ionic conductivity of potassium β-alumina is about an order of magnitude smaller than that of sodium β-alumina [4.65].

Raman scattering by ^{6}Li and ^{7}Li β-alumina crystals has been studied by *Kaneda* et al [4.66]. In spite of the small size of Li$^+$, the ionic conductivity of lithium β-alumina is considerably smaller than that of sodium or silver β-alumina. The Raman measurements show excitation of the Li$^+$ ions only in the 300–500 cm^{-1} range although a square-root-of-the-mass renormalisation of the results for sodium β-alumina suggests a frequency of ~ 110 cm^{-1}. From the high excitation frequency, it would appear that the Li$^+$ ions are located in positions different from the other alkali ions. *Kaneda* et al. [4.66] concluded that the Li$^+$ ions are displaced about 0.8 Å from the BR site along the c-axis, resulting in enhanced bonding to spinel-block oxygens and reduced mobility. Comparison of the Raman spectra of $Na_{1-x}Li_x$, $K_{1-x}Li_x$, and $K_{1-x}Sn_{x/2}$

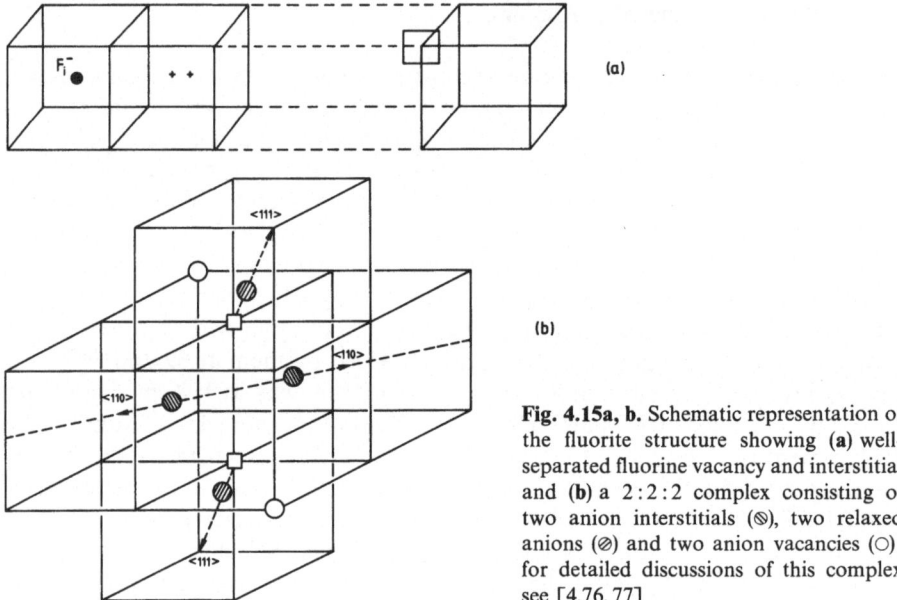

Fig. 4.15a, b. Schematic representation of the fluorite structure showing (**a**) well-separated fluorine vacancy and interstitial and (**b**) a $2:2:2$ complex consisting of two anion interstitials (⊗), two relaxed anions (⊘) and two anion vacancies (○); for detailed discussions of this complex see [4.76, 77]

β-alumina with ionic transport measurements in the mixed crystals [4.64] shows that when $Na^+ - Li^+$ and $K^+ - Li^+$ pairs are formed, the Na^+ and K^+ ions are trapped in the respective pairs because the activation energy for the dissociation of pairs by movement of Na^+ and K^+ ions is large. The activation energy for the movement of the Li^+ ions (~ 0.26 eV) seems to be a factor of about 3 smaller indicating that the Li^+ ions are the primary carriers of ionic current in the mixed crystals.

Finally, we mention that Raman studies have been carried out on ammonium β-alumina [4.67], on hydrogen β-alumina (for a list of references and a discussion of the spectroscopy of this material, see [4.68]) and on sodium β''-alumina [4.69]; the latter material has a structure closely related to that of sodium β-alumina [4.70] and a higher ionic conductivity [4.71]. However, work on the β'' compounds is still at a relatively early stage (see also [4.72]).

4.4 Crystals with Fluorite Structure

We pointed out in Sect. 4.1 that a heat-capacity anomaly occurred in some crystals with fluorite structure at a temperature T_c well below the melting temperature T_m (Table 4.1), and that above T_c, the crystals were extensively disordered, giving rise to a high ionic conductivity. This disorder occurs in the anion sublattice, resulting in the generation of anion Frenkel pairs at concentrations of up to ~ 10 mol %. The cation sublattice remains essentially intact. Below T_c, the anion interstitials are located at the centre of empty fluorine cubes (Fig. 4.15a, [4.73]). The first indications that the defect structure above T_c might be different from that below T_c came from molecular dynamics simulations of superionic $SrCl_2$ and CaF_2 [4.74, 75]. These studies found no evidence for the presence of cube-centre anion interstitials and led to the suggestion that cube-edge anion interstitials were involved. A subsequent theoretical study of the energetics of defect formation in CaF_2 [4.76] showed that the anion vacancy and cube-centre interstitial represented the lowest-energy configuration for small defect concentrations, but when the defect concentration increased to a level ($\leqslant 1\,mol\%$) at which strong defect interactions occurred, the so-called 2:2:2 complex was favoured. This aggregate consists of two anion interstitials associated with two relaxed anions and two anion vacancies (Fig. 4.15b) and is, of course, transient. It accounts quite well for much of the intensity observed in quasi-elastic neutron scattering [4.77].

The heat capacity anomaly suggests the onset of an order-disorder transition which is suppressed before massive interstitial disorder develops. The origin of the cooperative interaction between defects that gives rise to the anomaly was also discussed by *Catlow* et al. [4.78] who pointed out that there should be a net Coulomb attraction between defects up to a limiting concentration. This discussion is an extension of the Debye-Hückel concept which postulates that each charged defect is screened from defects of similar sign by a cloud of defects of opposite sign. They also considered the strain coupling between defects due to lattice distortion caused by individual defects but concluded that this is a smaller effect than the Coulomb interaction.

A study of the effect of anharmonicity ($T < T_c$) and lattice disorder ($T > T_c$) on the Raman spectrum of CaF_2, SrF_2, BaF_2, $SrCl_2$, and PbF_2 was carried out by *Elliott* et al. [4.79]. The first-order Raman spectrum of crystals with fluorite structure consists only of a phonon of T_{2g} symmetry in which the anion sublattices vibrate in antiphase and the cations are stationary. However, defects in the crystals can induce scattering in other symmetry configurations owing to the breakdown of wave vector conservation rules. Experiments were, therefore, carried out with XY, ZZ, and ZY scattering geometries corresponding to T_{2g}, $A_{1g}+4E_g$, and E_g symmetries.

We shall concentrate here on PbF_2 although similar data for the other fluorites were obtained [4.79]. The scattering $S(\omega)$ for PbF_2 was measured from 4 to 950 K (Fig. 4.16). At room temperature, the T_{2g} phonon at $257\,cm^{-1}$ is already broad (FWHM $= 50\,cm^{-1}$) and has an asymmetric shape. There is weak scattering in the $A_{1g}+4E_g$ and E_g symmetries which sharpens at very low temperatures giving structure similar to that found in the single-phonon density of states [4.80]. This scattering is due to defects present in as-grown crystals.

As the temperature is raised, the T_{2g} phonon continues to broaden and at T_c and above, an additional low-frequency scattering develops taking the form of an intense wing on the laser line (Fig. 4.16). The $A_{1g}+4E_g$ and E_g spectra also develop this low-frequency wing with increasing temperature and, at the highest temperatures, differ in appearance from the T_{2g} spectrum only to the extent that the latter retains some vestiges of the low-temperature T_{2g} phonon.

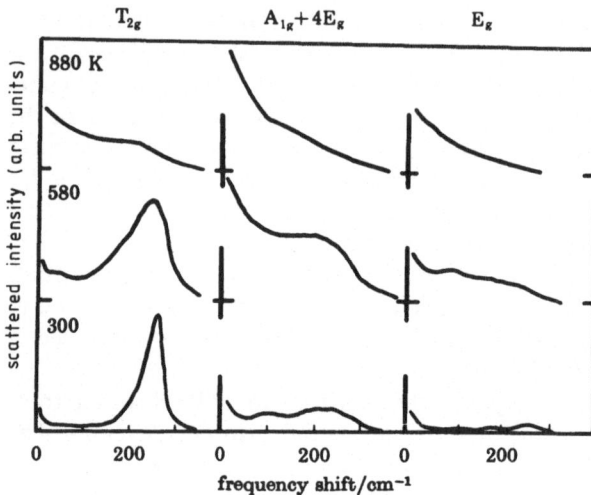

Fig. 4.16. Measured Raman scattering $S(\omega)$ of PbF_2 at three different temperatures for T_{2g}, $A_{1g}+4E_g$, and E_g symmetries [4.79]

Values of $|\chi'(\omega)|^2$ obtained from $S(\omega)$ with T_{2g} symmetry (Sect. 4.1) are shown in Fig. 4.17 for PbF_2 for four different temperatures, two below T_c and two above T_c. The position and shape of the bands below T_c were calculated [4.79] using third and fourth-order anharmonicity. The theory requires two parameters, B (third order) and C (fourth order) and the values required to give the calculated curves in Fig. 4.16a, b are comparable with values of third and fourth-order anharmonic parameters obtained by other experimental techniques. The additional scattering on the low energy side of the T_{2g} phonon at T_c and above is accounted for by a theory of defect-induced scattering which includes effects of both anion vacancies and interstitials ([4.79]; Fig. 6.17). It should be emphasised, however, that such calculations are not sensitive to the precise configuration of vacancies and interstitials. In the case of PbF_2, the defect-induced scattering almost develops into a subsidiary peak below the Raman peak because of a large difference between the Raman frequency and the peak in the smoothed-out, high-temperature, single-phonon density of states [4.80].

Figure 4.18 shows in more detail a comparison of the experimental and theoretical peak widths and peak positions for $|\chi'(\omega)|^2$ for PbF_2. It is interesting to note that the peak position first increases in energy with increasing T and then decreases, whereas, with all the other fluorites investigated, it deceases monotonically both in theory and experiment. This peculiarity arises from the detailed nature of the single-phonon density of states of PbF_2 [4.80].

The theory used above is not reliable at low frequencies since it ignores contributions at high temperatures due to hopping. On the experimental side, conventional Raman techniques have not proved adequate to study hopping. As in the case of β-alumina (Sect. 4.3.3), sophisticated Brillouin techniques will be required.

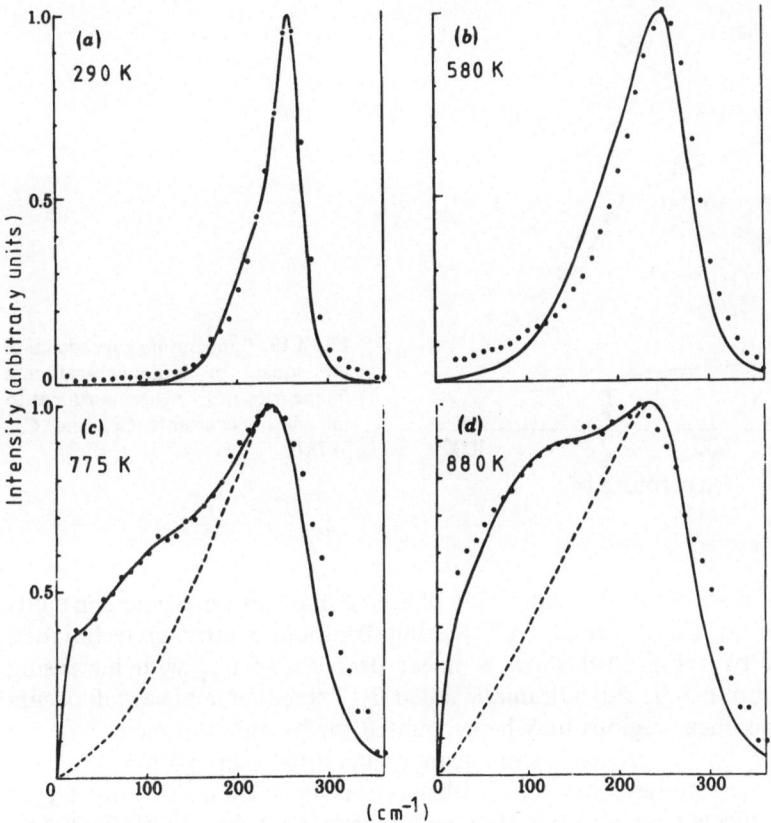

Fig. 4.17a–d. Measured (dots) and calculated (full lines) reduced Raman spectra of PbF_2 at (a) 290 K, (b) 580 K, (c) 775 K, (d) 880 K. The dashed lines in (c) and (d) represent the theoretical contribution of anharmonicity alone [4.79]

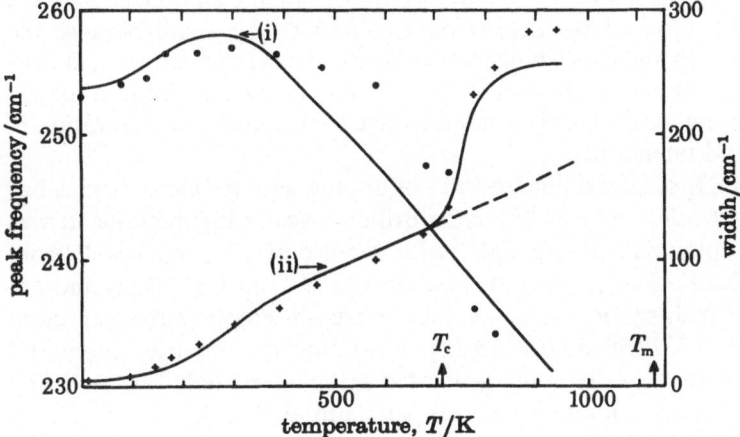

Fig. 4.18. Peak frequency (i) and linewidth (ii) of the squared Raman susceptibility of PbF_2 as a function of temperature. Full lines are the results of calculation. The dashed line is the result of the contribution of anharmonicity alone to the linewidth at high temperatures [4.79]

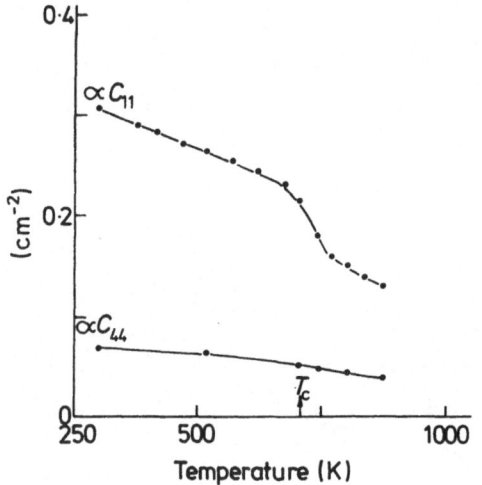

Fig. **4.19.** Temperature dependence of the square of Brillouin scattering frequencies in PbF_2, proportional to the elastic constants C_{11} and C_{44} [4.78]

Effects of disorder in the superionic phase of fluorites on elastic constants were studied by *Catlow* et al. [4.78] using Brillouin scattering techniques. Results for PbF_2 (Fig. 4.19) show a linear decrease of C_{11} with increasing temperature up to $\sim T_c$ and a dramatic fall at T_c. Thereafter, a linear fall occurs up to T_m. The linear regions may be accounted for by anharmonicity and the large fall at T_c by the generation of vacancy-interstitial pairs [4.78].

The Brillouin studies show that neither the elastic constant C_{12} nor C_{44} is appreciably affected by disorder (Fig. 4.19 shows the behavior of C_{44}). The different behavior of these elastic constants may be understood to some extent from the fact that the contributions to C_{11} from Coulomb and elastic forces have the same sign, whereas for C_{12} and C_{44} they have opposite signs [4.81]. It appears that the effect of the defects on C_{12} and C_{44} is small because the changes in Coulomb and short-range forces largely cancel each other. Again, as in the case of the Raman scattering [4.79], the theory used to account for the Brillouin-scattering results [4.78] is not sensitive to the precise configuration of the vacancies and interstitials.

The effect on T_c of superionic fluorites of doping with trivalent cations has been studied by *Catlow* et al. [4.82] using Brillouin-scattering methods. It was found, for example, that doping CaF_2 with 9 mol% of YF_3 reduces T_c from 1430 K to $\sim$ 1200 K. The yttrium ions dissolve as Y^{3+} in Ca^{2+} sites and are charge-compensated by F^- ions at the centre of empty cubes in close association with Y^{3+}. Calculation [4.82] shows that the impurity-interstitial complexes act as traps for thermally-generated anion interstitials, thus reducing the energy of formation of anion Frenkel pairs and also of T_c.

Finally, in this section we point out that superionics which conduct by oxygen-ion transport are useful for high-temperature fuel cells as well as for the measurement of thermodynamic and kinetic properties of systems containing oxygen. Most interest has centred on oxides with the fluorite structure, principally calcia stabilized zirconia [4.83]. ZrO_2 containing about 10 mol% of CaO has the fluorite structure although pure ZrO_2 does not. The doped material is a good ionic conductor above $\sim 700\,^\circ C$. ZrO_2 containing Y_2O_3 produces a similar result. The high conductivity arises from the presence of charge-compensating oxygen vacancies associated with the presence of aliovalent cations. The Raman spectrum of ZrO_2 containing 12 mol% of Y_2O_3 has been studied at room temperature by *Zherzdev* et al. [4.84] who found scattering corresponding to a disorder-induced single-phonon density of states (see Sect. 2.1.16 of [1.2]).

4.5 Conclusion

In the previous sections, we have touched on only a small section of the wide range of materials giving rise to superionic behavior [4.7], our choice reflecting the extent of light-scattering studies. We should, however, mention other important superionic conductors such as $Na_3Zr_2PSiO_2O_{12}$ [4.85], sometimes referred to as Nasicon. This compound consists of a combination of corner-shared octahedra and tetrahedra with some sodium ions octahedrally coordinated to oxygen at the intersection points of continuous networks of three-dimensional tunnels. Conductivity at room temperature is comparable to that of the best sodium β-alumina ($\sim 20\,Sm^{-1}$) but with a somewhat larger activation energy ($\sim 0.3\,eV$; see Sect. 4.3.2). A discussion of ionic transport in crystallographic tunnels has been given by *Flygare* and *Huggins* [4.86].

Studies of alkali ion conductors show that, in general, sodium compounds have substantially higher conductivity than lithium compounds. This result seems surprising at first glance because of the relatively small size of the Li^+ ion. However, the small Li^+ ions tend to move off-centre from open spaces, resulting in stronger bonding and reduced mobility (Sect. 4.3.4). Despite this, extensive studies of lithium-containing superionics have been made, in addition to those described for lithium β-alumina in Sect. 4.3.4, and many of these materials have useful applications [4.7]. Interesting examples of lithium superionics are the one-dimensional conductor β-eucryptite (β-$LiAlSiO_4$) [4.87], the two-dimensional conductor Li_3N [4.88] and also LiI [4.89]. The lattice dynamics of Li_3N have been thoroughly explored using infrared and Raman methods [4.90].

It will be apparent from the earlier sections that a variety of behaviors occurs in superionic materials:

a) a sublattice of the material is effectively totally disordered, with liquid-like behavior, as in α-AgI (Sect. 4.2.1);

b) a sublattice of the material is partly disordered, as in PbF_2 (Sect. 4.4). The disorder in **(a)** and **(b)** is intrinsic and is a cooperative, thermally-induced phenomenon.

c) The disorder is caused by impurities, as in $ZrO_2 : Y_2O_3$ (Sect. 4.4), and is therefore extrinsic.

The characteristic features of light scattering in the above three categories are a Raman spectrum corresponding to a defect-induced single-phonon density of states and a quasi-elastic peak associated with diffusive motion. However, the study of quasi-elastic light scattering to provide direct information about the dynamics of mobile ions is still at a relatively early stage.

Finally, we mention

d) nonstoichiometric materials, such as sodium β-alumina (Sect. 4.3.2), where the disorder is intrinsic, being present in as-grown crystals. The conducting ions in these materials give rise to relatively sharp vibrational Raman lines of the localised type, giving detailed information about the environment. Quasi-elastic light-scattering studies of these materials are only now beginning [4.91].

Acknowledgements. The author is indebted to J. F. Ryan for his comments on the manuscript.

References

4.1   W.Hayes: Cont. Physics **19**, 469 (1978)
4.2   N.F.Mott, R.W.Gurney: *Electronic Processes in Ionic Crystals* (Oxford University Press, Oxford 1953)
4.3   N.N.Greenwood: *Ionic Crystals, Lattice Defects and Nonstoichiometry* (Butterworths, London 1970)
4.4   A.D.Le Claire: In *Fast Ion Transport in Solids*, ed. by W. van Gool (Plenum Press, New York 1973) p. 51
4.5   G.D.Mahan, W.L.Roth (eds.): *Superionic Conductors* (Plenum Press, New York 1976)
4.6   A.Hooper: Cont. Phys. **19**, 147 (1978)
4.7   P.Vashishta, J.N.Mundy, G.K.Shenoy (eds.): *Fast Ion Transport in Solids* (North-Holland, Amsterdam 1979)
4.8   M.B.Salamon (ed.): *Physics of Superionic Conductors*, Topics Current Phys., Vol. 15 (Springer, Berlin, Heidelberg, New York 1979)
4.9   W.Hayes, R.Loudon: *Scattering of Light by Crystals* (Wiley, New York 1978)
4.10 M.J.Delaney, S.Ushioda: [Ref. 4.8, p. 111]
4.11 H.U.Beyeler, P.Brüesch, L.Pietronero, W.R.Schneider, S.Strässler, H.R.Zeller: [Ref. 4.8, p. 77]
4.12 W.Dieterich, P.Fulde, I.Peschel: Adv. in Phys. **29**, 527 (1980)
4.13 T.Geisel: [Ref. 4.8, p. 201]
4.13a S.Glassstone, K.Laidler, H.Eyring: *The Theory of Rate Processes* (McGraw-Hill, New York 1968)
4.14 R.J.Cava, F.Reidinger, B.J.Wuensch: Solid State Commun. **24**, 411 (1977)
4.15 J.B.Boyce, T.M.Hayes: [Ref. 4.8, p. 5]
4.16 G.L.Bottger, C.V.Damsgaard: J. Chem. Phys. **57**, 1215 (1972)
4.17 W.Bührer, P.Brüesch: Solid State Commun. **16**, 155 (1975)
4.18 R.C.Hanson, T.A.Fjeldly, H.D.Hochheimer: Phys. Stat. Sol. b **70**, 567 (1975)
4.19 M.J.Delaney, S.Ushioda: Phys. Rev. B**16**, 1410 (1977)
4.20 R.Alben, G.Burns: Phys. Rev. B**16**, 3746 (1977)
4.21 T.Geisel: Solid State Commun. **24**, 155 (1977)
4.22 G.Winterling, W.Senn, M.Grimsditch, R.Katiyar: *Lattice Dynamics*, ed. by M. Balkanski (Flammarion Press, Paris 1977) p. 553
4.23 A.Kvist, R.Tärneberg: Z. Naturforsch. **259**, 257 (1970)
4.24 G.Eckhold, K.Funke, J.Kalus, R.Lechner: J. Phys. Chem. Sol. **37**, 1097 (1976)
4.25 S.Geller: Science **157**, 310 (1967)

4.26 S.Geller: Phys. Rev. B**14**, 4345 (1976)
4.27 D.A.Gallagher, M.V.Klein: Phys. Rev. B**19**, 4282 (1979)
4.28 R.A.Field, D.A.Gallagher, M.V.Klein: Phys. Rev.B**18**, 2995 (1978)
4.29 G.G.Bentle: J. Appl. Phys. **39**, 4036 (1968)
4.30 T.Fukumoto, S.Namashima, K.Tabuchi, A.Mitsuishi: Phys. Stat. Sol. b**73**, 341 (1976)
4.31 G.Burns, F.H.Dacol, M.W.Shafer, R.Alben: Solid State Commun. **24**, 753 (1977)
4.32 R.J.Nemanich, J.C.Mikkelsen, Jr.: *Physics of Semiconductors*, ed. by B. L. H. Wilson, The Institute of Physics, Bristol (1978) p. 66
4.33 R.J.Nemanich, R.M.Martin, J.C.Mikkelsen, Jr.: [Ref. 4.7, p. 547]
4.34 R.J.Nemanich, R.M.Martin, J.C.Mikkelsen, Jr.: Solid State Commun. **32**, 79 (1979)
4.35 W.Dietrich, T.Geisel, I.Peschel: Z. Phys. B**29**, 5 (1978)
4.36 D.Grieg, D.F.Shriver, J.R.Ferraro: J. Chem. Phys. **66**, 5248 (1977)
4.37 A.Nitzan, M.A.Ratner, D.F.Shriver: J. Chem. Phys. **72**, 3320 (1980)
4.38 W.L.Bragg, C. Gottfried, J.West: Z. Krist. **77**, 255 (1931)
4.39 C.H.Beevers, M.A.S.Ross: Z. Krist. **97**, 59 (1937)
4.40 W.L.Roth, F.Reidinger, S.La Placa: [Ref. 4.5, p. 223]
4.41 J.C.Wang, M.Gaffari, Choi Sang-I: J. Chem. Phys. **63**, 772 (1975)
4.42 D.Wolf: J. Phys. Chem. Sol. **40**, 757 (1979)
4.43 Ph.Colomban, J.-P.Boilot, A.Kahn, G.Lucazeau: Nouveau J. Chim. **2**, 21 (1978)
4.44 W.Hayes, L.Holden, G.F.Hopper: J. Phys. C (Solid State Phys.) **13**, L317 (1980)
4.45 W.Hayes, L.Holden, B.C.Tolfield: Solid State Ionics, 1980, in course of publication
4.46 B.McWhan, S.J.Allen, J.P.Remeika, P.D.Dernier: Phys. Rev. Lett. **35**, 953 (1975)
4.47 W.Hayes, L.Holden, B.C.Tofield: J. Phys. C (Solid State Phys.) **14**, 511 (1981)
4.48 A.S.Barker, Jr., J.A.Ditzenberger, J.P.Remeika: Phys. Rev. B**14**, 386 (1976)
4.49 L.L.Chase, C.H.Hao, G.D.Mahan: Solid State Commun. **18**, 401 (1976)
4.50 C.H.Hao, L.L.Chase, G.D.Mahan: Phys. Rev. B**13**, 4306 (1976)
4.51 S.J.Allen, Jr., H.S.Cooper, F.de Rosa, J.P.Remeika, S.K.Ulasi: Phys. Rev. B**18**, 4031 (1978)
4.52 P.B.Klein, D.E.Schafer, U.Strom: Phys. Rev. B**18**, 4411 (1978)
4.53 W.Hayes, L.Holden: J. Phys. C (Solid State Phys.) **13**, L321 (1980)
4.54 D.B.McWhan, S.Shapiro, J.P.Remeika, G.Shirane: J. Phys. C (Solid State Phys.) **8**, L487 (1975)
4.55 R.E.Walstedt, R.Dupre, J.P.Remeika: [Ref. 4.5, p. 369]
4.56 P.M.Richards: Solid State Commun. **25**, 1019 (1978)
4.57 J.M.Newsam, B.C.Tofield: J. Phys. C (Solid State Phys.) (submitted for publication)
4.58 A.S.Barker, Jr., J.A.Ditzenberger, J.P. Remeika: Phys. Rev. B**14**, 4254 (1976)
4.59 P.Y.Le Cars, R.Comes, L.de Schamps, J.Thery: Acta Cryst. A**30**, 305 (1974)
4.60 J.P.Boilot, J.Thery, R.Collongues, R.Comes, A.Guinier: Acta Cryst. A**32**, 250 (1976)
4.61 D.B.McWhan, P.D.Dernier, C.Vettier, A.S.Cooper, J.P.Remeika: Phys. Rev. B**17**, 4043 (1978)
4.62 J.P.Boilot, Ph.Colomban, R.Collongues, G.Collin, R.Comes: Phys. Rev. Lett. **42**, 785 (1979)
4.63 K.B.Lyons, P.A.Fleury: J. Appl. Phys. **47**, 4898 (1976)
4.64 T.Kaneda, J.B.Bates, J.C.Wang, H.Engstrom: [Ref. 4.7, p. 371]
4.65 S.J.Allen, Jr., H.S.Cooper, F.de Rosa, J.P.Remeika, S.K.Ulasi: Phys. Rev. B**17**, 4031 (1978)
4.66 T.Kaneda, J.B.Bates, J.C.Wang: Solid State Commun. **28**, 469 (1978)
4.67 J.B.Bates, T.Kaneda, J.C.Wang, H.Engstrom: J. Chem. Phys. **73**, 1503 (1980)
4.68 W.Hayes, L.Holden, B.C.Tofield: J. Phys. C (Solid State Phys.) **13**, 4217 (1980)
4.69 J.B.Bates, T.Kaneda, W.E.Brundage, J.C.Wang, H.Engstrom: Solid State Commun. **32**, 261 (1979)
4.70 W.L.Roth, M.Anne, D.Tranqui, A.Heidemann: [Ref.4.7, p. 267]
4.71 J.T.Kummer: In *Prog. in Solid State Chem.* Vol. 1, ed. by H. Reiss, J. O. McCalden (Pergamon Press, New York 1972) p. 141
4.72 G.Burns, G.V.Chandrashekhar, F.H.Dacol, L.M.Foster, H.R.Chandrashekhar: Phys. Rev. B**22**, 1073 (1980)
4.73 A.B.Lidiard: In *Crystals with the Fluorite Structure*, ed. by W.Hayes (Clarendon Press, Oxford 1974) p. 101

4.74 M.Dixon, M.J.Gillan: J. Phys. C (Solid State Phys.) **11**, L165 (1978)
4.75 G.Jacucci, A.Rahman: J. Chem. Phys. **69**, 4117 (1978)
4.76 R.H.Bartram, C.R.A.Catlow, W.Hayes: J. Phys. C (Solid State Phys.) (to be published)
4.77 K.Clausen, W.Hayes, M.T.Hutchings, J.K.Kjems, P.Schnabel, C.Smith: Sol. St. Ionics **5**, 589 (1981)
4.78 C.R.A.Catlow, J.D.Comins, F.A.Germano, R.T.Harley, W.Hayes: J. Phys. C (Solid State Phys.) **11**, 3197 (1978)
4.79 R.J.Elliott, W.Hayes, W.G.Kleppmann, A.J.Rushworth, J.F.Ryan: Proc. Roy. Soc. A**360**, 317 (1978)
4.80 M.H.Dickens, M.T.Hutchings: Solid State Commun. **34**, 559 (1980)
4.81 M.M.Elcombe: J. Phys. C (Solid State Phys.) **5**, 2702 (1972)
4.82 C.R.A.Catlow, J.D.Comins, F.A.Germano, R.T.Harley, W.Hayes, I.B.Owen: J. Phys. C (Solid State Phys.) (to be published) (1981)
4.83 A.S.Nowick, D.S.Park: [Ref.4.5, p.395]
4.84 A.V.Zherzdev, Z.M.Khashkhozhev, A.A.Andreev, B.T.Melekh: Sov. Phys. Solid States **20**, 2031 (1978)
4.85 J.B.Goodenough, H.Y-P Hong, J.A.Kafalas: Mater. Res. Bull. **11**, 203 (1976)
4.86 W.H.Flygare, R.A.Huggins: J. Phys. and Chem. of Solids **34**, 1199 (1973)
4.87 W.Press, B.Renker, H.Schulz, H.Böhm: Phys. Rev. B**21**, 1250 (1980)
4.88 A.Rabenau: In *Festkörperprobleme*, Vol. 18, (Vieweg, Braunschweig 1978) p. 77
4.89 B.B.Owens, P.M.Skarstad: [Ref.4.7, p.61]
4.90 H.R.Chandrasekhar, G.Bhattacharya, R.Migoni, H.Bilz: Phys. Rev. B**17**, 884 (1978)
4.91 R.T.Harley, S.Andrews: Private communication

5. Raman Studies of Phonon Anomalies in Transition-Metal Compounds

M. V. Klein

With 21 Figures

Many of the interesting properties of transition metals and compounds are due to the high density of states of d electrons near the Fermi energy. Unusually strong electron–electron and electron–phonon scattering processes are often found in these materials. These processes are thought to be responsible for such diverse phenomena as spin density waves in chromium, high-temperature superconductivity in compounds such as V_3Si and in NbC, as well as anomalous phonon softening and related displacive phase transitions of the generalized charge-density-wave (CDW) type.

Evidence has been accumulating that Raman scattering by anomalous phonons is unusually strong in these materials. Some examples will be given in the first and third sections of this article. Section 5.2 deals with microscopic theories of phonon softening, and Sects. 5.4, 5 cover microscopic theories of Raman scattering. The unifying idea behind these theoretical discussions is that the same strong electron–phonon processes that produce phonon anomalies and phase transitions are responsible for the strong Raman scattering. This theory is quite new, and the development given here is detailed.

Section 5.5 is devoted to charge-density waves. Some ideas from mean-field theory are introduced in Sect. 5.5.1 and then applied to a theory of Raman scattering from CDW phonons in Sects. 5.5.3, 4.

The last section (Sect. 5.6) presents a brief discussion of experimental Raman results on a superconducting CDW system which show coupling between CDW phonons and electronic excitations across the superconducting gap.

5.1 Examples of Raman Spectra in Transition Metals

5.1.1 Two-Phonon Spectra of Cubic Metals

Many transition-metal carbides and nitrides form crystals with the rock-salt structure, which, when ideal, has no one-phonon Raman activity. They are usually nonstoichiometric and therefore can be described by the formulas MN_x or MC_x with $x < 1$. The vacancies on the nonmetal lattice induce a one-phonon Raman spectrum that essentially mirrors the phonon density of states weighted by a poorly known factor that characterizes the polarizability fluctuations. An example is given in Fig. 5.1, which presents results on ZrN obtained by *Spengler* and *Kaiser* [5.1]. The one-phonon peaks are labeled A and O for acoustic and optical branches. Their appearance is typical for transition-metal nitrides and

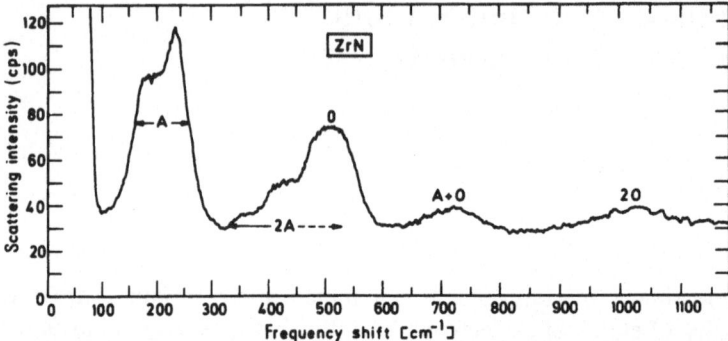

Fig. 5.1. Raman spectra of ZrN [5.1]. A and O label one-phonon peaks from acoustic and optic phonons; 2A, A+O, and 2O label two-phonon peaks

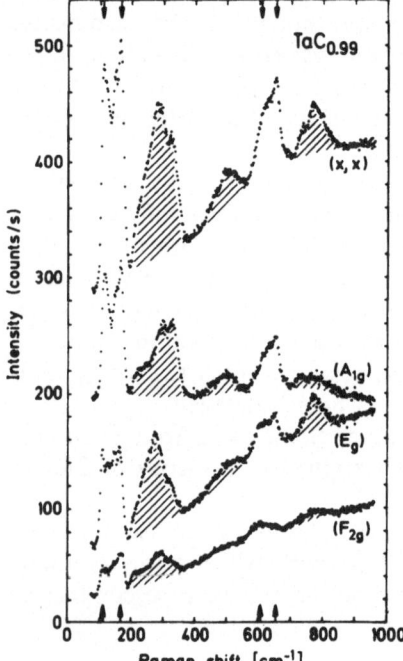

Fig. 5.2. Raman spectra of TaC [5.2]. Arrows label the one-phonon peaks. Shaded regions denote two-phonon spectra. The peak near $500\ cm^{-1}$ may contain contributions from 3A and from $O-A$ processes

carbides. The peaks labeled 2A, A+O, and 2O are two-phonon peaks. Their behavior is not typical of all these materials. They are absent, for example, in the group IVb carbides TiC, ZrC, and HfC, but are present in the group IVb nitrides TiN, ZrN and in the group Vb carbides NbC and TaC.

Figure 5.2 shows Raman results of *Wipf* et al. [5.2] for $TaC_{0.99}$. Arrows locate the positions of the one-phonon A and O peaks. The shaded areas are two-phonon features: 2A near $300\ cm^{-1}$, $O-A$ and perhaps 3A near $500\ cm^{-1}$, and $O+A$ near $800\ cm^{-1}$. Similar two-phonon results are seen in NbC_x, which is "isoelectronic" to ZrN_x. A requirement for strong two-phonon peaks is the presence of nine valence electrons per unit cell. Under the replacements NbC→ZrC or TaC→HfC, which reduce the number of valence electrons to eight, the two-phonon Raman peaks vanish. They also reduce

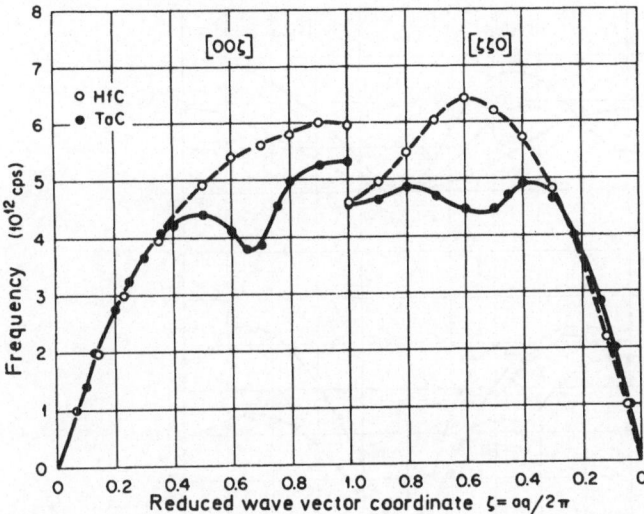

Fig. 5.3. Comparison of longitudinal acoustic phonons of "normal" HfC (○) with "anomalous" TaC (●) [5.3]

considerably in intensity when one increases the vacancy concentration from 2 % in $NbC_{0.98}$ to 13 % in $NbC_{0.87}$. This change reduces the number of valence electrons also.

It has been known for some time that cubic carbides and nitrides with nine valence electrons have anomalous phonon dispersion curves, whereas compounds with eight valence electrons do not. An example is the curve measured by *Smith* and *Gläser* [5.3] for the longitudinal acoustic (LA) modes of TaC and HfC shown in Fig. 5.3. The metals are in the same row of the periodic table and have practically identical atomic masses. The initial slopes of the dispersion curves are the same, but note the anomalies in the form of dips in the curves for the group Vb compound TaC at larger wave vector. The group IVb compound HfC has the usual normal behavior. A more complete set of dispersion curves for TaC is shown in Fig. 5.4 [5.4]. The anomalous acoustic phonon branches have frequencies "softer" by about 24 % than those of HfC. The optical branches are somewhat anomalous also, with frequencies 4 %–6 % lower than in HfC.

Two-phonon Raman spectra are difficult to analyze without detailed calculations of two-phonon densities of states weighted by matrix elements that obey the Raman selection rules. The fully symmetric A_{1g} spectrum is simplest to understand, since all overtones will have a component with A_{1g} symmetry. In comparing the Raman data from Fig. 5.2 with the dispersion curves in Fig. 5.4 we make tentative assignments as follows: The shoulder in the A_{1g} two-phonon spectrum at 220 cm^{-1} may be an overtone of TA at $q=0.5(111)$ or of $TA_{\parallel}$ near $q=0.5(110)$. The A_{1g} two-phonon peak at 300 cm^{-1} may be overtones of the anomalous LA modes near $q=0.5(111)$, $q=0.5(110)$, and $q=0.5(100)$. The E_g two-phonon peak at 270 cm$^{-1}=8.1$ THz could be mainly the contributions of anomalous $q=0.4(110)$ modes – $TA_{\parallel}$ at 3.2 THz plus LA at 4.9 THz are

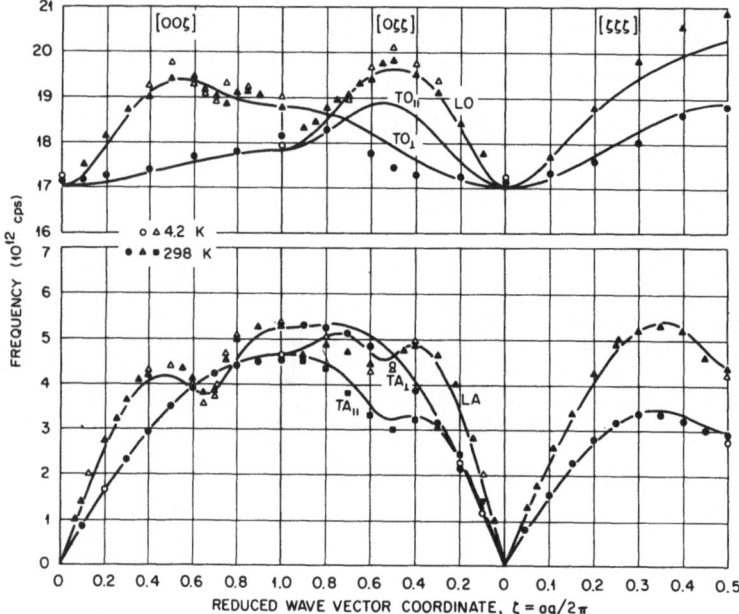

Fig. 5.4. Phonon dispersion curves for TaC [5.4]. The closed (open) symbols are the experimental data points at room temperature (4.2 K). The solid lines are calculations using a double-shell model

allowed by the selection rules [5.5]. Note that the middle-to-upper frequency range of the two-acoustic-phonon spectrum is enhanced. There are more anomalous phonons in this range than at lower frequencies. This suggests that the matrix elements for two-phonon Raman processes are enhanced when the phonons are anomalous.

Figure 5.5 shows a Raman spectrum for the bcc metal Ta [5.6]. This curve approximately matches the density of states calculated from an 11-parameter Born–von Karman fit to the neutron dispersion curves. Phonons in Ta have anomalous dispersion curves with a depression in the LA branch at $q = 0.7(100)$, where the frequency is 4.2 THz. The overtones of these phonons may produce the Raman peak at 300 cm^{-1}. The Raman peak at 190 cm^{-1} may be due to overtones of LA modes near $q = 0.7(111)$ with a frequency of 3 THz $= 100$ cm^{-1}.

5.1.2 Layered Compounds

Among the most spectacular two-phonon Raman spectra are those found in the 2 H polytype of the group Vb transition-metal dichalcogenides. Unlike the 1 T polytypes of the same materials, which have such strong phonon anomalies that "normal" phases do not exist, the 2 H polytypes are found in normal phases at room temperature and then suffer phase transitions of the charge-density-wave type at lower temperatures [5.7]. In 2 H–TaSe$_2$ and 2 H–NbSe$_2$ the transition is preceded by partial softening of LA phonons near a wave

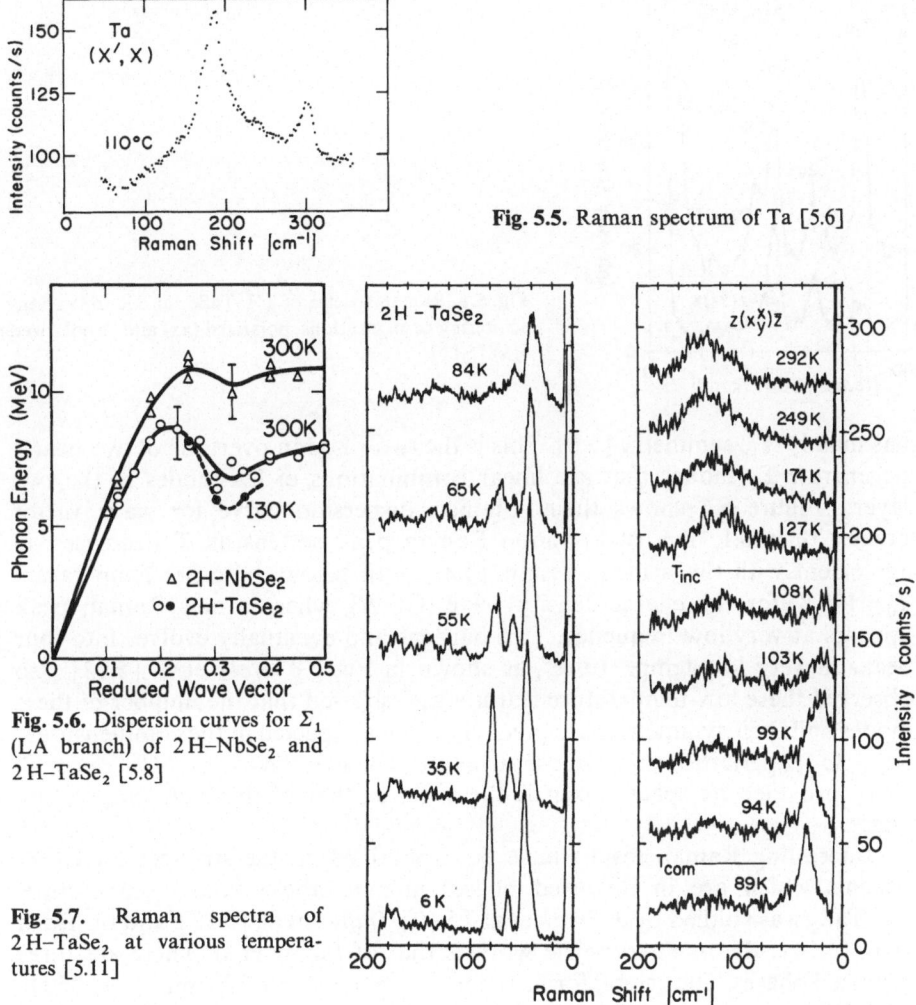

Fig. 5.5. Raman spectrum of Ta [5.6]

Fig. 5.6. Dispersion curves for Σ_1 (LA branch) of 2 H–NbSe$_2$ and 2 H–TaSe$_2$ [5.8]

Fig. 5.7. Raman spectra of 2 H–TaSe$_2$ at various temperatures [5.11]

vector $q = (2/3)\,\Gamma M$ in the basal plane of the hexagonal Brillouin zone and of a very similar phonon, technically an optic phonon, having the same q in the layer plane, but with a π phase shift between adjacent layers. Figure 5.6 shows the dispersion curves at room temperature, as measured by *Moncton* et al. [5.8]. That these curves are anomalous may been seen by comparing them with results on semiconducting MoS$_2$ by *Wakabayashi* et al. [5.9]. The lattices have the same structure, but there are no conduction electrons in MoS$_2$, and the dispersion curves are "ordinary" and can be understood using ordinary short-range force constants and atomic polarizability effects.

Room-temperature Raman data on 2 H–TaSe$_2$ were published by *Smith* et al. [5.10] and by *Steigmeier* et al. [5.11]. Results from the latter workers are shown in Fig. 5.7. The strongest feature is the broad peak near 135 cm^{-1}, which

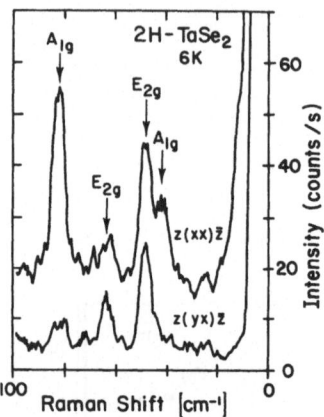

Fig. 5.8. Raman spectra of 2 H–TaSe₂ at 6 K for various scattering configurations, polarized (*xx*) and depolarized (*yx*) [5.11]

has mainly A_{1g} symmetry [5.6]. This is the two-phonon overtone of two nearly degenerate Σ_1 modes that are linear combinations of LA-modes in the two layers. Figure 5.6 shows their common dispersion curve for wave vector greater than 0.1. The two-phonon Raman peak softens as T decreases, in agreement with the neutron results [5.8], until below the onset temperature $T_0 = 120$ K for the charge-density wave (CDW), where a new Raman peak appears at very low frequency. This hardens and eventually evolves into four peaks at very low temperatures, as shown in Fig. 5.8. *Holy* et al. [5.12] also observed these low-temperature features and showed that the number of these modes and their symmetries are precisely what is expected in the commensurate $3a_0 \times 3a_0$ superlattice state, whose structure was investigated by *Moncton* et al. [5.8], provided its space group is the same as that of the high-temperature phase.

Interesting Raman results have been obtained on the stronger 1 T CDW systems which are in distorted phases already above room temperature. 1 T–TaS₂ was studied by *J. Smith* et al. [5.10], *Duffey* et al. [5.13], and by *Tsang* et al. [5.14]; the most complete work is that of *Duffey* et al. There are three distorted phases. The crystal forms in the incommensurate $1 T_1$ phase. At 350 K the CDW wave vectors rotate into another incommensurate $(1 T_2)$ phase, which finally becomes commensurate at 190–200 K $(1 T_3$ phase) with a $\sqrt{13}a_0 \times \sqrt{13}a_0 \times 13c_0$ superlattice containing 169 Ta atoms. The CDW Raman spectrum in the $1 T_3$ phase is extremely rich [5.13]. Elastic and inelastic neutron scattering measurements were done on $1 T_2$–TaS₂ by *Ziebeck* et al. [5.15]. Similar but less spectacular Raman spectra were observed in 1 T–TaSe₂ [5.10, 14].

The Raman activity of the new one-phonon peaks that appear below T_0 in 2 H–TaSe₂ (and in 2 H–NbSe₂ [5.16]) can be crudely understood as derived from that of the anomalously strong two-phonon peak in the normal phase. The two-phonon part of the Raman polarizability tensor can be written phenomenologically in the form

$$\delta\alpha = \sum_q \alpha^{(2)} u(q) u(-q), \tag{5.1}$$

where the $u(q)$'s are phonon coordinates. In the CDW phases u acquires a static or condensed part u_s and a dynamic part δu:

$$u(q_0) = u_s(q_0) + \delta u(q_0), \tag{5.2}$$

where q_0 is the wave vector of the condensed phonons. The fluctuating polarizability will then contain contributions of the form

$$\delta\alpha \sim \alpha^{(2)}[u_s(q_0)\delta u(-q_0) + \ldots] \tag{5.3}$$

which is first order in the $\delta u(-q_0)$, and the resulting intensity will be proportional to the square of the "order parameter" u_s. This argument is based on ideas from "mean-field theory," which may not accurately apply here; they furnish, however, a convenient starting point for discussion. These considerations imply that an approximate understanding of the Raman activity of the "new" modes can be simply based on an understanding of the strong two-phonon spectra in the normal phase (see Sect. 5.5.3 for a full discussion).

A similar strong two-phonon Raman spectrum is observed in the normal phase of $2\,H{-}NbSe_2$ [5.16, 17]. In this material the soft LA phonon condenses at 33 K to form an incommensurate CDW state that has an approximate $3a_0 \times 3a_0$ superlattice.

5.2 Microscopic Theories of Phonon Softening

A variety of theoretical work has been done to explain the origins of the phonon anomalies in transition-metal compounds. Some introduce the susceptibility of noninteracting Bloch electrons

$$\chi(q) = \sum_{ack} \frac{f_{ak} - f_{ck+q}}{E_{ak} - E_{ck+q}}. \tag{5.4}$$

Here f_{ak} is the Fermi function and E_{ak} the energy of a state with wave vector k and band a. The quantity $\chi(q)$ enters into simplified dielectric theories of screening that make the assumption of constant matrix elements. It will be large when Fermi surface "nesting" occurs. This happens when large pieces of the Fermi surface in band a at wave vector k and in band c at wave vector $k+q$ are parallel to one another. It is relatively easy to estimate $\chi(q)$ using results from band-structure calculations. *Gupta* and *Freeman* [5.18] and *Klein* et al. [5.19] performed such a calculation for carbides, and *Doran* et al. did one for $2\,H{-}NbSe_2$ [5.20]. They showed a gentle peaking of $\chi(q)$ for q's near the wave vectors of the anomalous phonons, but this effect is not large enough to produce phonon anomalies. More complicated models generalize $\chi(q)$ by introducing local field corrections, i.e., umklapp terms, to the dielectric constant [5.21, 22]. They may be formally correct, but they are often difficult to use for accurate numerical calculations.

The phonon anomalies have been successfully explained quantitatively by a microscopic theory developed by *Varma* and *Weber* [5.23] and applied by them to Nb and Mo and by *Weber* to NbC [5.24]. They employ a nonorthogonal tight-binding parameterization of the band structure to calculate phonon energies. In effect they calculate the phonon self-energy matrix within the random-phase approximation by evaluating the self-energy, or polarization "bubble" of Fig. 5.15a. They show that great computational simplification occurs when one renormalizes the two electron–phonon vertices by Coulomb interactions. To compensate for this double vertex correction, a term must be subtracted from the dynamical matrix which almost exactly cancels the ion-core – ion-core Coulomb interaction. Their expression for the dynamical matrix additively contains parts easily parameterized by simple short-range ion – ion interactions and a part, $D^{(2)}$, due mainly to the d electrons, which produces the anomalies. In fact, the anomalies result from the contribution to $D^{(2)}$ of d electrons that are within $1/2$ eV of the Fermi energy. $D^{(2)}$ is essentially the phonon self-energy due to the d electrons projected onto the space of ionic displacements.

The *Varma–Weber* theory as generalized to lattices with a basis by *Weber* introduces the form factor

$$g^{\lambda\beta}_{kc,\,k+qd}.$$

This is the renormalized matrix element for scattering of electrons from k, c to $k+q, d$ due to *a* "frozen" plane displacement wave (with wave vector q) in the βth Cartesian direction of the atoms in the λth sublattice. *Varma* and *Weber* [5.23] and *Weber* [5.24] gave explicit formulas for g in terms of gradients with respect to atomic position of self-consistent overlap integrals and tight-binding matrix elements. In terms of the g's $D^{(2)}$ is given by

$$D^{(2)}(\lambda\beta, \lambda'\beta'|q) = \sum_{kcd} \frac{(f_{dk+q} - f_{ck})}{(E_{dk+q} - E_{ck})} g^{\lambda'\beta'}_{kc,\,k+qd} g^{\lambda\beta}_{k+qd,\,kc}. \tag{5.5}$$

The total dynamical matrix is

$$D = D^{(0)} + D^{(1)} + D^{(2)}. \tag{5.6}$$

Varma and *Weber* stated that $D^{(0)}$ and $D^{(1)}$ do not give rise to phonon anomalies [5.23]. They showed by explicit calculation that $D^{(2)}$ has anomalies, essentially because there are rather localized regions in k space where the energies of the bands (c, k) and $(d, k+q)$ are not very different and where the form factors g are quite large. $D^{(2)}$ is essentially the contribution of the d electrons near the Fermi energy to the dynamical matrix.

The *Varma–Weber* theory has been applied to NbSe$_2$ by *Doran* [5.25], who showed that the calculated $D^{(2)}$ has the necessary properties to explain the observed LA phonon softening. A similar result was found by *Inglesfield* for $2\,H$–TaS$_2$ [5.26].

5.3 The Anomalous E_g Optical Phonon in A15 Compounds

The A15 class of intermetallic compounds is remarkable both for the high superconducting transition temperature T_c in some cases and for the occurrence in some of these compounds of a cubic-to-tetragonal martensitic transformation somewhat above T_c [5.27–29]. In this structure, the transition-metal atoms are closely bonded into chains along [100] directions. The transformation occurs with a softening to zero of the E_g component of the elastic constant tensor. The E_g strain component is allowed by symmetry to couple bilinearly in the free energy or Hamiltonian to an optical phonon mode of E_g symmetry. The eigenvector for this phonon describes a "dimerization" of the transition-metal atoms in the [100] chains. Primary phonon softening is thought to be driven by interband electronic transitions of the same general type discussed above for phonons of nonzero wave vector. For the E_g optical mode such coupling to electronic bands at the center of the Brillouin zone is equivalent to the dynamic Jahn–Teller effect [5.30], and if it occurs at the X-zone boundary point, it is similar to a Peierls distortion in a one-dimensional chain, as pointed out by *Gor'kov* [5.31]. For the E_g strain only Jahn–Teller-type coupling is possible [5.32].

The E_g Raman spectrum of V_3Si, which has $T_c = 17$ K, was found to be anomalous by *Wipf* et al. [5.33]. Their results are shown in Fig. 5.9. There are several anomalous features. The frequency is quite low compared to prior theoretical estimates [5.34]. The line shape is asymmetrical; the halfwidth on the high-energy side is about half that on the low-energy side, and the width decreases with increasing temperature. The solid lines in the figure represent a least-squares fit to a Lorentzian with a superimposed Fano-type antiresonance [Ref. 5.35, Chap. 4]. A possible theoretical explanation for this will be discussed in Sect. 5.4.2.

The E_g Raman spectrum in stoichiometric and nonstoichiometric V_3Si has also been observed by *Schicktanz* et al. [5.36]. In a later paper *Schicktanz* et al. [5.37] studied its temperature dependence in these crystals and in Nb_3Sn. Figure 5.10 shows their spectra for Nb_3Sn. The peak at 134 cm^{-1} has T_{2g} symmetry; that at 192 cm^{-1} has E_g symmetry. No other first-order peaks are expected or observed. The E_g peak is asymmetric in a similar way to that of V_3Si. Results for the temperature dependence of the E_g peak in Nb_3Sn are shown in Fig. 5.11. With decreasing temperature the width increases and the frequency decreases until somewhat above the temperature $T_m \approx 40$ K, where some Nb_3Sn samples transform; this is the minimum frequency reached, and an increase follows. This Nb_3Sn sample was nontransforming, however. In transforming stoichiometric V_3Si the width was found to reach a maximum and the frequency a minimum at 100 K, far above $T_m = 20$ K. At temperatures lower than 100 K the width decreased and the frequency increased. For a nonstoichiometric, nontransforming sample of V_3Si, frequency and width remained essentially constant at their respective minimum and maximum values for temperatures below 100 K.

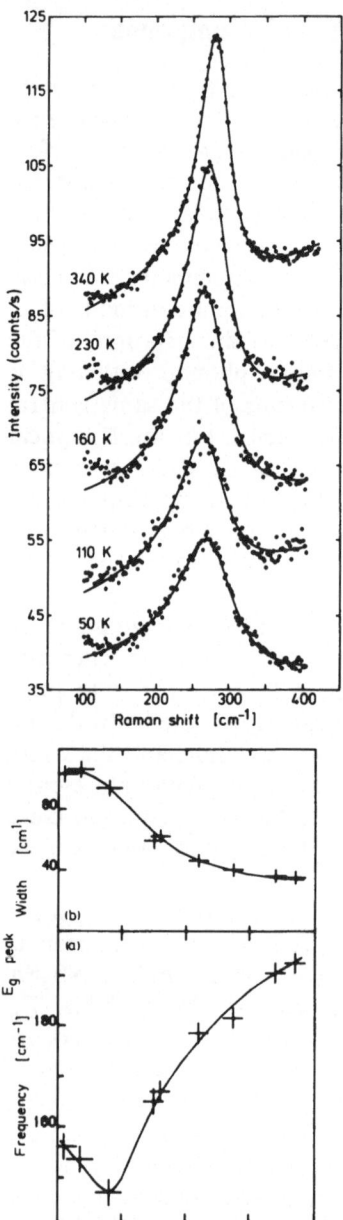

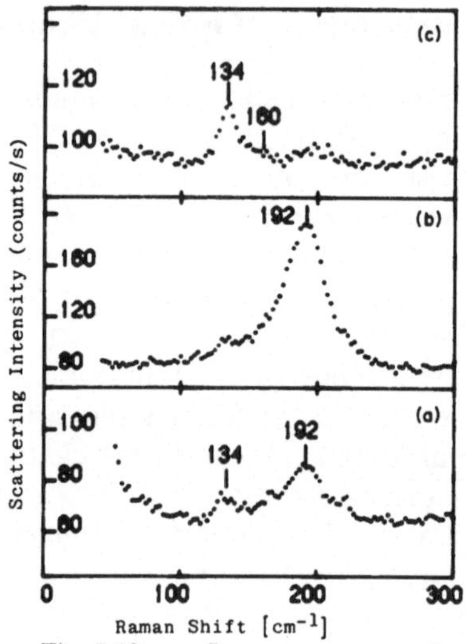

Fig. 5.10a–c. Raman spectra of nontransforming Nb$_3$Sn in three geometries [5.37]. **(a)** $A_{1g} + 1/3 E_g + T_{2g}$ geometry, **(b)** E_g geometry, **(c)** T_{2g} geometry

◀ **Fig. 5.9.** The E_g Raman spectrum of V$_3$Si at 5 temperatures [5.33]. Lines represent fits to a Lorentzian curve multiplied by an antiresonant factor

Fig. 5.11a, b. Temperature dependence of the frequency **(a)** and width **(b)** of the E_g Raman peak of nontransforming Nb$_3$Sn [5.37]

Schicktanz et al. [5.37] interpreted their results in terms of a microdomain model in which tetragonal domains exist even above T_m [5.38] and contribute a high-frequency component to the E_g Raman line, which therefore would have an inhomogeneous origin. This interesting suggestion needs to be developed more fully. A previous explanation for the line shape above T_m has been given by *Wipf* et al. [5.33], namely that it is homogeneous and due to a Fano-type interference. A general microscopic theory of such an effect will be discussed in Sect. 5.4.

5.4 Microscopic Theory of Raman Scattering in Metals

Light scattering in metals takes place in three steps [5.39]: 1) The incident photon field propagates from outside the sample. It is transmitted through the interface, and it attenuates strongly as it penetrates inside. 2) This photon is inelastically scattered by an excitation in the metal. 3) The scattered photon propagates back to the interface while being attenuated. It is transmitted through the interface, and propagates to the detector. Steps 1 and 3 can be treated by classical electromagnetic theory using Fresnel relations for the transmission coefficients at the interface. When a correction is made for the collection solid angle, steps 1 and 3 together introduce a coefficient that multiples the cross section by $(4/|1 + \sqrt{\varepsilon}|^2)^2$, where ε is the complex dielectric constant of the metal evaluated at an average photon frequency ω_l.

The component of the photon momentum q_z in the direction perpendicular to the interface is not conserved. Instead each photon acquires a Lorentzian distribution of this momentum peaked about zero with a width $1/\delta$ where δ is the optical penetration depth

$$\delta^{-1} = (2\omega_l/c)\,\mathrm{Im}\{\sqrt{\varepsilon}\}. \tag{5.7}$$

If both the matrix element for the scattering process in step 2 and the frequency (or density of states) of the created excitations are insensitive to the wave vectors of the two photons, one can take the zero wave-vector limit of these quantities. The spatial integration over the coordinate of the scattering event then introduces a factor of $(\delta/2)$.

The external photon scattering probability per unit solid angle (Ω_s) and frequency (ω_s) for the scattered photon is thus given for the 180° backscattering geometry normal to the surface by

$$\frac{d^2P}{d\omega_s d\Omega_s} = \left(\frac{4}{|1 + \sqrt{\varepsilon}|^2}\right)^2 \frac{\delta}{2V} \frac{d^2\sigma_{in}}{d\omega_s d\Omega_s'}, \tag{5.8}$$

where σ_{in} is the scattering cross section and Ω_s' the solid angle "inside" the sample, and where V is the quantization volume for the wave functions inside the sample, which is assumed to be uniformly irradiated on one surface by the incident light.

The entire scattering process has been given a Green's function treatment by *Kawabata* [5.40]. *Kawabata*'s formalism calls for the computation of an irreducible four-vertex function, which can be evaluated in various approxi-

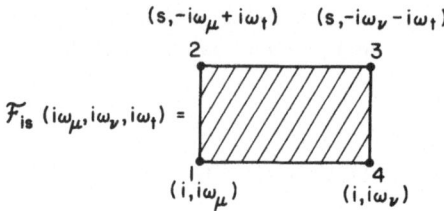

$$\mathcal{F}_{is}(i\omega_\mu, i\omega_\nu, i\omega_t) = $$

(s, $-i\omega_\mu + i\omega_t$)   (s, $-i\omega_\nu - i\omega_t$)

2   3

(i, $i\omega_\mu$)   (i, $i\omega_\nu$)

Fig. 5.12. The four-vertex function $\mathcal{F}$ for Raman scattering [5.40]

mations using the methods of diagrammatic perturbation theory [5.41]. In spite of their formal complexity, these methods are quite easy to use in practice, and they have several specific advantages for describing Raman scattering in metals with strong electron–phonon interactions, namely: 1) An easy connection can be made to the powerful Green's function description of electron–phonon interactions. 2) Ordinary multiple-order time-dependent perturbation theory generates many terms each with various orderings of electron–phonon and electron–photon operators and of energy denominators. Keeping track of all these terms becomes difficult. The Green's function method generates the equivalent of each term automatically. 3) The *Kawabata* formalism gives a clear prescription of how to treat the singularities that accompany processes of virtual photon absorption and emission. 4) This formalism uses the imaginary frequencies of Matsubara. Appropriate analytical continuation leads to results for finite temperatures with no more formal manipulation than is necessary at zero temperature. 5) Interferences of the Breit–Wigner–Fano type between two or more types of final states can be included with little formal difficulty.

In *Kawabata*'s formulation, what we have referred to above as steps 1 and 3 are described by one-photon Green's functions. Step 2 is described by the irreducible part of a four-vertex function $\mathcal{F}$ which has the structure shown in Fig. 5.12. At vertex 1 there is a matrix element of $(\mathbf{J} \cdot \mathbf{e}_i)$ where $\mathbf{J}$ is the current-density operator and $\mathbf{e}_i$ is a unit polarization vector for the incident photon. A Matsubara frequency $i\omega_\mu$ is added to the system at vertex 1. At vertex 2 $(\mathbf{J} \cdot \mathbf{e}_s)$ acts, where $\mathbf{e}_s$ is the scattered photon's polarization vector, and the added frequency is $-i\omega_\mu + i\omega_t$, etc. The frequencies ω_μ, ω_t, and ω_ν are equal to $2\pi T$ multiplied respectively by the integers μ, t, ν. The internal cross-section is then proportional to the following expression in which $\mathcal{F}$ is analytically continued as indicated:

$$[1 + n(\omega_0)] \frac{1}{2\pi i} \lim_{\varepsilon \to 0+} [\mathcal{F}_{is}(\omega_i + i\delta, -\omega_i + i\delta, \omega_0 + i\varepsilon)$$

$$- \mathcal{F}_{is}(\omega_i + i\delta, -\omega_i + i\delta, \omega_0 - i\varepsilon)].$$ (5.9)

Here δ is a fixed positive infinitesimal; ω_i and ω_s are the incident and scattered photon frequencies; $\omega_0 = \omega_i - \omega_s$ and $n(\omega) = [\exp(\omega/T) - 1]^{-1}$ is the Bose function. Units are used in which $\hbar$ and Boltzmann's constant are unity.

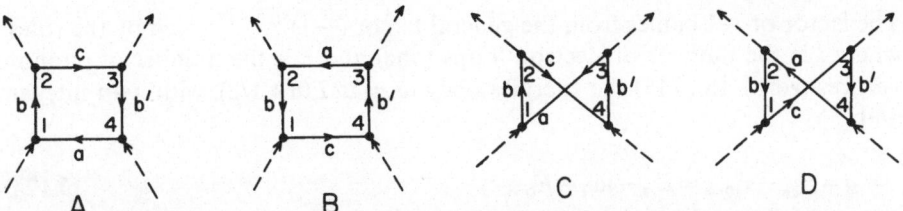

Fig. 5.13. Diagrams for $\mathscr{F}$ for interband electronic Raman scattering. The solid lines represent one-particle electronic Green's functions

Equation (5.9) is a generalization of the usual fluctuation-dissipation theorem for the Raman scattering cross section σ

$$d\sigma \propto [1 + n(\omega_0)] \, \mathrm{Im}\,\{R(\omega_0 + i\varepsilon)\},$$

where the "response function" R is the Fourier transform of a retarded Green's function that has the analytical properties of a one-particle Boson Green's function. R is obtained by analytical continuation of the Matsubara Green's function $R\,(i\omega_t)$.

5.4.1 Interband Electronic Raman Scattering

An application of the general theory was made by *Kawabata* to intraband electronic Raman scattering [5.40]. His result is altered slightly here to apply to interband scattering, which should be important to transition metals and compounds with complex band structure near the Fermi energy.

The Hamiltonian of the system is purely electronic; $H = H_e$ where

$$H_e = \sum_{ka} E_{ak} c_{ak}^+ c_{ak}. \tag{5.10}$$

In band a at wave vector k, $c_{ak}^+(c_{ak})$ creates (destroys) an electron of energy E_{ak}. Figure 5.13 shows the relevant Feynman diagrams. At vertices 1 and 4 one inserts matrix elements of the operator $p_i = (\boldsymbol{p} \cdot \boldsymbol{e}_i)$, where $\boldsymbol{p}$ is the momentum operator, and at vertices 2 and 3 the matrix elements are those of $p_s = (\boldsymbol{p} \cdot \boldsymbol{e}_s)$. Here $\boldsymbol{e}_i$ and $\boldsymbol{e}_s$ are photon polarization unit vectors. The solid lines represent one-electron Green's functions, which have the form $(i\omega_n + E_0 - E_{ak})^{-1}$ in the noninteracting limit for a line a. Here the notation is that of (5.4), i.e., a is a band index and k a wave vector. E_0 is the chemical potential. Using the rules of computation of diagrams we find that each diagram in Fig. 5.13 may be written in the form

$$(-1) \sum_{kabcb'} T \sum_n \frac{(p_i)(p_s)(p_i)'(p_s)'}{(i\omega_n - \underline{a})(i\omega_n - \underline{b})(i\omega_n - \underline{c})(i\omega_n - \underline{d})}. \tag{5.11}$$

The factor of -1 comes from the general factor $(-1)^{(1/2)v+l}$ given by the rules, where l is the number of electron loops (one) and v is the number of phonon vertices (zero). In (5.11) the energies obey $\omega_n = 2\pi T(n+1/2)$, with n an integer, and

$$\underline{a} = E_{ak} - E_0, \underline{c} = E_{ck} - i\omega_t - E_0, \tag{5.12a}$$

$$\left.\begin{aligned} \underline{b} &= E_{bk} - i\omega_\mu - E_0 && \text{for } A \text{ and } C \\ \underline{b} &= E_{bk} + i\omega_\mu - i\omega_t - E_0 && \text{for } B \text{ and } D, \end{aligned}\right\} \tag{5.12b}$$

$$\left.\begin{aligned} \underline{b}' &= E_{b'k'} - i\omega_v - i\omega_t - E_0 && \text{for } B \text{ and } C \\ \underline{b}' &= E_{b'k'} + i\omega_v - E_0 && \text{for } A \text{ and } D, \end{aligned}\right\} \tag{5.12c}$$

where

$$(p_i) = (bk|p_i|ak) \quad \text{and} \quad (p_s) = (ck|p_s|bk) \tag{5.12d}$$

for A and C, and

$$(p_i) = (ck|p_i|bk) \quad \text{and} \quad (p_s) = (bk|p_s|ak) \tag{5.12e}$$

for B and D, and where

$$(p_i)' = (ak'|p_i|b'k') \quad \text{and} \quad (p_s)' = (b'k'|p_s|ck') \tag{5.12f}$$

for A and D, and

$$(p_i)' = (b'k'|p_i|ck') \quad \text{and} \quad (p_s)' = (ak'|p_s|b'k') \tag{5.12g}$$

for B and C. To evaluate the sum over n we use the identity

$$T \sum_n H(i\omega_n) = \frac{1}{2\pi i} \int_c f(z) H(z) dz, \tag{5.13a}$$

where

$$f(z) = (e^{z/T} + 1)^{-1} \tag{5.13b}$$

and where the contour C goes counterclockwise around the poles of $H(z)$. This gives for (5.11)

$$- \sum_{kabcb'} (p_i)(p_s)(p_i)'(p_s)' \left(\frac{f_a}{(\underline{a}-\underline{b})(\underline{a}-\underline{c})(\underline{a}-\underline{b}')} + \frac{f_b}{(\underline{b}-\underline{c})(\underline{b}-\underline{d})(\underline{b}-\underline{d}')} \right.$$

$$\left. + \frac{f_c}{(\underline{c}-\underline{a})(\underline{c}-\underline{b})(\underline{c}-\underline{b}')} + \frac{f_{b'}}{(\underline{b}'-\underline{a})(\underline{b}'-\underline{b})(\underline{b}'-\underline{c})} \right), \tag{5.14a}$$

where, for example

$$f_a \equiv f(E_{ak} - E_0) \quad \text{and} \quad f_b \equiv f(E_{bk} - E_0 - i\omega_\mu) = f(E_{bk} - E_0). \tag{5.14b}$$

Of the four terms in square brackets in (5.14a) only the first and third contribute to (5.9), through the factor $(a - c)^{-1}$, giving

$$\frac{1}{2\pi i} \lim_{\varepsilon \to 0+} \left(\frac{1}{E_{ak} - E_{ck} + \omega_0 + i\varepsilon} - \frac{1}{E_{ak} - E_{ck} + \omega_0 - i\varepsilon} \right) = -\delta(E_{ak} - E_{ck} - \omega_0). \tag{5.15}$$

Step 2, as described in the beginning of Sect. 5.4, involves scattering of photons inside the metal. Making the analytical continuations $i\omega_\mu \to \omega_i + i\delta$, $i\omega_\nu \to -\omega_i + i\delta$ implied by (5.9), we find that the contribution of diagram A of Fig. 5.13 to the interior cross section is

$$\frac{d^2\sigma_{in}}{d\Omega'_s d\omega_s} = \frac{\omega_s}{\omega_i} \frac{r_0^2}{m^2} [1 + n(\omega_0)] \sum_{kac} |M(ac)|^2 [f(E_{ak} - E_0) - f(E_{ck} - E_0)]$$
$$\cdot \delta(\omega_0 + E_{ak} - E_{ck}), \tag{5.16}$$

where $r_0 = e^2/mc^2$ is the Thompson radius, m is the electron mass, and $M(ac)$ is given by

$$M_A(ac) = \sum_{b \neq a,c} \frac{(ck|p_s|bk)(bk|p_i|ak)}{E_{ak} - E_{bk} + \omega_i + i\delta}. \tag{5.17a}$$

The factor $r_0^2 \omega_s/(m^2\omega_i)$ in (5.16) is implicit in *Kawabata*'s expressions [5.40]. Alternatively, it may be justified by comparison with the results from use of ordinary perturbation theory. When diagrams B, C, and D are added, the same result (5.16) is obtained except that now

$$M(ac) = M_A(ac) + M_B(ac), \tag{5.17b}$$

with

$$M_B(ac) = \sum_{b \neq a,c} \frac{(ck|p_i|bk)(bk|p_s|ak)}{E_{ak} - E_{bk} - \omega_s - i\delta}. \tag{5.17c}$$

Broadening of E_{bk} can be treated by replacing it by a parameter E', by multiplying (5.17) by $A_{bk}(E')$, the spectral function for the state $|bk)$, and by integrating over E'. If $A_{bk}(E')$ is a Lorentzian with width γ_{bk}, then the result is simply to replace δ in (5.17) by γ_{bk}.

5.4.2 One-Phonon Scattering

To H_e (5.10) we now add phonon and electron–phonon interaction terms

$$H = H_e + H_p + H',$$ (5.18a)

where

$$H_p = \sum_{qj} \omega_{qj}^0 b_{qj}^+ b_{qj}$$ (5.18b)

and

$$H' = \sum_{kdaqj} (d, k+q|V_j(-q)|ak) c_{d,k+q}^+ c_{ak}(b_{qj} + b_{-qj}^+).$$ (5.18c)

In (5.18b, c) $b_{-qj}^+(b_{qj})$ creates a phonon of wave vector $-q$ and polarization j (annihilates a phonon qj) and frequency $\omega_{qj}^0 = \omega_{-qj}^0$; $(d, k+q|V_j(-q)|ak)$ is the matrix element for scattering of an electron from band a and wave vector k to band d and wave vector $k+q$. It takes the form

$$(d, k+q|V_j(-q)|ak) = (2\omega_{qj}^0)^{-1/2}(d, k+q|\partial H/\partial u_{-qj}|ak),$$ (5.18d)

where

$$\frac{\partial H}{\partial u_{-qj}} = \sum_{\lambda\beta l} \frac{\partial H}{\partial u_{\lambda\beta}(l)} \frac{\varepsilon^{\lambda\beta}(-qj)}{\sqrt{M_\lambda N}} e^{iq \cdot R_l}.$$ (5.18e)

In (5.18e) l labels the unit cell; λ labels the atoms in the basis; β is a Cartesian coordinate index; N is the number of unit cells; and M_λ is the mass of atom λ. The $\varepsilon^{\lambda\beta}(-qj)$ are the components of the phonon polarization vector for mode $(-qj)$, and $u_{\lambda\beta}(l)$ is the displacement in the βth direction of the atom at site λ in the lth unit cell. Using a tight-binding description of the electronic bands, $(d, k+q|\partial H/\partial u_{\lambda\beta}(l)|ak)$ may be parameterized (5.23, 24). It is closely related to the derivative of nearest-neighbor atomic overlap parameters with respect to displacement of one of the atoms, and has the order of magnitude of such a derivative. The matrix element $(d, k+q|V_j(-q)|ak)$ is of the order of $N^{-1/2}$ times that derivative times a root-mean-square vibrational amplitude.

The four-vertex function for one-phonon Raman scattering may be written as

$$\mathcal{F}(1\text{ph}) = - \sum_j (A+B) D_{0j}(i\omega_t)(A'+B').$$ (5.19a)

The diagram AD_jA' is shown in Fig. 5.14a. The other electron loops B and B' are shown in Fig. 5.14b. The phonon Green's function for mode j of zero wave

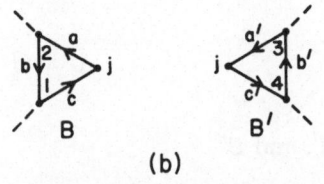

(c)

Fig. 5.14. (a) One possibility for $\mathscr{F}$ for one-phonon Raman scattering. **(b)** alternate possibilities for electron loops. **(c)** "Dyson equation" for dressed phonon Green's function (double wavy line) in terms of bare phonon Green's function (single wavy line)

vector is denoted by $D_{0j}(i\omega_t)$. It should be regarded as dressed and is obtained by formally summing the series shown in Fig. 5.14c, which gives

$$D_{0j}(i\omega_t) = \frac{2\omega_{0j}^0}{(i\omega_t)^2 - (\omega_{0j}^0)^2 - 2\omega_{0j}^0 \pi_{jj}(0, i\omega_t)} \tag{5.19b}$$

and where

$$2\omega_{0j}^0 \pi_{ij}(q=0, i\omega_t) = \sum_{kac} \frac{(f_a - f_c)|(a|V_j(0)|c)|^2 2\omega_{0j}^0}{E_a - E_c + i\omega_t} \tag{5.19c}$$

is the phonon self-energy (polarization). We neglect mixing between phonon modes j and j' of the same symmetry. We neglect vertex corrections due to the electron–phonon interaction. We do assume, however, that the electron–phonon matrix elements, which take the form $(ak|V_j|ck) \equiv (a|V_j|c)$, are fully renormalized with respect to Coulomb interactions. In final expressions these matrix elements always appear multiplied by $(2\omega_{0j}^0)^{1/2}$. The product is invariant under renormalization of the phonon frequency, (5.18d). In (5.19c) and in what follows we drop explicit reference to the wave vector k.

Diagram A gives the result

$$A = \sum_{abck} (a|V_j|c)(c|p_s|b)(b|p_i|a) F_A, \tag{5.20a}$$

where

$$F_A = T \sum_n [(i\omega_n - \underline{a})(i\omega_n - \underline{b})(i\omega_n - \underline{c})]^{-1} \tag{5.20b}$$

as given by (5.12a–c), and where $V_j \equiv V_j(0)$. Use of (5.13) to perform the sum over n allows F_A to be expressed as

$$F_A = F_A^{(1)} + F_A^{(2)}, \tag{5.21a}$$

where

$$F_A^{(1)} = \frac{f_a - f_c}{(E_a - E_c + i\omega_t)(E_a - E_b + \omega_i + i\gamma_b)}$$ (5.21b)

and

$$F_A^{(2)} = \frac{f_b - f_c}{(E_b - E_a - \omega_i - i\gamma_b)(E_b - E_c - \omega_s - i\gamma_b)}.$$ (5.21c)

Similar expressions are obtained for diagrams B, A', and B'.

One finds that as $i\omega_t \to \omega_0 \pm i\varepsilon$

$$A + B = R_j \mp i I_j$$ (5.22a)

and

$$A' + B' = R_j^* \mp i I_j^*,$$ (5.22b)

where

$$R_j = \sum_{abck} \left\{ \frac{(a|V_j|c)(c|p_s|b)(b|p_i|a)}{E_a - E_b + \omega_i + i\gamma_b} \left[\frac{f_b - f_c}{E_c - E_b + \omega_s + i\gamma_b} + \frac{f_a - f_c}{(E_a - E_c + \omega_0)_p} \right] \right.$$
$$\left. + \frac{(a|V_j|c)(c|p_i|b)(b|p_s|a)}{E_a - E_b - \omega_s - i\gamma_b} \left[\frac{f_b - f_c}{E_c - E_b - \omega_i - i\gamma_b} + \frac{f_a - f_c}{(E_a - E_c + \omega_0)_p} \right] \right\}$$ (5.22c)

and

$$I_j = \pi \sum_{ack} M(ac)(a|V_j|c)(f_a - f_c)\delta(E_a - E_c + \omega_0),$$ (5.22d)

with $M(ac)$ given by (5.17) with δ replaced by γ_b. The subscript p in (5.22c) denotes principal value.

The one-phonon contribution to the discontinuity (5.9) is

$$S_1 = \frac{1}{2\pi i} \lim_{\varepsilon \to 0+} [\mathscr{F}(\omega_0 + i\varepsilon) - \mathscr{F}(\omega_0 - i\varepsilon)]$$

$$= \pi^{-1} \sum_j [(|R_j|^2 - |I_j|^2)\text{Im}\{D_j\} + (R_j I_j^* + R_j^* I_j)\text{Re}\{D_j\}].$$ (5.23)

The last term in (5.23) describes electronic Raman scattering. It will interfere with (5.16). Using (5.19) we find that S_1 can be written

$$S_1 = \pi^{-1} \sum_j \sigma_{0j} \left(\frac{|Q_j + \varepsilon_j|^2}{|1 + \varepsilon_j|^2} - 1 \right),$$ (5.24a)

where

$$Q_j = R_j / I_j, \tag{5.24b}$$

$$\sigma_{0j} = -|I_j|^2 / \mathrm{Im}\{\pi_j\}, \tag{5.24c}$$

$$\varepsilon_j = \frac{\delta_j - \mathrm{Re}\{\pi_j\}}{(-\mathrm{Im}\{\pi_j\})}, \tag{5.24d}$$

$$\delta_j = (\omega_0^2 - \omega_j^{02})/(2\omega_j), \tag{5.24e}$$

$$-\mathrm{Im}\{\pi_j\} = \pi \sum_{kac} (f_a - f_c)|(a|V_j|c)|^2 \delta(\omega_0 + E_a - E_c), \tag{5.24f}$$

$$\mathrm{Re}\{\pi_j\} = \sum_{kac} \frac{(f_a - f_c)|(a|V_j|c)|^2}{(E_a - E_c + \omega_0)_p}. \tag{5.24g}$$

A result very similar to (5.24) was obtained by *Bechstedt* and *Peuker* [5.42] using a different Green's function technique.

If we denote the sum over k, a, and c in (5.16) by S_0, we find that the contribution of the electronic term (5.16) and the one-phonon term (5.23) and (5.24) to the cross section is proportional to $S_0 + S_1$ and equals

$$\frac{d^2 \sigma_{in}}{d\Omega_s' d\omega_s} = \frac{\omega_s r_0^2}{\omega_i m^2} [1 + n(\omega_0)] (\sigma_0 + \sigma_1 + \sigma_{1a}), \tag{5.25a}$$

where

$$\sigma_0 = S_0 + \sum_j \sigma_{0j} = \sum_j (-N_j)/\mathrm{Im}\{\pi_j\} \tag{5.25b}$$

with

$$N_j = \left[\sum_{ack} (f_a - f_c)|M_{ac}|^2 \delta(E_a - E_c + \omega_0)\right]$$
$$\cdot \left[\sum_{a'c'k'} (f_{a'} - f_{c'})\delta(E_{a'} - E_{c'} + \omega_0)|(a'|V_j|c')|^2\right]$$
$$- \left|\sum_{ack} (f_a - f_c)(a|V_j|c)M_{ac}\delta(E_a - E_c + \omega_0)\right|^2, \tag{5.25c}$$

where

$$\sigma_1 = \sum_j \sigma_{0j} \frac{(\mathrm{Im}\{Q_j\})^2}{1 + \varepsilon_j^2} \tag{5.25d}$$

and where

$$\sigma_{1a} = \sum_j \sigma_{0j} \frac{(\mathrm{Re}\{Q_j\} + \varepsilon_j)^2}{1 + \varepsilon_j^2}. \tag{5.25e}$$

By the triangle inequality N_j as given by (5.25c) is nonnegative. σ_0 describes an electronic Raman scattering background; σ_1 is essentially a sum of Lorentzians at the renormalized phonon frequencies; σ_{1a} has the Fano-form and represents interferences between phonon and electronic Raman scattering.

A result equivalent to (5.25) was derived by *Ipatova* and *Subashiev* [5.43] to explain microscopically the interference between phonon and interhole-band Raman scattering in degenerate *p*-type silicon. They used a three-band model for which the $(f_b - f_c)$ terms in (5.22c) are not present. They also omitted the damping γ_b in the intermediate electronic state. Their parameter Q was then real. This would give $\sigma_1 = 0$. By writing $Q_j = |Q_j| \exp(i\phi_j)$ we sum (5.25d, e) to give

$$\sigma_1 + \sigma_{1a} = \sum_j \sigma_{0j} \frac{|Q_j|^2 + 2\varepsilon_j |Q_j| \cos\phi_j + \varepsilon_j^2}{1 + \varepsilon_j^2}. \tag{5.26}$$

Bechstedt and *Peuker* show plots of a single term of (5.26) for various values of Q and ϕ [5.42].

The damping constant γ_b will be large in transition metals and compounds for any level not close to the Fermi level. If E_b is 2–3 eV above or below the Fermi energy, γ_b could be of the order of 0.2–0.5 eV based on values measured on lower density-of-states metals [5.44–46]. Such large values of γ_b in (5.22) will prevent sharp enhancements of the cross section by resonances between incident or scattered photons and real interband transitions. Note that if one of the $(f_b - f_c)$ terms in (5.22c) is in resonance, its denominator is imaginary, and it will be $\pm 90°$ out of phase with the $(f_a - f_c)$ term. Nevertheless, the latter term is likely to be stronger because its denominator can become considerably smaller than γ_b.

The solid lines in Fig. 5.9 represent a linear background plus one antiresonant term from the sum (5.25e). No attempt has yet been made to relate the parameters of the fit to microscopic band-structure parameters.

5.4.3 Intraband Electronic Processes

Thus far we have considered only interband electronic scattering processes. Intraband electronic Raman scattering has been discussed by *Kawabata* [5.40] for normal metals and by *Abrikosov* and *Genkin* [5.47] for superconductors. The results depend on q_z, the wave vector transferred during the scattering. As discussed in Sect. 5.4, in metals q_z has a distribution of values. Equations (5.16, 17) must be changed by replacing (ck) by $(a, k + q_z \hat{z})$ and then integrating over the distribution of q_z. The quantity $M(ac)$ in (5.16) needs to be evaluated for a single band (a, k) at the Fermi surface. Corrections for Coulomb interactions require that the average of M over the Fermi surface must be subtracted from M before it can be used to compute the cross section [5.40, 47]. The resulting expression for intraband Raman scattering in a normal metal will be approximately proportional to ω_0 for $0 < \omega_0 < v_f/\delta$ with a cutoff to zero as ω_0 increases beyond v_f/δ. For a superconductor at $T=0$ there will be no scattering until $\omega_0 = 2\Delta$ where Δ is the superconducting energy gap. Then there will be a rapid rise in the cross section, until the normal-metal value is reached [5.47].

The role of intraband electronic processes on the small wave-vector dispersion of optical phonons has been discussed by *Ipatova* and co-workers [5.48]. For $q \lesssim \omega_0/v_f$ there is a large nonadiabatic correction to the phonon frequency, one not containing the small parameter ω_0/E_F, which is not present for $q > \omega_0/v_f$. The transition value of $q \sim \omega_0/v_f$ is of the order of the reciprocal of the penetration depth δ. The integration over q_z referred to in the preceding paragraph will then give an asymmetric Raman line shape for the optical phonon. This dispersive effect is related to the closing of the intraband decay channel for $\omega_0 > q v_f$.

The transition metals and compounds have complicated band structures. The Fermi surface has a complicated multisheeted nature. Near points of degeneracy where the sheets intersect, zero wave-vector decay processes for optical phonons are possible, and there will be no anomalous small q phonon dispersion if these processes provide the dominant decay channels in these materials. If so, the interband electronic Raman processes discussed here are also likely to dominate over the intraband electronic scattering processes discussed by *Kawabata* [5.40], and by *Abrikosov* and *Genkin* [5.47].

5.4.4 Two-Phonon Raman Scattering

The aim of this section is to explain theoretically why two-phonon Raman scattering in transition metals and compounds seems to involve selectively those phonons that show anomalies. A detailed theory has been given by *Klein* [5.49] based on a short calculation for layered compounds by *Maldague* and *Tsang* [5.50]. The reason that Raman scattering is related to phonon anomalies is that both processes involve the same phonon-assisted scattering of d electrons close to the Fermi energy. *Maldague* and *Tsang* [5.50] pointed out that if one treats the momentum matrix elements and photon energy denominators as constant in k space, then the most important term in the two-phonon Raman amplitude is proportional to the phonon self-energy $\pi(q)$.

Figure 5.15 shows schematically that the phonon self-energy $\pi(q)$, which is due primarily to the d electrons, lowers the dispersion curve and makes it anomalous. The *Maldague–Tsang* approximation to the Raman intensity says that the two-phonon overtone Raman scattering intensity is proportional to $|\pi(q)|^2$. Hence the two-phonon spectrum will be weighted by how anomalous the phonons are, i.e., by how large $|\pi(q)|^2$ is.

The derivation of the microscopic theory is discussed in some detail on the next few pages. Some important steps in the derivation and important results are found in (5.29, 30, 37, 38, 43, 53, 54).

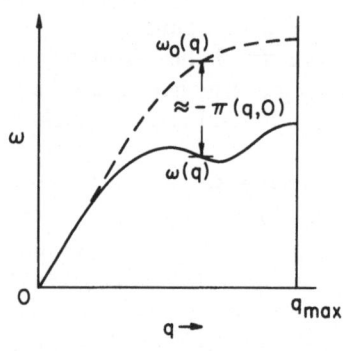

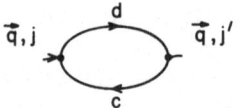

Fig. 5.15. (a) Diagram for self-energy $\pi_{ij'}(q, \omega)$. (b) Schematic dispersion curve for anomalous LA phonon. The dashed curve represents the effect of all interactions except that with d electrons near the Fermi energy. The solid line includes the d electrons

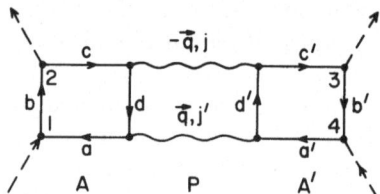

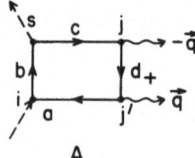

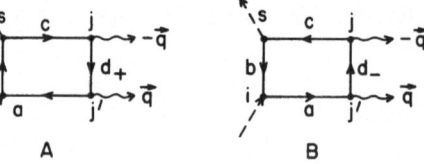

Fig. 5.16. One of the diagrams for the four-vertex function $\mathscr{F}$ for two-phonon Raman scattering [5.49]

Fig. 5.17. Diagrams representing all possible orderings of photon absorption and emission and phonon emission, under the assumption that state b does not couple to phonons [5.49]

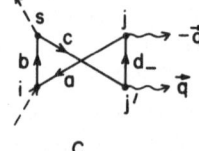

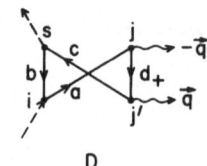

The two-phonon contribution to *Kawabata*'s four-vertex function [5.40] is from all distinct diagrams of the form shown in Fig. 5.16. Diagrams have been omitted that have phonon vertices inserted between the photon vertices 1 and 2 or 3 and 4 because it is assumed that the states b and b' do not strongly couple to phonons, whereas states a, c, and d (and a', c', and d') are assumed to be close to the Fermi energy and to couple strongly to phonons. Figure 5.17 shows four electron loops, A–D; there are four other electron loops, A'–D', that involve vertices 3 and 4; loop A' is shown in Fig. 5.16. The pair of phonon lines will be denoted by P. There are many equivalent combinations of diagrams, for instance $CPC' = APA'$. The distinct combinations can be shown to give for the 4-vertex function of Fig. 5.12

$$\mathscr{F}_{\mathrm{is}}(i\omega_\mu, i\omega_\nu, i\omega_t) = (A + B + C + D) P(A' + B')$$
$$= \tfrac{1}{2}(A + B + C + D) P(A' + B' + C' + D'). \tag{5.27}$$

According to (5.9), we must take the discontinuity of (5.27) as $i\omega_t$ is analytically continued across the real axis. For a diagram such as APA' (Fig. 5.16), this gives energy-conserving delta functions for various possible final states. These can be represented by "cuts" in the diagrams. The vertical cut through both phonon lines produces two phonons in the final state and is the one we want. Among other cuts are those that pass vertically through the electron loops, A or A', say. These correspond to electronic Raman scattering. They must be considered if the simple electronic scattering processes of Fig. 5.13 are also considered, and they will produce a Fano-type interference between electronic and two-phonon Raman scattering. We simplify the discussion and eliminate the interference formally by assuming that bands a and c (and a' and c') are identical. Recall that we have already formally eliminated intraband electronic processes by going to the limit of zero momentum transfer. In future work one might have to consider these omitted processes.

Other cuts will produce a final state consisting of one phonon plus an electron-hole pair. They will contribute a broad background to the two-phonon Raman spectrum. Another contribution to the background comes from cuts that lead to a two-pair final state. We do not wish to consider such final states and will therefore interpret the energy denominators whose imaginary parts would give them as principal values. Then we have $A = (A')^*$, $B = (B')^*$, $C = (C')^*$, and $D = (D')^*$.

We consider only the contribution of two-phonon creation processes to the two-phonon spectrum. Using

$$D_{-qj}(i\omega_n) = \frac{2\omega_{qj}^0}{(i\omega_n)^2 - \omega_{qj}^2} \tag{5.28a}$$

for the renormalized phonon Green's function, we find that the frequency integration for the phonon loop $P(i\omega_l)$ leads to this result before summation over q, j, and j':

$$\frac{1}{2\pi i}[P(\omega_0 + i\varepsilon) - P(\omega_0 - i\varepsilon)] = [1 + n(\omega_{-qj}) + n(\omega_{qj'})]\delta(\omega_0 - \omega_{-qj} - \omega_{qj'})$$

$$\cdot \left(\frac{\omega_{qj}^0 \omega_{qj'}^0}{\omega_{qj}\omega_{qj'}}\right), \tag{5.28b}$$

where ω_{qj} is the renormalized phonon frequency and ω_{qj}^0 is the bare frequency. Defining

$$M_{is}(-qj, \omega_{-qj}; qj', \omega_{qj'}) = (A + B + C + D)/\sqrt{2}, \tag{5.29}$$

where A–D are the contributions of the diagrams in Fig. 5.17 gives for the Raman cross section inside the sample

$$\frac{d^2\sigma_{in}}{d\Omega_s' d\omega_s} = \frac{\omega_s r_0^2}{\omega_i m^2} \sum_{qjj'} |M_{is}(-qj, \omega_{-qj}; qj', \omega_{qj'})|^2 \delta(\omega_{-qj} + \omega_{qj'} - \omega_0)$$

$$\cdot [1 + n(\omega_{-qj})][1 + n(\omega_{qj'})]. \tag{5.30a}$$

We have eliminated the last factor in (5.28b) by renormalizing the electron–phonon matrix element. Instead of (5.18d) we are to use

$$(d, k+q|V_j(-q)|ak) = (2\omega_{qj})^{-1/2}(d, k+q|\partial H/\partial u_{-qj}|ak) \tag{5.30b}$$

in calculating M_{is}.

For each diagram in Fig. 5.17 we evaluate the frequency summation and obtain results similar to (5.11–14) (but with different matrix elements and energies). We then let $E_c \rightarrow E_a$. Each diagram can be shown to take the form

$$\sum_{bcdk} (\text{Matrix elements}) F, \tag{5.31}$$

where

$$F = \frac{f_b - f_d}{(\underline{b}-\underline{a})(\underline{b}-\underline{c})(\underline{b}-\underline{d})} + \frac{(f_d-f_c)(\underline{a}+\underline{c}-\underline{b}-\underline{d})}{(\underline{b}-\underline{a})(\underline{b}-\underline{c})(\underline{d}-\underline{a})(\underline{d}-\underline{c})}. \tag{5.32}$$

For diagram A the energies in (5.32) are

$$\underline{a} = E_{ck} - E_0,$$
$$\underline{b} = E_{bk} - \omega_i - i\gamma_b - E_0, \tag{5.33}$$
$$\underline{c} = E_{ck} - \omega_0 - E_0,$$

and

$$\underline{d} = E_{d,k+q} - \omega_0 + \omega_{-qj} - E_0.$$

For diagrams B, C, and D expressions similar to (5.33) hold.

Since the width γ_b of level b is expected to be much larger than either or both of the phonon frequencies, it is a good approximation to neglect the frequencies ω_0, $\omega_{-qj'}$ and $\omega_{qj'}$ in the photon energy denominators. This is done symmetrically in (5.32) by the replacement

$$(\underline{b}-\underline{a})(\underline{b}-\underline{c}) \rightarrow [\underline{b} - \tfrac{1}{2}(\underline{a}+\underline{c})]^2. \tag{5.34}$$

The first term in (5.32), which will be denoted F_3, is, for diagram A,

$$F_3^A = \frac{f_b - f_{d+}}{(E_b - E_c - \omega_l - i\gamma_b)^2 (E_b - E_{d+} - \omega_l \mp i\gamma_b)}, \tag{5.35a}$$

where

$$\omega_l = \tfrac{1}{2}(\omega_i + \omega_s). \tag{5.35b}$$

In this and in subsequent expressions $d\pm$ denotes the Bloch state $(d, k\pm q)$. For diagrams A and B, F_3 is triply resonant when bands c and d are near the Fermi energy E_0 and when $E_b \approx \omega_i + E_0$. Such a term gives the most-resonant contribution to two-phonon scattering in insulators or semiconductors. In the present case, the largeness of γ_b prevents strong resonance enhancements, and other terms in F may be larger. For diagrams C and D, F_3 is triply resonant for $E_b \approx -\omega_i + E_0$.

The second term in (5.32) has parts with double and single photon resonances, denoted by F_2 and F_1 respectively. For diagram A, we have

$$F_2^A = \frac{-\frac{1}{2}(f_{d+} - f_c)}{(E_b - E_c - \omega_l - i\gamma_b)^2} \left[\frac{1}{(E_{d+} - E_c - \omega_{qj'})_p} + \frac{1}{(E_{d+} - E_c + \omega_{-qj})_p} \right] \quad (5.36a)$$

and

$$F_1^A = \frac{-\frac{1}{2}(f_{d+} - f_c)}{(E_b - E_c - \omega_l - i\gamma_b)(E_{d+} - E_c - \omega_{qj'})_p(E_{d+} - E_c + \omega_{-qj})_p}. \quad (5.36b)$$

Similar expressions exist for the contributions of diagrams B, C, and D to F_1 and F_2. In (5.36) the subscript p again denotes "principal value." Use has been made of phonon energy conservation, i.e., the delta function (5.28b).

During the summation over the wave vector k in (5.31) the energies of bands d and c will rise and fall through the Fermi energy, and $(f_{d+} - f_c)$ will alternate in sign, producing a tendency for F_1 to integrate to zero. In the following F_1 will be neglected.

The contribution of F_2 to the matrix element M in (5.29) is then

$$M_{is}^{(2)}(-qj, \omega_{-qj}; qj', \omega_{qj'})$$

$$= -\frac{1}{2\sqrt{2}} \sum_{kcd} \sum_{b \neq c} \left[\frac{(c|p_s|b)(b|p_i|c)}{(E_b - E_c - \omega_l - i\gamma_b)^2} + \frac{(c|p_i|b)(b|p_s|c)}{(E_b - E_c + \omega_l + i\gamma_b)^2} \right]$$

$$\cdot [\Delta_{jj'}(-q, -\omega_{-qj}; k, cd) + \Delta_{jj'}(-q, \omega_{qj'}; k, cd)$$

$$+ \Delta_{j'j}(q, -\omega_{qj'}; k, cd) + \Delta_{j'j}(q, \omega_{-qj}, k, cd)] \quad (5.37a)$$

where

$$\Delta_{jj'}(\pm q, \omega; k, cd) \equiv \frac{(c|V_{j'}(\mp q)|d\mp)(d\mp|V_j(\pm q)|c)(f_{d\mp} - f_c)}{(E_{d\mp} - E_c - \omega)_p}. \quad (5.37b)$$

Equation (5.37a) gives what is expected to be the most important contribution to the two-phonon Raman matrix element as a sum over the product of momentum matrix elements and photon energy denominators (first square bracket) multiplied by a function $\Delta_{jj'}(q, \omega; k, cd)$ which sums to give the "real part" of the self-energy matrix:

$$\pi_{jj'}(q, \omega) = P\tilde{\pi}_{jj'}(q, \omega) = \sum_{kcd} \Delta_{jj'}(q, \omega, k, cd), \quad (5.38)$$

where "P" means that the energy denominator of π is a principal value.

In the static ($\omega = 0$) limit the self-energy is related to the $D^{(2)}$ matrix of *Varma* and *Weber* [5.23], see (5.5), as follows:

$$2\sqrt{\omega_{-qj'}\omega_{qj}}\,\pi_{jj'}(q,0) = \sum_{\substack{\lambda\lambda' \\ \beta\beta'}} \frac{D^{(2)}(\lambda\beta, \lambda'\beta'| - q)\varepsilon^{\lambda\beta}(qj)\varepsilon^{\lambda'\beta'}(-qj')}{\sqrt{M_\lambda M_{\lambda'}}}, \qquad (5.39)$$

where M_λ is the mass of the atom on sublattice λ, and where the $\varepsilon^{\lambda\beta}(-qj)$ are the components of the eigenvector for phonon $(-qj)$. These are assumed to be orthonormal and to obey

$$\varepsilon^{\lambda\beta}(-qj) = \varepsilon^{\lambda\beta}(qj)^*. \qquad (5.40)$$

The quantity $\varepsilon^{\lambda\beta}(qj)/\sqrt{M_\lambda}$ is the eigenvector of the total D matrix, not necessarily of $D^{(2)}$ alone. Mode mixing (off-diagonal elements of $\pi_{jj'}$) may occur if allowed by symmetry. For overtones, where $j=j'$, we have the simple result

$$\pi_{jj}(q,0) = [(\omega_{qj})^2 - (\omega_{qj}^0)^2]/[2\omega_{qj}], \qquad (5.41)$$

where the "bare" frequency ω_{qj}^0 is the contribution of $D^{(0)} + D^{(1)}$ to the frequency of mode (qj).

The matrix element M_{is} becomes symmetric in the indices i and s if two more approximations are valid, namely, 1) neglect of spin, so that by time-reversal symmetry

$$|-kc) = |kc)^* \qquad (5.42)$$

and 2) frozen phonon approximation where one sets $\omega_{qj'} = \omega_{-qj} = 0$ in the expressions for $\Delta_{jj'}$ in (5.37). With these assumptions one can obtain the result:

$$M_{is}^{(2)}(-qj,0;qj',0) = -\frac{1}{\sqrt{2}}\sum_{kcd}$$

$$\cdot \left\{ \sum_{b\neq c} [(c|p_s|b)(b|p_i|c) + (c|p_i|b)(b|p_s|c)] \right.$$

$$\cdot \left. [(E_b - E_c - \omega_l - i\gamma_b)^{-2} + (E_b - E_c + \omega_l + i\gamma_b)^{-2}] \right\} \Delta_{jj'}(-q,0;k,cd). \qquad (5.43)$$

5.4.5 Symmetries, Simplifications, and Selection Rules for the Two-Phonon Case

Because a sum over all phonon wave vectors q and mode indices j,j' is taken in (5.30), we can symmetrize $|M_{is}|^2$ for a given q with respect to all other wave vectors in the "star of q." First, we introduce Cartesian components $e_{i\alpha}$ and $e_{s\beta}$ of the photon polarization vectors. Then

$$M_{is}((q)) = \sum_{\alpha\beta} M_{\alpha\beta}((q))e_{i\alpha}e_{s\beta}, \qquad (5.44a)$$

where $((q))$ denotes

$$((q)) \equiv (-qj, \omega_{-qj}; qj', \omega_{qj'}).$$ (5.44b)

To calculate the average of

$$|M_{is}((q))|^2 = \sum_{\substack{\alpha\beta \\ \alpha'\beta'}} M_{\alpha\beta}((q)) M_{\alpha'\beta'}((q))^* e_{i\alpha} e_{s\beta} e_{i\alpha'} e_{s\beta'}$$ (5.45)

we need the following average over the star of q, denoted by $\langle ... \rangle$:

$$\langle M_{\alpha\beta}((q)) M_{\alpha'\beta'}((q))^* \rangle = \sum_l (\alpha\beta, \alpha'\beta'|l)|M_l((q))|^2.$$ (5.46)

In (5.46) l labels the irreducible representation of the point group of the crystal. The coefficients $(\alpha\beta, \alpha'\beta'|l)$ depend only on the symmetry properties of the crystal. $|M_l((q))|^2$ is the same for any q in the star of q [5.49]. Explicit formulas will be given in two cases [5.49]. For layered compounds, such as $2\,H-TaSe_2$, photon polarizations are in the layer plane, and taking the x and y directions in this plane

$$|M_{A_{1g}}((q))|^2 = \tfrac{1}{2}|M_{xx}((q)) + M_{yy}((q))|^2$$ (5.47a)

for the fully symmetric A_{1g} mode and

$$|M_{E_{2g}}((q))|^2 = \tfrac{1}{4}[|M_{xx}((q)) - M_{yy}((q))|^2 + |M_{xy}((q)) + M_{yx}((q))|^2]$$ (5.47b)

for the doubly degenerate E_{2g} mode. The coefficients $(\alpha\beta, \alpha'\beta'|l)$ are

$$(xx, xx|A_{1g}) = (yy, yy|A_{1g}) = (xx, yy|A_{1g}) = \tfrac{1}{2},$$ (5.48a)

$$(xx, xx|E_{2g}) = (yy, yy|E_{2g}) = -(xx, yy|E_{2g}) = \tfrac{1}{2},$$ (5.48b)

and

$$(xy, xy|E_{2g}) = (yx, yx|E_{2g}) = \tfrac{1}{2}.$$ (5.48c)

For cubic crystals the results are

$$|M_{A_1}((q))|^2 = |\text{Trace}\, M_{\alpha\beta}((q))|^2/3$$ (5.49a)

for the fully symmetric A_1 mode,

$$|M_E((q))|^2 = \tfrac{1}{6}[|M_{xx}((q)) - M_{yy}((q))|^2 + |M_{yy}((q)) - M_{zz}((q))|^2 \\ + |M_{zz}((q)) - M_{xx}((q))|^2]$$ (5.49b)

for the doubly degenerate E mode, and

$$|M_{T_2}((q))|^2 = \tfrac{1}{6}[|M_{xy}((q)) + M_{yx}((q))|^2 + |M_{xz}((q)) + M_{zx}((q))|^2$$
$$+ |M_{yz}((q)) + M_{zy}((q))|^2] \tag{5.49c}$$

for the triply degenerate T_2 mode. The coefficients for the cubic case are

$$(xx, xx|A_1) = (xx, yy|A_1) = \tfrac{1}{3}, \tag{5.50a}$$

$$(xx, xx|E) = \tfrac{2}{3}, \tag{5.50b}$$

$$(xx, yy|E) = -\tfrac{1}{3}, \tag{5.50c}$$

and

$$(xy, xy|T_2) = \tfrac{1}{2}. \tag{5.50d}$$

Equations (5.44–50) were concerned with the extraction of as much relevant information as possible from the set of tensor components $M_{\alpha\beta}((q))$ for a given pair of phonon modes $(-qj)$ and (qj'). The selection rules for $M_{\alpha\beta}((q))$ are the usual ones for two-phonon Raman scattering. They may be derived from an explicit expression for the matrix element $M_{\alpha\beta}$ using arguments applied by *Hulin* [5.51] to infrared absorption [5.49]. In the integrand for $M_{\alpha\beta}$ one replaces the wave vector k by Rk and averages over all symmetry elements R in G_q, the group of the wave vector q. This leads to the statement that if phonons $(-qj)$ and (qj') belong to representations Γ_j and $\Gamma_{j'}$ of G_q, then a necessary condition for two-phonon Raman activity is that the product representation $\Gamma_j \times \Gamma_{j'}$ contain one of the irreducible tensor representations Γ_t of G_q. For overtones $(j = j')$ an additional condition is necessary. It can be stated as a sum over products of characters χ:

$$\sum_{R \in G_q} \{\chi_{t'}(R^{-1})\chi_j(R)\chi_j(J^{-1}RJ) + \chi_t[(JR)^{-1}]\chi_j[(JR)^2]\} = 2ng_q, \tag{5.51}$$

where the sum is over the g_q symmetry elements R in the group G_q, and where n is a positive integer. J in (5.51) is an operation of the space group which takes q into $-q$.

The averaging over Rk mentioned above is also useful in reducing the complexity of explicit expressions. For example, for the LA, $q = q(100)$ mode in bcc metals like Nb it is easy to see that in the static approximation for overtone scattering the matrix element with the double photon energy denominator is

$$M_{\alpha\beta}^{(2)}[-q(100)L, 0; q(100)L, 0] = -\sqrt{2}\delta_{\alpha\beta}\sum_{kcd}\left\{\sum_{b \neq c} \langle|(c|p_\alpha|b)|^2\rangle_k\right.$$

$$\left. \cdot [(E_b - E_c - \omega_l - i\gamma_b)^{-2} + (E_b - E_c + \omega_l + i\gamma_b)^{-2}]\right\} \Delta_{LL}[-q(100), 0; k, cd],$$
$$\tag{5.52a}$$

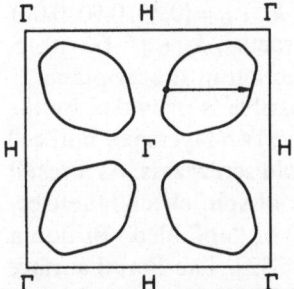

Fig. 5.18. The Fermi surface of Nb in the $k_z = 0.06$ plane [5.23]. The arrow denotes the electron scattering process that softens the $q = 0.7$ (100) LA phonon

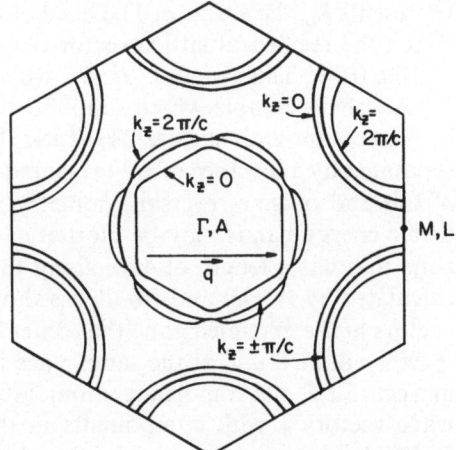

Fig. 5.19. Fermi surface of $2\,H$–TaSe$_2$, redrawn from results of [5.52] (Copyright, The Institute of Physics.) The arrow denotes a wave vector q of $2/3$ ΓM

where

$$\langle|(c|p_\alpha|b)|^2\rangle_k \equiv \frac{1}{g_q} \sum_{R\in G_q} |(cRk|p_\alpha|bRk)|^2$$

$$= \frac{1}{g_q} \sum_{R\in G_q} |(ck|(Rp)_\alpha|bk)|^2 . \tag{5.52b}$$

Suppose there is a region about k_0 in k space that makes a major contribution to (5.52a). According to (5.37b) the state (ck) is then strongly coupled to $(d, k_0 + q)$. We should then examine (5.52) for k near $k_0 + q$. This will have already been included in the case $k \sim k_0$, provided that k_0 and $k_0 + q$ are "equivalent," which means that to within a reciprocal lattice vector $k_0 + q$ equals one of the wave vectors Rk_0 or $-Rk_0$, where R is a symmetry element of G_q. In such a case (5.52b) is needed only at k_0, and one finds for the overtone of this mode

$$|M_{T_2}((q))|^2 = 0 , \tag{5.53a}$$

$$|M_A^{(2)}((q))|^2 = 2|\pi_{LL}[q(100), \omega = 0]|^2 \tfrac{1}{3}|P_{xx} + P_{yy} + P_{zz}|^2 , \tag{5.53b}$$

$$|M_E^{(2)}((q))|^2 = 2|\pi_{LL}[q(100), \omega = 0]|^2 \tfrac{1}{3}|P_{xx} - \tfrac{1}{2}P_{yy} - \tfrac{1}{2}P_{zz}|^2 , \tag{5.53c}$$

where

$$P_{\alpha\alpha} = \sum_{b \neq c} \langle|(c|p_\alpha|b)|^2\rangle_{k_0}[(E_b - E_c - \omega_l - i\gamma_b)^{-2} + (E_b - E_c + \omega_l + i\gamma_b)^{-2}]_{k_0} \tag{5.53d}$$

For Nb, $k_0 = (0.15, 0.50, 0.06)$ may be a good example for the LA phonon with $q = 0.7(1, 0, 0)$ [5.23, 49]. This is shown in Fig. 5.18. The closed circle represents k_0, the arrow q. The head of the arrow is at $k_0 + q = (0.85, 0.50, 0.06)$. When the reciprocal lattice vector $G = (1, 1, 0)$ is subtracted, $k_0 + q - G = (0.15, -0.50, 0.06)$. This equals $-\sigma_{xy}k_0$, where σ_{xy} is a reflection in the xy plane.

Another example where simplification may be possible is provided by the layered compounds such as 2 H–TaSe$_2$ [5.7]. They have two layers per unit cell separated by a distance of $c/2$ measured along the sixfold screw axis. As a result of this and of time-reversal symmetry in the absence of spin effects, the electronic energy bands may be plotted in an "extended" or "unfolded" Brillouin zone that has a length of $4\pi/c$ along the screw axis [5.52]. The Fermi surface calculated by *Wexler* and *Woolley* is shown in Fig. 5.19 [5.52]. States with wave vectors in the extended zone that differ by components along this axis equal to $\pm 2\pi/c$ will be found at the same place in the smaller, true zone. As mentioned in Sect. 5.1.2, the important anomalous phonons in TaSe$_2$ and NbSe$_2$ have wave vectors q with components in the plane of the layers (xy plane) near $(2/3)\Gamma M$. There are two nearly degenerate Σ_1 modes that are linear combinations of LA-modes in each of the two layers. We assume exact degeneracy. If the unfolded zone is used, the symmetric linear combination has $q_z = 0$, and the anti-symmetric linear combination has $q_z = \pm 2\pi/c$. The symmetric mode induces electronic transitions within the same band at constant k_z. The anti-symmetric mode causes interband electronic scattering in the true small zone, but, to a good approximation, only intraband scattering occurs (with $k_z \rightarrow k_z \pm 2\pi/c$) in the extended zone. One would hope that expressions like (5.53d) would be relatively insensitive to the z component of k_0 when evaluated in the extended-zone scheme. If true, then the following approximate results should hold for overtone scattering:

$$|M_A^{(2)}(q)|^2 = 2|\pi(q)|^2 \tfrac{1}{2}|P_{xx} + P_{yy}|^2 , \qquad (5.54a)$$

$$|M_E^{(2)}(q)|^2 = 2|\pi(q)|^2 \tfrac{1}{2}|P_{xx} - P_{yy}|^2 , \qquad (5.54b)$$

where P_{xx} and P_{yy} obey (5.53d) and where $q \cong \tfrac{2}{3}\Gamma M$ for the true LA mode and $q \cong \tfrac{2}{3}\Gamma M + 2\pi\hat{c}/c$ for the LA-like interlayer mode. The x (or y) axis in (5.54) must be along ΓM.

Equations (5.53, 54) are explicit realizations of the approximation of *Maldague* and *Tsang* [5.50] which says that the Raman intensity is proportional to the square of the phonon self-energy. It is hoped that this theory will lead to numerical calculations of the two-phonon Raman spectrum for model systems.

5.5 Charge-Density-Wave State

We have seen in Sect. 5.4 that two-phonon Raman scattering is closely related to the self-energy of the phonons due to d electrons. If the self-energy matrix $\pi_{jj'}(q, 0)$ of (5.39) becomes sufficiently negative, the lattice may become unstable, and a phase transition will occur. The nature of the transition and the phonon dynamics on both sides of the transition temperature may be strongly influenced by anharmonic coupling terms that can either be properties of the "bare" system or be mediated by the d electrons.

5.5.1 Mean-Field Theory

Below the transition temperature new phonon modes become strongly Raman active. The microscopic mechanism for the Raman activity of these modes is closely related to the electronic processes that within mean-field theory (MFT) explain their "condensation." Although anharmonicity may make MFT an inaccurate predictor of critical behavior and phonon dynamics, we discuss some of its concepts here for later use in a theory of Raman scattering. A convenient description is that of *Kotani* [5.53], which we generalize to three dimensions.

We return to the Hamiltonian of (5.10, 18)

$$H = H_0 + H', \tag{5.55a}$$

where

$$H_0 = H_e + H_p = \sum_{ka} E_{ak} c^+_{ak} c_{ak} + \sum_{qj} \omega^0_{qj} b^+_{qj} b_{qj} \tag{5.55b}$$

and

$$H' = \sum_{kdaqj} (d, k+q|V_j(-q)|ak) c^+_{d,k+q} c_{ak} (b_{qj} + b^+_{-qj}). \tag{5.55c}$$

In a simple single-band MFT there is a static charge-density wave (CDW) described by the expectation value

$$\langle \varrho_q \rangle = \sum_k \langle c^+_{k+q} c_k \rangle. \tag{5.56}$$

We make the more general assumption that

$$\langle c^+_{d,k+q} c_{ak} \rangle \neq 0 \tag{5.57}$$

for a few bands d, a (the d electron bands close to the Fermi energy) and for certain wave vectors q. For layered compounds the "primary" values of q are $q_1, q_2,$ and q_3, three equivalent vectors in the basal plane that meet at angles of $\pm 120°$. Harmonics are possible of the form

$$q = n_1 q_1 + n_2 q_2 + n_3 q_3, \tag{5.58}$$

where n_j are integers. For a *commensurate* distortion, integers m_i exist for which

$$m_i q_i = G, \tag{5.59a}$$

where G is a reciprocal lattice vector. For an incommensurate distortion, (5.59a) is not true for at least one i. In the commensurate case only a finite number of q's satisfy (5.58), and MFT gives a finite number of coupled integral equations. In the incommensurate case one truncates the coupled equations. This is the equivalent of using a finite number of harmonics (5.58). Umklapp terms are also important in this case [5.8], which means that q can obey

$$q = \sum n_i q_i - G. \tag{5.59b}$$

We write H' in mean-field theory as

$$H' = \delta H_e + \delta H_p + H_{int}, \tag{5.60a}$$

where

$$\delta H_e = \sum_{kda} \left[\sum_{qj} (d, k+q|V_j(-q)|ak) \langle b_{qj} + b^+_{-qj} \rangle \right] c^+_{d,k+q} c_{ak}, \tag{5.60b}$$

$$\delta H_p = \sum_{qj} \left[\sum_{kda} (d, k+q|V_j(-q)|ak) \langle c^+_{d,k+q} c_{ak} \rangle \right] (b_{qj} + b^+_{-qj}), \tag{5.60c}$$

and where H_{int} involves only "fluctuations":

$$H_{int} = \sum_{kdaqj} (d, k+q|V_j(-q)|ak) c^+_{d,k+q} c_{ak} (b_{qj} + b^+_{-qj}). \tag{5.60d}$$

We determine $\langle b_{qj} \rangle$ and $\langle b^+_{-qj} \rangle$ by requiring the mean time derivatives of the fluctuating quantities b_{qj} and b^+_{-qj} to vanish:

$$0 = \langle [b_{-qj}, (H_p + \delta H_p)] \rangle \tag{5.61a}$$

and

$$0 = \langle [b^+_{-qj}, (H_p + \delta H_p)] \rangle. \tag{5.61b}$$

Thus

$$\langle b^+_{-qj} \rangle = \langle b_{qj} \rangle = -\frac{1}{\omega^0_{qj}} \sum_{kda} (d, k-q|V_j(q)|ak) \langle c^+_{d,k-q} c_{ak} \rangle. \tag{5.61c}$$

Hence

$$\delta H_e = \sum_{kdaq} (d, k+q|V_s(-q)|ak) c^+_{d,k+q} c_{ak}, \tag{5.61d}$$

where

$$(d, k+q|V_s(-q)|ak) \equiv -2\sum_j \frac{(d, k+q|V_j(-q)|ak)}{\omega_{qj}^0} \sum_{k'd'a'} (d', k'-q|V_j(q)|a'k')$$
$$\cdot \langle c_{d',k'-q}^+ c_{a'k'} \rangle \tag{5.61e}$$

is the matrix element for electron scattering from the static potential of the CDW.

The presence of the mean-field potential $V_s(-q)$ makes the new electronic Green's function off diagonal in k. This function is defined by

$$G_{ek;d,k+q}(i\omega_n) = -\frac{1}{2} \int_{-1/T}^{1/T} e^{i\omega_n \tau} \langle T_\tau (c_{k+q,d}(\tau) c_{ke}^+(0))) \rangle d\tau, \tag{5.62a}$$

where T_τ is the time-ordering operator for time τ. It obeys the "Dyson" equation

$$G_{ek;d,k+q}(i\omega_n) = \frac{1}{i\omega_n - E_{ek} + E_0} \left[\delta_{q0}\delta_{de} + \sum_{q'e'} G_{e',k+q';d,k+q}(i\omega_n) \right.$$
$$\left. \cdot \langle k+q', e'|V_s(-q')|ke \rangle \right], \tag{5.62b}$$

where δ_{q0} and δ_{de} are Kronecker deltas.

The chemical potential E_0 is determined from

$$n_{\text{tot}} = T \sum_{nke} G_{ek,ek}(i\omega_n), \tag{5.63a}$$

where n_{tot} is the total number of electrons.

The self-consistency condition follows by manipulating (5.62a):

$$\langle c_{d',k'-q}^+ c_{a'k'} \rangle = T \sum_n G_{d',k'-q;a'k'}(i\omega_n). \tag{5.63b}$$

In lowest order in V_s we have

$$G_{ek,ek}^{(0)}(i\omega_n) = \frac{1}{i\omega_n - E_{ek} + E_0}. \tag{5.64a}$$

From (5.62) we find to first order in V_s for $q \neq 0$:

(set $q' = q, e' = d$)

$$G_{ek;d,k+q}^{(1)}(i\omega_n) = \frac{\langle d, k+q|V_s(-q)|ek \rangle}{(i\omega_n - E_{ek} + E_0)(i\omega_n - E_{d,k+q} + E_0)}. \tag{5.64b}$$

With (5.64b), (5.63b) gives

$$\langle c_{d',k'-q}^{+} c_{a',k'} \rangle_{(1)} = \frac{f_{d',k'-q} - f_{a',k'}}{E_{d',k'-q} - E_{a',k'}} \langle a'k'|V_s(-q)|d',k'-q \rangle. \tag{5.65}$$

When substituted in (5.61c, e), (5.65) gives a homogeneous linear equation for the matrix elements of $V_s(-q)$. A nonzero solution exists when

$$\det|\delta_{jj'} + \frac{2}{\omega_{qj}^0} \pi_{jj'}(q, \omega=0)| = 0. \tag{5.66}$$

Equation (5.66) determines the phase-transition temperature in mean-field theory. It is equivalent to the condition that in the high-temperature phase there exists a phonon of wave vector q and renormalized frequency equal to zero. It is expected that as the temperature is lowered, (5.66) is satisfied first for a q equal to any of the primary wave vectors q_i in (5.58).

5.5.2 Static Distortion

From the general expression for the displacements $u_{\lambda\beta l}(n)$ of the λth atom in the nth unit cell of layer l one finds

$$\langle u_{\lambda\beta l}(n) \rangle = \frac{1}{2} \sum_{1/2q} [e_{\lambda\beta l}(q) e^{iq\cdot R_n} + e_{\lambda\beta l}(q)^* e^{-iq\cdot R_n}] \tag{5.67a}$$

where

$$\frac{1}{2} e_{\lambda\beta l}(q) = \sum_j \frac{\varepsilon^{\lambda\beta l}(qj)(\langle b_{qj} \rangle + \langle b_{-qj}^+ \rangle)}{\sqrt{2NM_\lambda \omega_{qj}^0}} \tag{5.67b}$$

and where the sum in (5.67a) is over half of all the q's that contribute to the distortion such that the pair $(q, -q)$ occurs only once.

For 1-T type materials there is a single layer per unit cell in the undistorted phase. We drop the index l in (5.67a, b) and write

$$e_{\lambda\beta}(q) = \mathscr{E}_{\lambda\beta}(q) e^{i\phi_{\lambda\beta}} \tag{5.68}$$

where the amplitudes $\mathscr{E}_{\lambda\beta}(q)$ and phases $\phi_{\lambda\beta}$ are real. One may invert (5.67a, b) to obtain $\langle b_{qj} \rangle$ and then obtain the matrix elements of the static potential $V_s(-q)$

$$(d, k+q|V_s(-q)|ak) = \frac{1}{2}\sqrt{N} \sum_j (d, k+q|V_j(-q)|ak)\sqrt{2\omega_{qj}^0}$$
$$\cdot \sum_{\beta\lambda} \sqrt{M_\lambda} \mathscr{E}_{\lambda\beta}(q)\varepsilon^{\lambda\beta}(qj)^* e^{i\phi_{\lambda\beta}}. \tag{5.69}$$

The CDW itself is often described by parameters that represent Fourier components of electronic charge density [5.54–60]. Ionic displacements are then essentially given by the gradient of the local charge density. This reduces the number of parameters in (5.69) to those for an LA phonon. Thus (5.69) may be approximated by

$$(d, k+q|V_s(-q)|ak) = \tfrac{1}{2}\sqrt{2N\omega_q^0 M^*}\,(d, k+q|V_{LA}(-q)|ak)\mathscr{E}\,e^{i\phi} \tag{5.70}$$

where $\mathscr{E}$, ϕ are close to $\mathscr{E}_{\lambda\beta}$ and $\phi_{\lambda\beta}$ for λ and β corresponding to displacements of the metal atom along q. M^* is an effective mass that is not too different from the metal-atom mass.

The undistorted structure of 2H–TaSe$_2$ has two TaSe$_2$ layers per unit cell with a center of inversion half way between the two Ta-atoms. Lattice vectors R_n in the basal plane will locate these centers of inversion. The layers will be labelled by $l=1, 2$. The inversion operation takes site λ in layer 1 into site λ in layer 2. Equation (5.68) now becomes

$$e_{\lambda\beta l}(q) = \mathscr{E}_{\lambda\beta}(q)\exp\{i[(-1)^{l+1}\phi_{\lambda\beta l}(q) + \phi_0]\}\,. \tag{5.71}$$

Inversion symmetry is maintained in the distorted phase if $\phi_0 = \pm\pi/2$ and if

$$\phi_{\lambda\beta 1}(q) = \phi_{\lambda\beta 2}(q) + 2q\cdot d\,,\ \textit{modulo } 2\pi\,, \tag{5.72a}$$

where the inversion center is now at $R_n = d$. Now introduce three nearest-neighbor lattice vectors in the basal plane, a_1, a_2, and a_3, making angles of 120° with one another, and making 30° angles with q_1, q_2, and q_3, respectively, so that

$$a_i\cdot q_i = 2\pi/3\,. \tag{5.72b}$$

Then if d is written

$$d = \sum_i n_i a_i\,, \tag{5.72c}$$

where $n_i = 0, 1, 2$ we find for (5.72a)

$$\phi_{\lambda\beta 1}(q) = \phi_{\lambda\beta 2}(q) + 4\pi(n_1 - n_3)/3\,,\ \textit{modulo } 2\pi,\ \text{plus cyclic permutations}\,. \tag{5.73d}$$

In the spirit of (5.70) we now consider a phonon "basis" consisting of LA modes in each layer. These will be assumed to be degenerate with frequency ω_q^0. This gives

$$(d, k+q|V_s(-q)|ak) = \tfrac{1}{2}\sqrt{2N\omega_q^0 M^*}\,\mathscr{E}\,e^{i\phi_0}$$
$$\cdot\sum (d, k+q|V_l(-q)|ak)\exp[i(-1)^{l+1}\phi_{ql}]\,, \tag{5.73}$$

where $V_l(-q)$ is the electron-phonon perturbation for an LA mode in the layer l.

We keep a q-dependence and a l-dependence to the phases ϕ_{ql} in (5.73) because the most general Landau theory allowed by symmetry, namely that of *Jacobs* and *Walker* [5.60], allows for them. Recent electron-diffraction experiments performed on a cold-stage in an electron microscope by *Fung* et al. [5.61] show that the low-temperature commensurate CDW of 2 H–TaSe$_2$ is orthorhombic, rather than hexagonal, as thought previously. The space group is either Cmc2$_1$ or Cmcm, which is favored and which has a center of inversion. Recent theoretical work [5.62–64] has shown that Cmcm structures can be obtained for certain values of the Landau parameters. Such theories can also explain the existence of the "striped" phase first seen in the x-ray diffraction studies of *Fleming* et al. [5.65] and then in electron microscope images by *Chen* et al. [5.66], and by *Fung* et al. [5.61]. This phase has one commensurate and two incommensurate wave vectors. These theories also seem able to explain the re-entrant nature of the pressure-dependence of the commensurate states discovered by *McWhan* et al. [5.67].

5.5.3 Theory of Raman Scattering from Weak CDW Phonons

Since the static potential $V_s(-q)$ makes the electronic Green's function $G_{k,k+q}(i\omega_n)$ off diagonal in wave vector, the electron–phonon interaction H_{int} of (5.60) mixes $\pm q$ in the phonon Green's function, which now becomes

$$D_{qj,q'j'}(i\omega_n) = \langle\langle (b_{q'j'} + b^+_{-q'j'}) ; (b^+_{qj} + b_{-qj}) \rangle\rangle$$

$$\equiv - \int_0^{1/T} e^{i\omega_n\tau} \langle T_\tau(b_{q'j'}(\tau) + b_{-q'j'}(\tau)^+)(b_{qj}(0)^+ b_{-qj}(0)) \rangle \, d\tau. \qquad (5.74)$$

For q equal to one of the primary q_j, examination of the Dyson equation for D [5.68] reveals that strong coupling via $G_{k,k+q}$ occurs only when $q' = \pm q$ in (5.74). (We do, however, allow for anharmonic mixing of different q_j in D.)

The important contributions to the four-vertex function are of the form shown in Fig. 5.14:

$$\mathscr{F} = - \sum_{jj'qq'} (A_{qj} + B_{qj}) D_{qj;q'j'}(i\omega_t)(A'_{q'j'} + B'_{q'j'}). \qquad (5.75)$$

In the sum q and q' range over $\pm$ the three primary q's. $A_{qj}(B_{qj})$ is an A-type (B-type) electron loop with a matrix element of $V_j(q)$ at the phonon vertex. We assume that band b couples neither to the phonons nor to V_s and find [5.69, 70]

$$A_{qj} = T \sum_n \sum_{kbcd} \frac{(ck|p_s|bk)(bk|p_i|ck)}{(i\omega_n + i\omega_\mu - E_{bk} + E_0)}$$

$$\cdot [G_{ck;d,k+q}(i\omega_n + i\omega_t)(ck|V_j(q)|d, k+q)G_{ck,ck}(i\omega_n)$$

$$+ G_{ck,ck}(i\omega_n + i\omega_t)(d, k-q|V_j(q)|ck)G_{d,k-q;ck}(i\omega_n)]. \qquad (5.76)$$

For a relatively weak static potential V_s, we use (5.64a, b) in (5.76) and find in the approximation which sets phonon frequencies equal to zero that A_{qj} equals the sum of diagrams A and C in Fig. 5.17, where in those diagrams $V_s(-q)$ acts at the vertex labeled j and $V_j(q)$ acts at vertex j'. Similarly A_{-qj} equals the sum of A and C in Fig. 5.17, where $V_s(q)$ acts at vertex j' and $V_j(-q)$ acts at j; B_{qj} equals the sum of B and D with $V_s(-q)$ at vertex j and $V_j(q)$ at vertex j'; and B_{-qj} equals the sum of B and D with $V_s(q)$ at vertex j'.

We now make the approximation that terms with double photon resonances dominate. Earlier this led to our use of only the F_2 part of (5.32) and gave (5.43) for two-phonon scattering in the static approximation. Now this approximation gives [dropping explicit reference to the wave-vector k and denoting the state $(d, k+q)$ by $(d+)$]:

$$A_{qj} + B_{qj} = - \sum_{kcd} P_{is}(ck)(c|V_j(q)|d+)(d+|V_s(-q)|c)\frac{(f_{d+} - f_c)}{(E_{d+} - E_c)}, \qquad (5.77a)$$

where

$$P_{is}(ck) = \sum_{b \neq c} [(c|p_s|b)(b|p_i|c) + (c|p_i|b)(b|p_s|c)]$$
$$\cdot [(E_b - E_c - \omega_l - i\gamma_b)^{-2} + (E_b - E_c + \omega_l + i\gamma_b)^{-2}] \qquad (5.77b)$$

and

$$A_{-qj} + B_{-qj} = \sum_{kcd} P_{is}(ck)(c|V_s(q)|d+)(d+|V_j(-q)|c)\frac{(f_{d+} - f_c)}{(E_{d+} - E_c)}. \qquad (5.77c)$$

For 1T-type samples where (5.70b) holds for a single condition band c, we have

$$A_{qj} + B_{qj} = \tfrac{1}{2}\sqrt{N} e^{i\phi}(\mathscr{E}/u_0)N_{is}(q, \text{LA}), \qquad (5.78a)$$

where

$$u_0 = (2\omega_q^0 M^*)^{-1/2} \qquad (5.78b)$$

is a root-mean-square amplitude for the LA mode, and where N_{is} obeys

$$N_{is}(q, \text{LA}) = - \sum_k P_{is}(ck)|(c|V_{\text{LA}}(q)|c+)|^2 \frac{(f_{c+} - f_c)}{(E_{c+} - E_c)}. \qquad (5.78c)$$

We also have

$$A_{-qj} + B_{-qj} = \tfrac{1}{2}\sqrt{N} e^{-i\phi}(\mathscr{E}/u_0)N_{is}(q, \text{LA}). \qquad (5.78d)$$

Note that $N_{is}(q, \text{LA})$ in (5.78c) equals $\sqrt{2}M_{is}^{(2)}(-q, \text{LA}, 0; q, \text{LA}, 0)$, the matrix element for LA-overtone two-phonon scattering in (5.43).

For each primary q we make a unitary transformation to new variables

$$b_\eta^q \propto \delta \mathscr{E}_q,$$ (5.79a)

$$b_\phi^q \propto \mathscr{E} \delta \phi_q.$$ (5.79b)

Explicitly

$$\sqrt{2} b_\eta^q = b_q e^{-i\phi} + b_{-q} e^{i\phi},$$ (5.80a)

$$i\sqrt{2} b_\phi^q = b_q e^{-i\phi} - b_{-q} e^{i\phi}.$$ (5.80b)

The Raman intensity is then proportional to the imaginary part as $i\omega_t \to \omega_0 + i\varepsilon$ of

$$\mathscr{F}(i\omega_t) = -\tfrac{1}{2} N(\mathscr{E}/u_0)^2 \sum_{qq'} N_{is}(q, \mathrm{LA}) N_{is}(q', \mathrm{LA})^* D_{\eta\eta}^{qq'}(i\omega_t),$$ (5.81)

where

$$D_{\eta\eta}^{qq'}(i\omega_t) = \langle\!\langle [b_\eta^{q'} + (b_\eta^{q'})^+]; [(b_\eta^q)^+ + b_\eta^q] \rangle\!\rangle$$

$$\equiv - \int_0^{1/T} e^{i\omega_t \tau} \langle T_\tau [b_\eta^{q'}(\tau) + b_\eta^{q'}(\tau)^+] [b_\eta^q(0)^+ + b_\eta^q(0)] \rangle \, d\tau.$$ (5.82)

Anharmonic coupling will mix the b_η^q (and b_ϕ^q) for different q's and produce uncoupled amplitude (and phase) modes of A_{1g} and E_g symmetry [5.12, 57]. The transformations are

$$b_\eta^{q3} = \frac{1}{\sqrt{3}} b_\eta^A + \frac{2}{\sqrt{6}} b_\eta^{E2},$$ (5.83a)

$$b_\eta^{q1} = \frac{1}{\sqrt{3}} b_\eta^A + \frac{1}{\sqrt{2}} b_\eta^{E1} - \frac{1}{\sqrt{6}} b_\eta^{E2},$$ (5.83b)

$$b_\eta^{q2} = \frac{1}{\sqrt{3}} b_\eta^A - \frac{1}{\sqrt{2}} b_\eta^{E1} - \frac{1}{\sqrt{6}} b_\eta^{E2}.$$ (5.83c)

Then (5.81) becomes

$$\mathscr{F}(i\omega_t) = - N(\mathscr{E}/u_0)^2 \left[\tfrac{1}{6} | \sum_q{}' N_{is}(q)|^2 D_{\eta\eta}^{AA} \right.$$

$$+ \tfrac{1}{12} D_{\eta\eta}^{E2E2} |2N_{is}(q_3) - N_{is}(q_1) - N_{is}(q_2)|^2$$

$$\left. + \tfrac{1}{4} D_{\eta\eta}^{E1E1} |N_{is}(q_1) - N_{is}(q_2)|^2 \right].$$ (5.84a)

Since the E modes are degenerate, their Green's functions equal a common $D_{\eta\eta}^{EE}$. (5.84a) can then be written after use of symmetry arguments

$$\mathscr{F}(\mathrm{i}\omega_t) = -\tfrac{3}{8}N(\mathscr{E}/u_0)^2 \{[|N_{xx}(q_1) + N_{yy}(q_1)|^2 D_{\eta\eta}^{AA}$$
$$+ |N_{xx}(q_1) - N_{yy}(q_1)|^2 D_{\eta\eta}^{EE}](e_i \cdot e_s)^2$$
$$+ |N_{xx}(q_1) - N_{yy}(q_1)|^2 D_{\eta\eta}^{EE}|e_i \times e_s|^2\}, \tag{5.84b}$$

where e_i and e_s are photon polarization vectors lying in the basal plane, and where $N_{xx}(q_1)$ and $N_{yy}(q_1)$ are calculated with the x-axis parallel to the primary q-vector q_1.

Inversion symmetry holds in each layer of the 1-T polytypes. Since amplitude and phase modes have opposite parity, the amplitude-mode Green's functions $D_{\eta\eta}^{AA}$ and $D_{\eta\eta}^{EE}$ appearing in (5.84b) will show no mixing from phase modes, even with anharmonicity present. If damping is neglected, the imaginary parts of $D_{\eta\eta}^{AA}$ and $D_{\eta\eta}^{EE}$ for $\mathrm{i}\omega_t \to \omega_0 + \mathrm{i}\varepsilon$ will give $2\omega_{q\mathrm{LA}}\,\delta(\omega_0 - \omega_\eta^A)$ and $2\omega_{q\mathrm{LA}}\,\delta(\omega_0 - \omega_\eta^E)$, respectively.

For 2 H polytypes we use (5.73) in (5.77a, c) to obtain

$$L_{\pm q1} = \tfrac{1}{2}\sqrt{N}(\mathscr{E}/u_0)\,\mathrm{e}^{\pm \mathrm{i}\phi_0}[\mathrm{e}^{\mp \mathrm{i}\phi_{q1}}N_{\mathrm{is}}(q, 11) + \mathrm{e}^{\pm \mathrm{i}\phi_{q2}}N_{\mathrm{is}}(q, 12)], \tag{5.85a}$$

$$L_{\pm q2} = \tfrac{1}{2}\sqrt{N}(\mathscr{E}/u_0)\,\mathrm{e}^{\pm \mathrm{i}\phi_0}[\mathrm{e}^{\mp \mathrm{i}\phi_{q1}}N_{\mathrm{is}}(q, 21) + \mathrm{e}^{\pm \mathrm{i}\phi_{q2}}N_{\mathrm{is}}(q, 22)], \tag{5.85b}$$

where

$$N_{\mathrm{is}}(q, ll') = -\sum_{kcd} P_{\mathrm{is}}(ck)\,(c|V_l(q)|d+)\,(d+|V_{l'}'(-q)|c)\left(\frac{f_{d+} - f_c}{E_{d+} - E_c}\right). \tag{5.85c}$$

The inversion symmetry of the 2 H structure may be used to show that

$$N_{\mathrm{is}}(q, 11) = N_{\mathrm{is}}(q, 22) \tag{5.85d}$$

and

$$N_{\mathrm{is}}(q, 12) = N_{\mathrm{is}}(q, 21) \tag{5.85e}$$

Equations (5.80a, b) then make it possible to write the 4-vertex function in the form

$$\mathscr{F} = -\langle\!\langle Q; Q \rangle\!\rangle, \tag{5.86a}$$

where

$$Q = \sum_{ql}{}' [L_{ql}(b_{ql}^+ + b_{-ql}) + L_{-ql}(b_{-ql}^+ + b_{ql})]. \tag{5.86b}$$

We introduce operators for amplitude $(b_{\eta l}^q)$ and phase modes $(b_{\phi l}^q)$ for each layer by the transformations:

$$e^{-i(\phi_0-\phi_{q1})}b_{q1} = \tfrac{1}{2}(b_{\eta 1}^q - ib_{\phi 1}^q),$$ (5.87a)

$$e^{-i(\phi_0+\phi_{q2})}b_{q2} = \tfrac{1}{2}(b_{\eta 2}^q + ib_{\phi 2}^q),$$ (5.87b)

$$e^{i(\phi_0-\phi_{q1})}b_{-q1} = \tfrac{1}{2}(b_{\eta 1}^q + ib_{\phi 1}^q),$$ (5.87c)

$$e^{i(\phi_0+\phi_{q2})}b_{-q2} = \tfrac{1}{2}(b_{\eta 2}^q - ib_{\phi 2}^q).$$ (5.87d)

This gives

$$Q = \sqrt{\frac{N}{2}\frac{\mathscr{E}}{u_0}} \sum_q{}' \{N_{is}(q,11)(b_{\eta 1}^q + b_{\eta 1}^{q+} + b_{\eta 2}^q + b_{\eta 2}^{q+})$$
$$+ N_{is}(q,12)[(b_{\eta 1}^q + b_{\eta 1}^{q+} + b_{\eta 2}^q + b_{\eta 2}^{q+})\cos(\phi_{q1}+\phi_{q2})$$
$$- (b_{\phi 1}^q + b_{\phi 1}^{q+} + b_{\phi 2}^q + b_{\phi 2}^{q+})\sin(\phi_{q1}+\phi_{q2})]\}.$$ (5.88)

Interlayer electronic couplings, which produce a non-zero value for $N_{is}(q,12)$, are seen from (5.88) to make the phase mode operators $b_{\phi 1}^q$ and $b_{\phi 2}^q$ Raman-active. We can estimate the order of magnitude of $|N_{is}(q,12)|$ relative to $|N_{is}(q,11)|$ by comparing (5.85c) with (5.37b) and (5.38). Thus

$$\left|\frac{N_{is}(q,12)}{N_{is}(q,11)}\right| \sim \left|\frac{\pi_{12}(q,\omega=0)}{\pi_{11}(q,\omega=0)}\right| \approx \frac{1}{2}\frac{(\Delta\omega^2)_{\text{IL}}}{(\Delta\omega^2)_r}.$$ (5.89a)

In (5.89a) π_{12} is the interlayer term in the phonon self-energy due to the d-electrons, and π_{11} is the interlayer term in the self-energy. We estimate π_{12} by assuming that it is solely responsible for the observed interlayer phonon splitting $(\Delta\omega^2)_{\text{IL}}$ of $(24\,\text{cm}^{-1})^2$ [5.12], and we estimate π_{11} to be $(\Delta\omega^2)_r \sim (96\,\text{cm}^{-1})^2$ from the size of the anomaly in the room temperature dispersion curve [5.8]. Thus we estimate

$$\left|\frac{N_{is}(q,12)}{N_{is}(q,11)}\right| \sim 0.03$$ (5.89b)

and find that the intrinsic Raman activity of the phase modes, as represented by the $\sin(\phi_{q1}+\phi_{q2})$ terms in (5.88) would give an intensity 0.1% or less that from the amplitude modes. Thus it is probably correct to approximate Q in (5.86b) by

$$Q = \sqrt{\frac{N}{2}\frac{\mathscr{E}}{u_0}} \sum_q{}' N_{is}(q)(b_{\eta 1}^q + b_{\eta 1}^{q+} + b_{\eta 2}^q + b_{\eta 2}^{q+}),$$ (5.90a)

where

$$N_{is}(q) = N_{is}(q,11)$$ (5.90b)

as given by (5.85c) for a single layer. Transformations like (5.83) may then be made to A and E amplitude mode coordinates for each layer. (5.90a) then becomes

$$Q = \sqrt{\frac{N}{2}\frac{\mathscr{E}}{u_0}}\left\{\frac{1}{\sqrt{3}}\left[N_{is}(q_1)+N_{is}(q_2)+N_{is}(q_3)\right](b_{\eta1}^A+b_{\eta1}^{A+}+b_{\eta2}^A+b_{\eta2}^{A+})\right.$$

$$+\frac{1}{\sqrt{6}}\left[2N_{is}(q_3)-N_{is}(q_1)-N_{is}(q_2)\right](b_{\eta1}^{E2}+b_{\eta1}^{E2+}+b_{\eta2}^{E2}+b_{\eta2}^{E2+})$$

$$\left.+\frac{1}{\sqrt{2}}\left[N_{is}(q_1)-N_{is}(q_2)\right](b_{\eta1}^{E1}+b_{\eta1}^{E1+}+b_{\eta2}^{E1}+b_{\eta2}^{E1+})\right\}. \tag{5.91}$$

How then do we explain the experimental observation of four strong CDW phonon modes ($2A$ plus $2E$) at low temperatures in $2\,H\text{–}TaSe_2$? Third and higher-order anharmonic couplings act within each layer to split each triply-degenerate set of modes into A and E amplitude modes and A and E phase modes. Each layer lacks inversion symmetry. These perturbations will also couple E amplitude modes with E phase modes [5.70, 71]. One can show using Landau theories such as that of *Jacobs* and *Walker* [5.60] that the terms that couple amplitude and phase modes are comparable with those that give the splitting between amplitude and phase modes. Furthermore, the ratio of coupling to splitting remains finite as the CDW amplitude $\mathscr{E}$ tends to zero. Since this mixing of modes of opposite parity is large, the new coupled modes will both be rather strongly Raman-active, but the stronger modes will be those with the greater amplitude-mode character, which we denote "amplitude-like".

The coupling terms can be shown to be real. Introducing annihilation operators $B_{\eta l}^\gamma$ and $B_{\phi l}^\gamma$ for amplitude-like and phase-like modes of symmetry γ in layer l, we can write the transformation in the form

$$b_{\eta l}^\gamma = \cos\theta_\gamma\, B_{\eta l}^\gamma + \sin\theta_\gamma\, B_{\phi l}^\gamma\,, \tag{5.92a}$$

$$b_{\phi l}^\gamma = -\sin\theta_\gamma\, B_{\eta l}^\gamma + \cos\theta_\gamma\, B_{\phi l}^\gamma\,, \tag{5.92b}$$

where $|\theta_\gamma|\leqq\pi/4$ and where $\theta_{E1}=\theta_{E2}=\theta_E$. Thus far the degeneracy of $E1$ and $E2$ is maintained.

We now consider the effects of the weak interlayer coupling. Its static effect is extremely important. As recent theoretical work has shown [5.62–64], it causes the re-entrant lock-in transition under pressure [5.67], and its competition with intralayer terms causes the stripes and textures in the incommensurate states [5.61, 65, 66]. A dynamical effect of interlayer coupling is the splitting of the degeneracy of CDW modes in layers 1 and 2. If the low temperature phase has inversion symmetry, the number of Raman-active modes does not change, but for $\gamma=A$, $E1$, and $E1$ there is a splitting into

Raman-active symmetric (s) and infrared-active antisymmetric (a) states whose annihilation operators obey

$$B^{\gamma}_{\eta s,a}= \frac{1}{\sqrt{2}}(B^{\gamma}_{\eta 1} \pm B^{\gamma}_{\eta 2}), \tag{5.92c}$$

$$B_{\phi s,a}= \frac{1}{\sqrt{2}}(B^{\gamma}_{\phi 1} \pm B^{\gamma}_{\phi 2}). \tag{5.92d}$$

If the inversion center is at $d=0$, see (5.72a, c, d), the resulting symmetry is that of the high-temperature phase, D_{6h}, but if $d \neq 0$, the structure is ortho-rhombic (D_{2h}). Suppose d is perpendicular to q_3. Then d and q_3 give the orthorhombic axes in the basal plane. Then the E2 modes of a single layer become A_g plus A_u of D_{2h}, the E1 modes become B_{1g} plus B_{1u}, and the A modes (actually, A_1 modes) become A_g plus A_u.

In the D_{2h} case the interlayer coupling will cause θ_{E1} to differ slightly from θ_{E2} in (5.92a, b). This coupling will also cause weak mixing of $B^{E2}_{\eta s}$ with $B^A_{\eta s}$ and of $B^{E2}_{\phi s}$ with $B^A_{\phi s}$. If the latter mixing is neglected, the intensity of the Raman scattering from the three modes can be obtained from the square modulus of the coefficients of $B^{\gamma}_{\eta s}$ and $B^{\gamma}_{\phi s}$ after (5.92a–c) are substituted in (5.91). The orthorhombic phase shows multiple twinning on a scale of 1 μm or less [5.61]. The resulting expression should therefore be averaged over all permutations of q_1, q_2, and q_3 for fixed photon polarization vectors e_i and e_s. When this is done, we recover (5.84b) with

$$D^{EE}_{\eta\eta} \text{ replaced by } (\cos^2\theta_{E1} D^{E1E1}_{\eta s \eta s} + \sin^2\theta_{E1} D^{E1E1}_{\phi s\phi s}+\cos^2\theta_{E2} D^{E2E2}_{\eta s \eta s}$$
$$+\sin^2\theta_{E2} D^{E2E2}_{\phi s\phi s}). \tag{5.93}$$

Thus far we have discussed no mechanism to produce the observed strong Raman coupling to the A phase mode. That requires terms in the Raman amplitude that are of second or higher order in $\mathscr{E}$ in the CDW state [5.70]. Such terms should scale with the third and fourth order anharmonic terms in a Landau expansion of the free energy [5.70, 49].

The above considerations about intensity of the coupled amplitude and phase modes suggest that the data of Fig. 5.8 on 2 H–TaSe$_2$ at low temperature be interpreted as follows: The 80 cm^{-1} A peak and the 50 cm^{-1} E peak correspond to amplitude-like modes, and the 40 cm^{-1} A peak and the 63 cm^{-1} E peaks correspond to phase-like modes. The recent data of *Sugai* et al. [5.72, 73] support this assignment, which is different from the original one made by *Holy* et al. [5.12]; they assigned the higher-frequency mode of each symmetry to the amplitude mode. *Sooryakumar* et al. [5.74] and *Sugai* et al. [5.72, 73] studied the evolution of the CDW Raman modes as the temperature is raised. There is considerable broadening and some softening. The A modes seem to merge together, and perhaps the E modes also, as the transition temperature to the striped phase is approached from below.

5.5.4 Raman Scattering in the Strong CDW Case

The electronic Green's function G that appears in (5.76) cannot be approximated by (5.64a,b) for a strong static perturbation $V_s(q)$. An exact formal solution for $G_{k,k+q}$ can be obtained for a single electronic band coupled to a single primary q. Thus k will couple only to $k+q$ if k is in the "left" half of the Brillouin zone near $-q/2$. The approximation is exact if $2q$ is a reciprocal lattice vector [5.68, 53].

For simplicity we let

$$V_q = (k+q|V_{LA}(-q)|k) \tag{5.94a}$$

be real and independent of k, and set the phase of the CDW equal to zero. Then by (5.70b, 78b) we have

$$(k+q|V_s(-q)|k) = \sqrt{N}\,(\mathscr{E}/u_0)\,V_q \equiv \Delta. \tag{5.94b}$$

Equation (5.62b) has the solution

$$G_{kk}(i\omega_n) = \frac{i\omega_n - E_{k+q}}{\mathscr{D}(i\omega_n)}, \tag{5.95a}$$

$$G_{k+q,k+q}(i\omega_n) = \frac{i\omega_n - E_k}{\mathscr{D}(i\omega_n)}, \tag{5.95b}$$

$$G_{k,k+q}(i\omega_n) = G_{k+q,k}(i\omega_n) = \Delta/\mathscr{D}(i\omega_n) \tag{5.95c}$$

where

$$\mathscr{D}(i\omega_n) = (i\omega_n - E_k)(i\omega_n - E_{k+q}) - \Delta^2. \tag{5.95d}$$

This can also be written

$$\mathscr{D}(i\omega_n) = (i\omega_n - E_1)(i\omega_n - E_2), \tag{5.96a}$$

where

$$E_{1,2} = E_{1,2}(k, k+q) = \frac{E_k + E_{k+q}}{2} \pm \sqrt{\left(\frac{E_k - E_{k+q}}{2}\right)^2 + \Delta^2} \tag{5.96b}$$

are the energies of the new coupled electronic states. Note that E_1 is greater than E_2 by at least the CDW "gap" 2Δ.

We evaluate the energy E_{bk} and the momentum matrix elements in (5.76) for $k = \pm q/2$. We assume that these matrix elements are real. We rewrite the second term in (5.76) by making the replacement of $k+q$ for k. This gives

$$A_q = A_{-q} = V_q \Delta \sum_k \sum_b (q/2|p_i|b, q/2)(b, q/2|p_s|q/2)$$

$$\cdot T \sum_n \frac{2i\omega_n - E_1 - E_2 + i\omega_t}{\mathscr{D}(i\omega_n + i\omega_t)\mathscr{D}(i\omega_n)(i\omega_n - E_{b,q/2} - i\omega_\mu - E_0)}. \tag{5.97}$$

The sum over n in (5.97) can be written as two terms of the form

$$F = T \sum_n \frac{1}{(i\omega_n - \underline{b})(i\omega_n - \underline{a})(i\omega_n - \underline{c})(i\omega_n - \underline{d})} . \tag{5.98}$$

For both terms

$$\underline{b} = E_{b, q/2} - i\omega_\mu - E_0 .$$

For one of them $\underline{a} = E_2$, $\underline{c} = E_2 - i\omega_t$, and $\underline{d} = E_1 - i\omega_t$. For the other term $\underline{a} = E_1$, $\underline{c} = E_1 - i\omega_t$, and $\underline{d} = E_2$. F then obeys the identity (5.32). Extracting the part with two photon energy denominators, (F_2) gives a total contribution from both F terms of

$$F_{2\,\text{tot}} = \frac{-(f_1 - f_2)}{E_1 - E_2} \left[\frac{1}{(\underline{b} - E_1)^2} + \frac{1}{(\underline{b} - E_2)^2} \right], \tag{5.99a}$$

where

$$f_{1,2} = f(E_{1,2} - E_0) \tag{5.99b}$$

[see (5.13b)]. Or

$$F_{2,\,\text{tot}} \approx \frac{-2(f_1 - f_2)}{(E_1 - E_2)(\underline{b} - E_{\text{av}})^2} , \tag{5.100a}$$

where

$$2E_{\text{av}} = E_1 + E_2 = E_k + E_{k+q} . \tag{5.100b}$$

After analytic continuation we set

$$\underline{b} \rightarrow E_b - \omega_l - i\gamma_b \tag{5.100c}$$

where (5.35b)

$$\omega_l = \tfrac{1}{2}(\omega_i + \omega_s) \tag{5.100d}$$

and

$$E_b = E_{b, q/2} . \tag{5.100e}$$

A similar expression holds for $B_q = B_{-q}$.

Thus

$$A_{qj} + B_{qj} \cong - V_q \Delta P_{is} \sum_k \frac{f_1 - f_2}{E_1 - E_2}, \tag{5.101a}$$

where $E_{1,2}$ obey (5.96b), and where

$$P_{is} = 2 \sum_b (q|2|p_i|b, q/2)(b, q/2|p_s|q/2)$$
$$\cdot [(E_b - E_{av} - \omega_l - i\gamma_b)^{-2} + (E_b - E_{av} + \omega_l + i\gamma_b)^{-2}]. \tag{5.101b}$$

Upon comparison of (5.101) with the one-band weak CDW result (5.78, 77b), we see that the main change is the substitution of

$$\frac{f_1 - f_2}{E_1 - E_2} \tag{5.101c}$$

for

$$\frac{f_k - f_{k+q}}{E_k - E_{k+q}}. \tag{5.101d}$$

In the case of the 2 H layered compounds we must use a two-band, two-phonon-mode model, and there are four strongly coupled components of the electron Green's function for a single q, but only two of them are strongly coupled if interlayer interactions are weak.

Within mean-field theory the self-consistency condition (5.63b) gives, using (5.61e, 94), the "gap" equation

$$1 = \frac{-2V_q^2}{\omega_q^0} \sum_k \frac{f_1 - f_2}{E_1 - E_2}. \tag{5.102a}$$

In one dimension this is BCS-like and gives for linear $E(k)$ near the Fermi energy

$$\Delta(T=0) = \Delta_0 = E_B e^{-\omega_q^0 E_B/g^2}, \tag{5.102b}$$

where E_B (assumed $\gg \Delta_0$) is the width of the band and where

$$g = \sqrt{N} V_q. \tag{5.102c}$$

Integrating (5.101a) in the one-dimensional case gives

$$A_{qj} + B_{qj} = - \frac{\Delta_0 P_{is} \sqrt{N}}{2} \left(\frac{g}{E_B} \ln \frac{E_B}{\Delta_0} \right). \tag{5.103}$$

If the mean-field gap equation (5.102a) holds, then for any dimension

$$A_{qj} + B_{qj} = - \frac{\Delta_0 P_{is} \sqrt{N}}{2} \frac{\omega_q^0}{g}. \tag{5.104}$$

The gap equation (5.102a) no longer holds in the presence of anharmonicity if the coherence length $\delta_0 = (\Delta q)^{-1}$ is small. Here (Δq) is the size of the region in q space where the phonons are anomalous, and this is large for layered dichalcogenides, as can be seen from Fig. 5.6. *McMillan* [5.57] showed that under these conditions phonon entropy dominates electronic entropy (due to excitations across the CDW gap) in determining Δ self-consistently. Numerical calculations for $2\,H\text{–}TaSe_2$ give a Δ about six times larger than implied by (5.102b) [5.57].

The sixfold gain in Δ_0 more than offsets a weaker decrease in

$$\frac{1}{N} \sum_k \frac{f_1 - f_2}{E_1 - E_2} = \frac{1}{E_B} \ln(E_B/\Delta_0) \tag{5.105}$$

due to the larger gap. In higher dimension this effect of the gap in reducing (5.105) will be weaker than the $\ln E_B/\Delta_0$ factor predicts, unless there is very good Fermi surface "nesting".

Such one-dimensional nesting is possible in principle even in three-dimensional materials [5.75], but it is not expected in the $2\,H$ polytype layered compounds [5.52].

The phonon entropy model [5.57] was inspired in part by infrared data that suggest a large CDW gap 2Δ of about $0.25\,eV$ for $2\,H\text{–}TaSe_2$ [5.76]. Other optical data and calculations support a gap of this magnitude [5.77]. This interpretation of the data has not been universally accepted [5.78].

5.6 The Superconducting Gap and Its Coupling to Charge-Density-Wave Phonons

At $33\,K$ $2\,H\text{–}NbSe_2$ distorts into an incommensurate CDW state similar to that of $2\,H\text{–}TaSe_2$ [5.8]. Raman spectra from this phase are shown in Fig. 5.20 [5.79]. The strong low-frequency peaks of A_{1g} and E_{2g} symmetry obey selection rules for a hexagonal phase, suggesting that the CDW is in the "anharmonic" or "soliton" limit described by hexagonal commensurate "domains" separated by

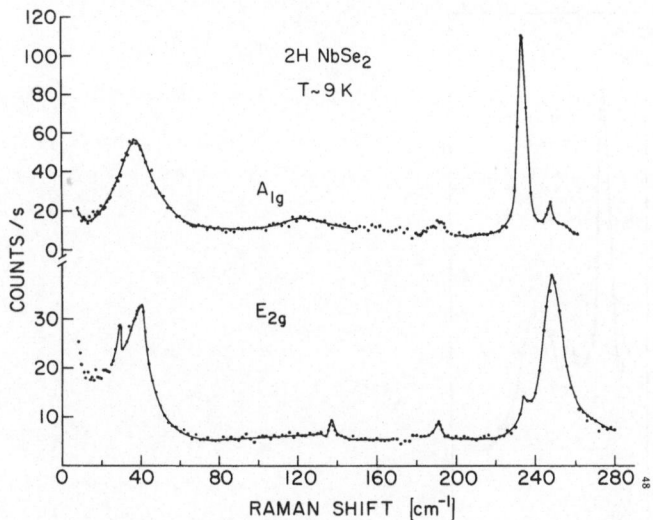

Fig. 5.20. Raman spectrum of 2 H–NbSe$_2$ in the CDW state [5.79]

"discommensurations" [5.54, 56, 58, 59]. The strong peaks are presumably amplitude modes. The "relative phase" modes are not visible.

Below 7.2 K, 2 H–NbSe$_2$ becomes a highly anisotropic type II superconductor [5.80]. Raman spectra taken by *Sooryakumar* [5.79] and *Sooryakumar* and *Klein* [5.81] when the sample of Fig. 5.20 is immersed in superfluid helium are shown in Fig. 5.21. Two new peaks are shown, one of A_{1g} symmetry at 19 cm^{-1} and an E_{2g} peak at 15.5 cm^{-1}. Their average position agrees with the superconducting gap $2\Delta_s$, as determined by far-infrared transmission [5.82]. The new Raman peaks weaken in a magnetic field and have lost all their strength when $H \approx H_{c2}/3$ [5.81, 83]. The missing strength is transferred to the CDW peaks. Raman spectra have been taken on a variety of samples with different CDW properties. CDW's are sensitive to small amounts of impurities and can be suppressed completely by an impurity concentration that hardly effects the superconductivity. In such samples the low-frequency Raman spectra show no CDW phonons and no peaks at the gap $2\Delta_s$ [5.17].

Electronic Raman scattering should reveal the gap by suppression of intraband electronic scattering below $\omega = 2\Delta_s$ [5.47, 84–86]. This effect has not been clearly seen in any superconductor. The new gap peaks in Fig. 5.21 are due to coupling of electronic excitations across the gap to the CDW amplitude modes. Two microscopic theories have been proposed for this coupling. Both give similar results for the renormalized CDW phonon spectral function $\text{Im}\{D_{\eta\eta}(\omega_0 + i\varepsilon)\}$.

The first calculation was done by *Balseiro* and *Falicov* who took an electron–phonon Hamiltonian like (5.55c) for a single band in the $q \to 0$ limit

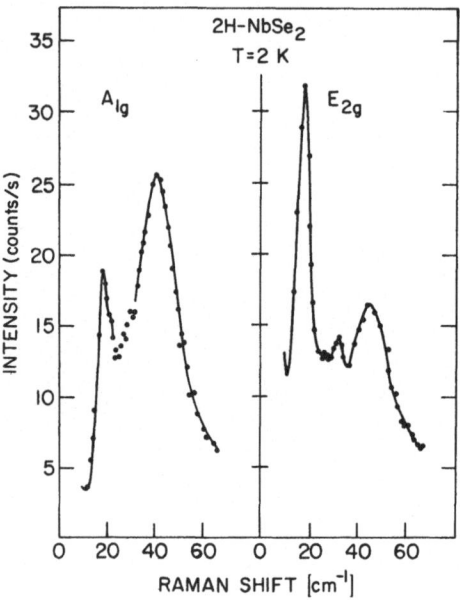

Fig. 5.21. Raman spectrum of 2 H–NbSe$_2$ in the superconducting state at 2 K [5.79, 81]

[5.87]. They calculated the phonon self-energy $\pi\,(q\approx 0,\,\omega)$ using the Bardeen–Cooper–Schrieffer [5.88] theory to calculate the electron Green's functions used in the integral for $\pi(q,\omega)$. They find a pole (bound state) in D for all values of coupling g with a strength and a distance below $2\Delta_s$ that grows with increasing $|g|^2$.

A second calculation has been done by *Littlewood* and *Varma* [5.89]. They argued that the self-energy calculated by *Balseiro* and *Falicov* is not gauge invariant, and that when it is made gauge invariant it will only contain features for ω near the plasma frequency. This second calculation starts with the observation that the CDW amplitude mode varies the electronic density of states $N(0)$ and hence the superconducting gap parameter Δ_s. Since Δ_s is a measure of the correlation between two electrons or two holes with opposite spin, a modulation of Δ_s by a phonon of wave vector $q \approx 0$ will be equivalent to a coupling term where the phonon decays by emitting two electrons or two holes of opposite spin and total wave vector q [instead of emitting an electron-hole pair of the same spin as with (5.55a)]. With such a coupling the self-energy can be made gauge invariant by a vertex correction [5.89, 90]. The resulting phonon spectral function is qualitatively similar to that calculated by *Balseiro* and *Falicov* [5.87].

No theory presently exists to describe how impurities modify the spectral function of the coupled system of phonon and superconducting electrons as measured by *Sooryakumar* et al. [5.17]. Nor does a theory exist to describe the coupled system in the vortex state in an applied magnetic field [5.81].

5.7 Conclusions

Many of the strong features observed in the Raman spectra of transition metals are doubly due to the *d* electrons near the Fermi energy. The *d* electrons are responsible both for the phonon anomalies and lattice instabilities and for the strong Raman matrix elements for scattering from the anomalies. As a result of these *d* electron effects, the Raman spectra can be qualitatively different from those found in insulators or semiconductors, where the strongest contributions often come from the most-resonant factors omitted here.

Acknowledgements. The author thanks many colleagues for collaboration and discussions: D.G.Bruns, S.B.Dierker, J.A.Holy, N.Nagaosa, W.L.McMillan, R.Merlin, R.Sooryakumar, B.H.Suits, and H.Wipf.

This work was supported by the NSF through the MRL grant DMR-80-20250.

References

5.1   W.Spengler, R.Kaiser: Solid State Commun. **18**, 881 (1976)
5.2   H.Wipf, W.S.Williams, M.V.Klein: Phys. Stat. Sol. (b) **108**, 489 (1981)
5.3   H.G.Smith, W.Gläser: Phys. Rev. Lett. **25**, 1611 (1970)
5.4   D.H.Douglass (ed.): *Superconductivity in d- and f-Band Metals*, (Plenum Press, New York 1976) see p. 223 by H.G.Smith, N.Wakabayashi, M.Mostoller
5.5   E.Burstein, F.A.Johnson, R.Loudon: Phys. Rev. **139**A, 1239 (1965)
5.6   J.A.Holy: Ph. D. dissertation, University of Illinois, Urbana (1977) (unpublished)
5.7   J.A.Wilson, F.J.DiSalvo, S.Mahajan: Adv. Phys. **24**, 117 (1975)
5.8   D.E.Moncton, J.D.Axe, F.J.DiSalvo: Phys. Rev. Lett. **34**, 734 (1975); Phys. Rev. B**16**, 801 (1977)
5.9   N.Wakabayashi, H.G.Smith, R.M.Nicklow: Phys. Rev. B**12**, 659 (1975)
5.10 J.E.Smith, J.C.Tsang, M.W.Shafer: Solid State Commun. **19**, 283 (1976)
5.11 E.F.Steigmeier, G.Harbeke, H.Auderset, F.J.SiSalvo: Solid State Commun. **20**, 667 (1976)
5.12 J.A.Holy, M.V.Klein, W.L.McMillan, S.F.Meyer: Phys. Rev. Lett. **37**, 1145 (1976)
5.13 J.R.Duffey, R.D.Kirby, R.V.Coleman: Solid State Commun. **20**, 617 (1976)
5.14 J.C.Tsang, J.E.Smith, M.W.Shafer, S.F.Meyer: Phys. Rev. B**16**, 4239 (1977)
5.15 K.R.A.Ziebeck, B.Dorner, W.G.Stirling, R.Schöllhorn: J. Phys. F. **7**, 1139 (1977)
5.16 J.C.Tsang, J.E.Smith, M.W.Shafer: Phys. Rev. Lett. **37**, 1407 (1976)
5.17 R.Sooryakumar, M.V.Klein, R.F.Frindt: Phys. Rev. B**23**, 3222 (1981)
5.18 M.Gupta, A.J.Freeman: Phys. Rev. B**14**, 5205 (1976)
5.19 B.Klein, D.A.Papaconstantopoulos, L.L.Boyer: [5.4], Solid State Commun. **20**, 937 (1976)
5.20 N.J.Doran, B.Ricco, M.Schreiber, D.Titterington, G.Wexler: J. Phys. C**11**, 699 (1978)
5.21 S.K.Sinha, B.H.Harmon: Phys. Rev. Lett. **35**, 1515 (1975)
5.22 W.Hanke, J.Hafner, H.Bilz: Phys. Rev. Lett. **37**, 1560 (1976)
5.23 C.M.Varma, W.Weber: Phys. Rev. B**19**, 6142 (1979)
5.24 W.Weber: In *Superconductivity in d- and f-Band Metals*, ed. by H.Suhl and M.B.Maple (Academic Press, New York 1980) p. 131
5.25 N.J.Doran: J. Phys. C**11**, L959 (1978)
5.26 J.E.Inglesfield: J. Phys. C**13**, 17 (1980)
5.27 Yu.A.Izyumov, Z.Z.Kurmaev: Usp. Fiz. Nauk **113**, 193 (1974) [Transl.: Sov. Phys.-Usp. **17**, 356 (1974)]
5.28 L.R.Testardi: Rev. Mod. Phys. **47**, 637 (1975)

5.29 J.Muller: Rep. Prog. Phys. **43**, 641 (1980)
5.30 J.Labbe, J.J.Friedel: J. Phys. Radium **27**, 708 (1966)
5.31 L.P.Gor'kov: Zh. Eksp. Teor. Fiz. **65**, 1658 (1973) [Transl.: Sov. Phys. JETP **38**, 830 (1974)]
     L.P.Gor'kov, O.N.Dorokhov: J. Low Temp. Phys. **22**, 1 (1976)
5.32 R.N.Bhatt: Phys. Rev. B**16**, 1915 (1977)
5.33 H.Wipf, M.V.Klein, B.S.Chandrasekhar, T.H.Geballe, J.H.Wernick: Phys. Rev. Lett. **41**, 1752 (1978)
5.34 N.N.Achar, G.R.Barsch: Phys. Stat. Sol. (b) **76**, 133, 677 (1976)
5.35 M.Cardona, G.Güntherodt (eds.): *Light Scattering in Solids* II, Topics Appl. Phys., Vol. 50 (Springer, Berlin, Heidelberg, New York 1982)
5.36 H.Schicktanz, R.Kaiser, W.Spengler, B.Seeber: Solid State Commun. **28**, 935 (1978)
5.37 S.Schicktanz, R.Kaiser, E.Schneider, W.Gläser: Phys. Rev. B**22**, 2386 (1980)
5.38 J.B.Hastings, G.Shirane, S.J.Williamson: Phys. Rev. Lett. **43**, 1249 (1979)
5.39 D.L.Mills, A.A.Maradudin, E.Burstein: Annals of Physics **56**, 504 (1970)
5.40 A.Kawabata: J. Phys. Soc. Japan **30**, 68 (1971)
5.41 A.A.Abrikosov, L.P.Gor'kov, I.E.Dzyaloshinski: *Methods of Quantum Field Theory in Statistical Physics* (Dover, New York 1963)
5.42 F.Bechstedt, K.Peuker: Phys. Stat. Sol. (b) **72**, 743 (1975)
5.43 I.P.Ipatova, A.V.Subashiev: Fiz. Tverd. Tela **18**, 2145 (1976)
     [Transl.: Sov. Phys. Solid State **18**, 1251 (1976)]
5.44 P.Thiry, D.Chandesris, J.Lecante, C.Guillot, R.Pinchaux, Y.Petroff: Phys. Rev. Lett. **43**, 82 (1979)
5.45 J.A.Knapp, F.J.Himpsel, D.E.Eastman: Phys. Rev. B**19**, 4952 (1979)
5.46 F.Sacchetti: J. Phys. F**10**, L231 (1980)
5.47 A.A.Abrikosov, V.M.Genkin: Zh. Eksp. Teor. Fiz. **65**, 842 (1973)
     [Sov. Phys. JETP **38**, 417 (1974)]
5.48 I.P.Ipatova, A.V.Subashiev, A.A.Maradudin: In *Light Scattering in Solids*, ed. by M.Balkanski (Flammarian, Paris 1971) p. 86
     I.P.Ipatova, A.V.Subashiev: Zh. Eksp. Teor. Fiz. **66**, 722 (1974) [Sov. Phys. JETP **39**, 349 (1974)]
5.49 M.V.Klein: Phys. Rev. B**24**, 4208 (1981)
5.50 P.F.Maldague, J.C.Tsang: *In Proc. Int. Conf. Lattice Dynamics*, ed. by M.Balkanski (Flammarion, Paris 1978)
5.51 M.Hulin: Phys. Stat. Sol. **21**, 607 (1967)
5.52 G.Wexler, A.M.Woolley: J. Phys. C**9**, 1185 (1976)
5.53 A.Kotani: J. Phys. Soc. Japan **42**, 408 (1977)
5.54 W.L.McMillan: Phys. Rev. B**12**, 1187, 1197 (1975)
5.55 K.Nakanishi, H.Takatera, Y.Yamada, H.Shiba: J. Phys. Soc. Japan **43**, 1509 (1977)
5.56 K.Nakanishi, H.Shiba: J. Phys. Soc. Japan **43**, 1839 (1977)
5.57 W.L.McMillan: Phys. Rev. B**16**, 643 (1977)
5.58 K.Nakanishi, H.Shiba: J. Phys. Soc. Japan **44**, 1465 (1978)
5.59 S.A.Jackson, P.A.Lee, T.M.Rice: Phys. Rev. B**17**, 3611 (1978)
5.60 A.E.Jacobs, M.B.Walker: Phys. Rev. B**21**, 4132 (1980)
5.61 K.K.Fung, S.McKernan, J.W.Steeds, J.A.Wilson: J. Phys. C**14**, 5417 (1981)
5.62 M.B.Walker, A.E.Jacobs: Phys. Rev. B**24**, 6770 (1981); Phys. Rev. B**25**, 4856 (1982)
5.63 P.B.Littlewood, T.M.Rice: Phys. Rev. Lett. **48**, 27 (1982)
5.64 W.L.McMillan: Unpublished work
5.65 R.M.Fleming, D.E.Moncton, D.B.McWhan, F.J.DiSalvo: Phys. Rev. Lett. **45**, 576 (1980)
5.66 C.H.Chen, J.M.Gibson, R.M.Fleming: Phys. Lett. **47**, 733 (1981)
5.67 D.B.McWhan, R.M.Fleming, D.E.Moncton, F.J.DiSalvo: Phys. Rev. Lett. **45**, 269 (1980)
5.68 P.A.Lee, T.M.Rice, P.W.Anderson: Solid State Commun. **14**, 703 (1974)
5.69 E.Hanamura, N.Nagaosa: Physica **105**B, 400 (1981)
5.70 M.V.Klein: Phys. Rev. B **25** (June 15, 1982)
5.71 N.Nagaosa, E.Hanamura: Solid State Commun. **41**, 809 (1982)
5.72 S.Sugai, K.Murase, S.Uchida, S.Tanaka: Physica **105**B, 405 (1981); and unpublished work

5.73 S. Sugai, K. Murase: Phys. Rev. B **25**, 2418 (1982)
5.74 R. Sooryakumar, D. G. Bruns, M. V. Klein: *2nd Joint US-USSR Symposium*, ed. by J. L. Birman, H. Z. Cummins, K. K. Rebane (Plenum Press, New York 1979) p. 347
5.75 B. Horovitz, H. Gutfreund, M. Weger: Phys. Rev. B **12**, 3174 (1975)
5.76 A. S. Barker, Jr., J. A. Ditzenberger, F. J. DiSalvo: Phys. Rev. B **12**, 2049 (1975)
5.77 G. Campagnoli, A. Gustinetti, A. Stella, E. Tosatti: Phys. Rev. Lett. **38**, 95 (1977); Nuovo Cimento **38** B, 562 (1977); Phys. Rev. B **20**, 2217 (1979)
5.78 J. A. Wilson: Phys. Rev. B **12**, 5748 (1977)
5.79 R. Sooryakumar: Ph. D. Thesis, University of Illinois (1980) (unpublished)
5.80 P. DeTrey, Suso Gygax, J. P. Jan: J. Low Temp. Phys. **11**, 421 (1973)
5.81 R. Sooryakumar, M. V. Klein: Phys. Rev. Lett. **45**, 660 (1980)
5.82 B. P. Clayman, R. F. Frindt: Solid State Commun. **9**, 1881 (1971)
5.83 R. Sooryakumar, M. V. Klein: Phys. Rev. B **23**, 3213 (1981)
5.84 S. Y. Tong, A. A. Maradudin: Mat. Res. Bulletin **4**, 563 (1969)
5.85 D. R. Tilley: Z. Phys. **254**, 71 (1972); J. Phys. F **38**, 417 (1973)
5.86 C. B. Cuden: Phys. Rev. B **13**, 1993 (1976); Phys. Rev. B **18**, 3156 (1978)
5.87 C. A. Balseiro, L. M. Falicov: Phys. Rev. Lett. **45**, 662 (1980)
5.88 J. Bardeen, L. N. Cooper, J. R. Schrieffer: Phys. Rev. **108**, 1175 (1957)
5.89 P. B. Littlewood, C. M. Varma: Phys. Rev. Lett. **47**, 811 (1981)
5.90 Y. Nambu: Phys. Rev. **117**, 648 (1960)

6. Trends in Brillouin Scattering: Studies of Opaque Materials, Supported Films, and Central Modes

J. R. Sandercock

With 14 Figures

The interaction between light and thermally excited acoustic excitations in solids was first investigated early this century by *Brillouin* [6.1] and independently by *Mandelshtam* [6.2]. Although some aspects of the theory were later verified using Raman scattering techniques, the subject generally received little attention until the advent of the laser. Then it was shown in some elegant experiments that Brillouin scattering is an excellent means of investigating the elastic and photo-elastic properties of materials [6.3], particularly in the neighbourhood of phase transitions [6.4, 5]. A step forward occurred with the introduction of the high-contrast spectrometer [6.6], allowing measurements on small and even opaque samples. Several reviews of the field have appeared [6.7–10].

More recently, however, there have been further advances resulting in the observation of Brillouin scattering from both phonons and magnons in metals. In metals, the light scattering takes place of necessity near the surface and yields, therefore, predominantly information about surface excitations, although particularly in the case of magnons, considerable information relating to the bulk excitations is still present in the spectra.

Diffuse type excitations or entropy fluctuations have long been known to give rise to a central mode in the spectrum of the scattered light from liquids [6.11]. Theoretical predictions for solids have recently been partially verified, opening up an interesting and challenging new field of study.

It is the purpose of this chapter to describe these latest developments. In order that the theory may be cogently presented to include the effects of surfaces and high opacity, the basic theory for transparent materials is first summarised in Sect. 6.1. The latest experimental techniques are described in Sect. 6.2 and this is followed in Sect. 6.3 with a step-by-step development of the theory of scattering from phonons in opaque materials. This is interlaced with experimental results to illustrate the various aspects as they are introduced. Section 6.4 covers the scattering from magnons in opaque materials in a similar manner. Finally in Sect. 6.5, a discussion of scattering from entropy fluctuations in solids is given.

6.1 Concept of Brillouin Scattering

A sound wave of wave vector q and frequency ω propagating through a medium of dielectric constant ε sets up a modulation in ε from which light may be scattered. Incident light of wave vector k^i and frequency ω_i is scattered into a state k^s, ω_s such that

$$k^s - k^i = \pm q, \tag{6.1}$$

where the $+$ sign refers to absorption (anti-Stokes process) and the $-$ sign to emission (Stokes process) of the phonon. Correspondingly, the frequencies are related by

$$\omega_s - \omega_i = \pm \omega. \tag{6.2}$$

Since the velocity of sound, v, is very much less than that of light, c, it is a very good approximation to take $|k^s| = |k^i|$ and so

$$2|k^s| \sin \varphi/2 = |q|, \tag{6.3}$$

where φ is the angle between k^s and k^i, known as the scattering angle. Typical phonons q lie near the centre of the Brillouin zone with a maximum frequency of about 150 GHz for visible incident light.

The theory of Brillouin scattering in transparent solids has been excellently presented by many authors, notably *Fabelinskii* [6.7], *Landau* and *Lifshitz* [6.12], and *Benedek* and *Fritsch* [6.3] for isotropic and cubic systems. For lower symmetries, the theory becomes tedious. Many useful results for lower symmetry systems have been tabulated by *Vacher* and *Boyer* [6.13]. Note that (6.1) may not always be reduced to the simple form of (6.3). Thus, in an optically anisotropic medium where the refractive index for the scattered light is different from that for the incident light, (6.3) no longer applies. This more complex situation is discussed in the literature [6.14–16] but has no bearing on the results to be presented here.

In this chapter, only the basis of the theory for isotropic systems is presented. The effects of reflection and refraction of the light at the sample surface are ignored since these can always be treated independently and would unnecessarily complicate the results.

6.1.1 Brillouin Scattering from Phonons in Transparent Isotropic Solids

Light is scattered from fluctuations in the dielectric constant. While the mean dielectric constant ε is considered isotropic, the fluctuations $\delta\varepsilon$ produced by an excitation of wave vector q will in general be anisotropic.

The electric field of the incident light produces a polarisation in the medium. The excess polarisation δP associated with the fluctuations radiates a scattered field E^s at point R_0. Following the treatment by *Landau* and *Lifshitz* [6.12]

$$E^s = -\frac{[\exp(i k^s \cdot R_0)]}{4\pi\varepsilon R_0} \cdot k^s \times (k^s \times G), \tag{6.4}$$

where the scattering vector G has components

$$G_i = \int \delta\varepsilon_{ik} \cdot E_k^i \exp[-i(k^s - k^i) \cdot r] dV. \tag{6.5}$$

The vector r scans the scattering volume V, and the subscripts i, k, l refer to Cartesian axes.

The fluctuation $\delta\varepsilon$ can be related to the local strain u, which is a symmetric combination of displacement gradients. (This is true only for optically isotropic materials – generally the antisymmetric combination must also be included as pointed out be *Nelson* and *Lax* [6.17a]). Let us consider the case of an isotropic material. The general case was discussed in [6.17b]. The fluctuation $\delta\varepsilon$ in the isotropic case:

$$\delta\varepsilon_{ik} = -\varepsilon^2[(p_{11} - p_{12})u_{ik} + p_{12}u_{ll}\delta_{ik}], \tag{6.6}$$

where the p's are the appropriate Pockels coefficients. The strain u associated with the excitation of wave vector q has components

$$u_{ik} = \tfrac{1}{2}i(e_i q_k + e_k q_i)\exp(iq \cdot r), \tag{6.7}$$

where e is the local displacement.

The volume integral I_v contained within G_i then becomes

$$I_v = \int u_{ik}\exp[-i(k^s - k^i) \cdot r]dV = \tfrac{1}{2}i V(e_i q_k + e_k q_i)\delta_{k^s, k^i + q}, \tag{6.8}$$

provided that the dimensions of the scattering volume V are large compared to q^{-1}. Thus, only those excitations with wave vector q satisfying (6.1) may scatter light. It will be shown below that in opaque materials where one dimension of the scattering volume is no longer large compared to q^{-1}, the δ-function becomes modified, relaxing the wave vector conservation requirement of (6.1).

The total power dP scattered into the solid angle $d\Omega$ is given by

$$\frac{1}{P_0}\left(\frac{dP}{d\Omega}\right) = \frac{LR_0^2}{V|E^i|^2} \cdot |E^s|^2, \tag{6.9}$$

where P_0 is the incident power and L is the length of the interaction volume. The scattered field E^s is given by (6.4–6.8).

The averaged scattered power is thus obtained in terms of the mean square displacement $\overline{|e|^2}$ which in turn is obtained by equating the mean kinetic energy per mode q to $k_B T$:

$$\overline{|e|^2} = \frac{4k_B T}{V \varrho v^2 q^2}, \tag{6.10}$$

where ϱ is the density and v the phase velocity of mode q. Finally, the power scattered by transverse phonons, $e \| (k^i \times k^s)$, is given by

$$\frac{1}{LP_0}\left(\frac{dP}{d\Omega}\right) = \frac{\pi^2 \varepsilon^4 (p_{11} - p_{12})^2}{4\lambda_0^4 \varrho v_t^2} \cdot k_B T \cdot \cos^2 \frac{\varphi}{2}, \tag{6.11}$$

where λ_0 is the vacuum wavelength of the light and v_t the transverse phonon velocity.

For scattering by longitudinal phonons with E^i, $E^s \| (k^i \times k^s)$,

$$\frac{1}{LP_0}\left(\frac{dP}{d\Omega}\right) = \frac{\pi^2 \varepsilon^4 (p_{12})^2}{\lambda_0^4 \varrho v_l^2} \cdot k_B T \tag{6.12}$$

and with E^i, $E^s \perp (k^i \times k^s)$,

$$\frac{1}{LP_0}\left(\frac{dP}{d\Omega}\right) = \frac{\pi^2 \varepsilon^4 \left[(p_{11} - p_{12}) \cos^2 \dfrac{\varphi}{2} + p_{12} \cos\varphi\right]^2}{\lambda_0^4 \varrho v_l^2} \cdot k_B T. \tag{6.13}$$

The important conclusions to be drawn from (6.11–13) are:

a) the light scattered by transverse phonons is completely depolarised;
b) there is no scattering by transverse phonons polarised in the scattering plane;
c) the intensity of scattering from transverse phonons falls to zero in back scattering ($\varphi = 180°$);
d) scattering by longitudinal phonons is fully polarised.

While the above statements apply strictly only to *isotropic materials*, they provide useful guidelines to what may be expected in materials of lower symmetry. Exact results for lower symmetry situations may be obtained from the literature [6.13].

For an excitation whose amplitude is damped as $\exp(-\gamma_0 t)$, the intensity I_0 of the scattered light has a frequency distribution

$$dI = \frac{I_0 \gamma_0 / \pi}{(\omega_s - \omega_i \pm \omega)^2 + \gamma_0^2} \cdot d\omega \tag{6.14}$$

which is seen to be normalised such that

$$\int_{\omega=-\infty}^{\omega=+\infty} \frac{dI}{I_0} = 1. \tag{6.15}$$

A differential scattered power in a frequency interval $d\omega$ can then be written as

$$\frac{d^2P}{d\Omega d\omega} = \frac{\gamma_0/\pi}{(\omega_s - \omega_i \pm \omega)^2 + \gamma_0^2} \cdot \frac{dP}{d\Omega}. \tag{6.16}$$

6.1.2 Spin Waves

A particularly important difference between spin waves and phonons lies in the wave vector dependence of their frequencies. Due to the linear wave vector – frequency dependence of phonons near the zone centre, Brillouin scattering measurements at different wave vectors normally yield no new information about the phonon system. On the other hand, the exchange coupling of spins produces a nonlinear wave vector dependence of the spin-wave energy, even near the zone centre in the region accessible to light scattering, so that important information on the strength of the exchange may be deduced by varying the wave vector in a light-scattering measurement. Furthermore, values for the gyromagnetic ratio, magnetisation, anisotropy and magneto-optic interactions may be obtained.

The basic dispersion relations for spin waves in an isotropic ferromagnet are summarised below. The spin-wave dispersion plays an important role in determining the spectrum of the scattered light. Demagnetisation and anisotropy fields are ignored for the sake of simplicity.

a) Bulk Spin Waves

The frequency ω_B of a bulk spin wave is given by [6.17c]

$$\omega_B^2 = \gamma^2 (H + Dq^2)(H + 4\pi M_s \sin^2\theta + Dq^2), \tag{6.17}$$

where γ is the gyromagnetic ratio (containing the g factor), H is the external applied magnetic field, D is the exchange constant, M_s the saturation magnetisation and θ the angle between q and M_s.

As $q \to 0$, ω_B has the limiting values

$$\left.\begin{array}{ll} \omega_B = \gamma H & \text{for} \quad q \| M_s \\ \omega_B = \gamma (HB)^{1/2} & \text{for} \quad q \perp M_s \end{array}\right\}, \tag{6.18}$$

where $B = H + 4\pi M_s$. For other values of θ, a spin-wave band is defined as shown by the shaded area in Fig. 6.1a.

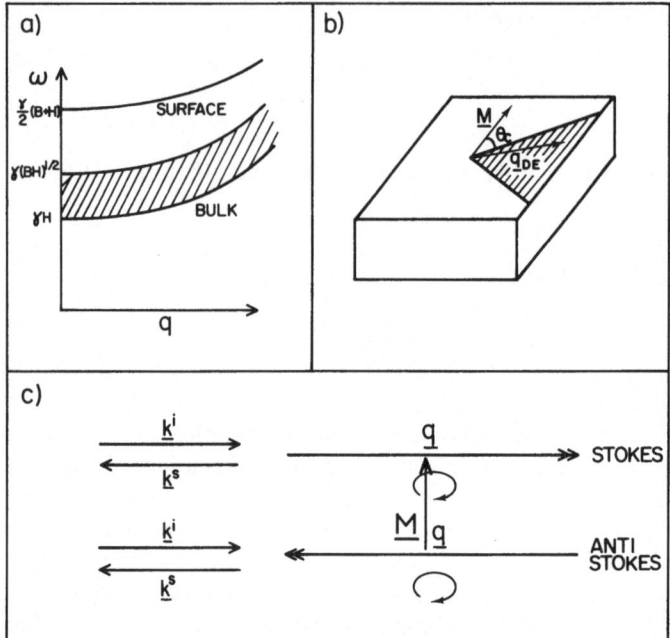

Fig. 6.1. (a) The dispersion relations for bulk and surface magnons; (b) range of propagation directions for Damon–Eshbach surface magnon; (c) wave vectors involved in Stokes and anti-Stokes scattering processes. Precession direction is the same for both, thereby destroying the time reversal symmetry

b) Surface Spin Waves

The *Damon-Eshbach* [6.18] surface spin wave, in the absence of exchange, has the energy

$$\omega_{DE} = \frac{\gamma}{2}(H/\sin\theta + B\sin\theta). \tag{6.19}$$

As θ increases or decreases from 90°, ω_{DE} decreases and eventually becomes degenerate with the top of the bulk spin-wave band at a critical angle θ_c. Beyond this angle, the Damon-Eshbach mode decays spontaneously into bulk spin waves.

$$\theta_c = \sin^{-1}(H/B)^{1/2} \tag{6.20}$$

corresponding to $\omega_{DE} = \gamma(HB)^{1/2}$.

The interesting property of the Damon-Eshbach mode is the nonreciprocal propagation. Propagation is only possible for the range of directions shown in Fig. 6.1b.

The effects of exchange are more complicated than for the bulk modes, but one may write approximately

$$\omega(q) = \omega(0) + \gamma D' q^2, \tag{6.21}$$

where $D' \sim 1.76 D$ for $\theta = 90°$.

The above dispersion formulae apply equally to metallic and insulating ferromagnets. The effect of the free electrons is merely to introduce an eddy current damping of the spin waves.

c) Light Scattering from Spin Waves

It might be expected that light would be scattered from spin waves due to a direct interaction between the spin fluctuation and the magnetic vector of the light. It has been shown, however, [6.19] that this mechanism is in general very weak and that the dominant interaction is via fluctuations produced in the dielectric constant ε by a spin-orbit coupling to the spin fluctuations. The scattered intensity can then be derived exactly analogous to that from phonons, as described in Sect. 6.2.1.

Following *Wettling* et al. [6.20], the scattered power may be expressed as

$$\frac{1}{LP_0}\left(\frac{dP}{d\Omega}\right) = \frac{k^{s^4}}{64\varepsilon^2\pi^2} \cdot V \cdot \overline{|\Delta\varepsilon|^2}. \tag{6.22}$$

The fluctuation $\Delta\varepsilon$ depends on the transverse component of the magnetisation M_t, i.e., the magnetisation component associated with the spin precession. It was pointed out by *Le Gall* et al. [6.21] that the coupling between $\Delta\varepsilon$ and M_t has terms both linear and quadratic in magnetisation (though linear in M_t):

$$\Delta\varepsilon = KM_t + GM_sM_t,$$

where K and G are tensor quantities.

The transverse magnetisation component M_t is obtained by equating the magnon energy per mode to $k_B T$ so that

$$M_t^2 = \frac{1}{V} \cdot \frac{2\gamma M_s}{\omega} \cdot k_B T \tag{6.23}$$

and the scattered power may then be obtained in the form

$$\frac{1}{LP_0}\left(\frac{dP}{d\Omega}\right) = \frac{k^{s^4}}{32\varepsilon^2\pi^2} \cdot \frac{\gamma M_s}{\omega} \cdot k_B T \cdot |GM_s \pm K|^2. \tag{6.24}$$

The $\pm$ signs refer to Stokes and anti-Stokes intensities. The difference arises because the linear and quadratic contributions are $90°$ out of phase (in

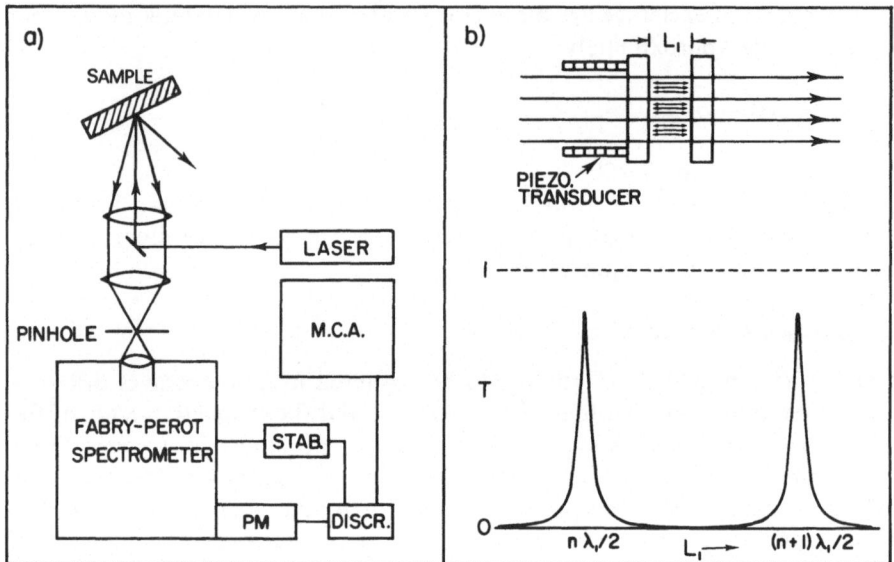

Fig. 6.2. (a) Experimental set-up for backscattering with a Fabry–Perot spectrometer; (b) transmission of monochromatic light by scanning Fabry–Perot interferometer

transparent materials K is imaginary and G real) and are therefore sensitive to the relative precession direction of the spins. This is illustrated in Fig. 6.1c – note the lack of time reversal symmetry between Stokes and anti-Stokes events.

The tensor K and G may be related to the appropriate magneto-optic constants.

6.2 Experimental Techniques

When studying transparent materials, all scattering configurations between forward scattering and 180° backscattering are possible, allowing the study of excitations with wave vectors q in the range 0 to $2k^i$ ($k^i = nk_0$ where n is the refractive index and k_0 the vacuum wave vector). When studying opaque materials, this freedom no longer exists–and generally one is limited to a scattering arrangement from one surface of the sample only. The minimum bulk wave vector that can be observed in this case is $2nk_0\sqrt{1 - 1/n^2}$, corresponding to incident and scattered light at grazing incidence and in the same direction. For the typically large values of refractive index n in opaque materials, this minimum wave vector does not differ significantly from the value $2nk_0$ for 180° backscattering. Since the wave vector spread due to the finite collection aperture is minimised for 180° scattering, it is convenient to use this geometry for all measurements where it is not absolutely necessary to have incident and scattered light in different directions. A suitable optical arrangement is indicated in Fig. 6.2a.

In order to achieve sufficiently high resolution for a Brillouin scattering measurement, a Fabry–Perot interferometer is used as a scanning spectrometer. In most applications, a plane parallel Fabry–Perot (PFP) is used, as illustrated schematically in Fig. 6.2b.

The PFP consists of two very flat mirrors mounted accurately parallel to each other with a spacing L_1 which may be varied. The instrument will transmit light of wavelength λ_1 if the spacing L_1 is such that

$$L_1 = \frac{m\lambda_1}{2}, \tag{6.25}$$

where m is an integer. The instrument acts as a tunable filter whose peak transmission is close to unity over a narrow spectral interval, falling to a very low value outside this interval. Two incident signals of wavelength λ_1 and $\lambda_1 + \Delta\lambda$ will be simultaneously transmitted if

$$m\lambda_1 = (m-1)(\lambda_1 + \Delta\lambda), \tag{6.26}$$

or in other words, neighboring orders of interference are separated in frequency by $1/2L\,\mathrm{cm}^{-1}$. This interorder spacing is called the free spectral range (FSR). The width of the transmission peak determines the resolution of the instrument. The ratio of FSR to width is known as the finesse F.

The finesse is primarily a function of mirror reflectivity, although instrumental aperture and mirror flatness are also important parameters. For a discussion of the Fabry–Perot, see, for example, the review article by *Jacquinot* [6.22]. In practice, the finesse is limited to values less than about 100 and this places an upper limit on the possible contrast, where the contrast C is the ratio of maximum to minimum transmission given by

$$C = 1 + \frac{4F^2}{\pi^2} \simeq \frac{4F^2}{\pi^2} \lesssim 10^4. \tag{6.27}$$

It is apparent that this contrast will be insufficient for measuring in situations where the elastically scattered component of the scattered light exceeds the intensity of the Brillouin component by more than a factor of 10^4 to 10^5. For backscattering measurements on opaque materials, this is generally the case and so some means of increasing the spectral contrast is essential.

6.2.1 Improving the Spectral Contrast

a) High-Contrast Multipass Interferometer

The contrast of an interferometer may be increased by placing two or more interferometers in series, but it is not straightforward to synchronise the scans of the separate units. An equivalent and more elegant technique is to pass the light two or more times through the same interferometer – in this case, provided that the mirrors remain parallel, all passes are identical and so the problem of synchronisation is obviated.

The design aspects of a multipass interferometer have been discussed previously [6.23–25] and will not be presented here. It has been demonstrated that a five pass interferometer can achieve a contrast of greater than 10^9, five or six orders of magnitude greater than that of a single interferometer. At the same time, the peak transmission ($\sim 50\%$) and finesse (50–100) stay comparable. The only possible disadvantage is the reduced beam diameter which may reduce the light throughput. However, for scattering from opaque materials where the scattering volume has the dimensions of the focussed laser beam on the sample surface, all the scattered light may still be collected within the available aperture and so the light throughput of the multipass system is not reduced with respect to that of single-pass interferometer.

The introduction of the multipass interferometer opened the way for Brillouin scattering experiments in opaque materials. While the resolution of this instrument is adequate for many experiments, for those experiments where the highest resolution is required, it becomes necessary to use a spherical Fabry–Perot. This interferometer has been well described by *Hercher* [6.26]. A resolution as high as 1 MHz is possible with high light throughput. The limitations on contrast are similar to those on the plane Fabry–Perot described above. The instrument, furthermore, does not lend itself to multipassing, although a two pass system may be operated [6.27]. If high resolution is required combined with high contrast, for example to analyse the linewidth of a narrow Brillouin component, a tandem arrangement of a multipass interferometer and a spherical one may be successfully employed although the stability problems inherent in a tandem system are inevitably present [6.28, 29].

b) Rejection Filters

In a Brillouin spectrum it is the elastically scattered peak which is intense. If this peak were not present, a low contrast spectrometer would be adequate for resolving the spectrum. The alternative approach to improving the spectral contrast would therefore be the use of a notch filter to remove the elastic peak (width of peak ~ 1 to 10 MHz). Unfortunately, no notch filter exists which is narrow enough for this purpose. The best that can be achieved is a comb filter which of course removes slices from throughout the spectrum.

Interferometric rejection comb filters have been proposed [6.30] but these have poor rejection and above all, an aperture which is far too small for inclusion in a Brillouin scattering experiment.

Somewhat more feasible is the iodine filter which has many molecular absorption bands, one of which fortuitously coincides with the 5145 Å line of the argon laser. A rejection greater than 10^7 can be achieved [6.31]. This filter suffers from the severe disadvantage that the comb is irregular and shows considerable absorption even in between the absorption bands. Since the absorption bands are broader than typical Brillouin peak widths, it cannot be used in normal Brillouin scattering investigations [6.32]. Measurements of broad central modes have been performed using a computer normalisation technique [6.33, 96] to compensate for the messy transmission characteristics of the iodine filter. That this normalisation is only partially successful can be seen from the published data [6.33, 34, 96] and Fig. 6.13. There is no situation in which the iodine filter can achieve better results than the multipass Fabry–Perot. Its use is not to be recommended.

6.2.2 Stabilising the Fabry–Perot

There are two types of scanning interferometers in common use which differ in their scanning means. Pressure scanned [6.35] instruments use a slow isothermal pressure change to vary the optical spacing of the mirrors, while piezoelectric scanning [6.36] uses a transducer to move one of the mirrors. To scan through a transmission peak of Fig. 6.2 requires a change of only 50 Å in the mirror spacing. Mechanical stability is clearly of the utmost importance. Stability of better than 5 Å is required over the whole recording time which may be several hours when recording weak signals.

Pressure scanning instruments can only be passively stabilised (thermally stable environment and good mechanical construction) and are generally only suitable for shorter recording times. Piezoelectric scanning, on the other hand, lends itself to dynamic stabilisation whereby the instrument is rapidly scanned (a few scans per second) and continously realigned to maintain correct mirror spacing and parallelness. A multichannel storage device is used to sum the spectrum over successive scans. In order for a given spectral feature to occur always at the same point in the scan, a small correction is made to the scan voltage offset [6.37, 38]. The mirror parallelness is maintained by making small alignment corrections so that a dominant spectral feature has either maximum amplitude [6.38, 39] or, equivalently, minimum linewidth [6.40, 41]. These above corrections are performed automatically in a feedback loop.

6.2.3 Increasing the FSR

It is well known that, at least in principle, the FSR of an interferometer may be extended by combining two interferometers of unequal spacing. The first interferometer of spacing L_1 transmits wavelengths

$$\lambda_1 = \frac{2L_1}{m_1} \quad \text{for integral } m_1, \tag{6.28}$$

while the second interferometer of spacing L_2 transmits wavelengths

$$\lambda_2 = \frac{2L_2}{m_2} \quad \text{for integral } m_2. \tag{6.29}$$

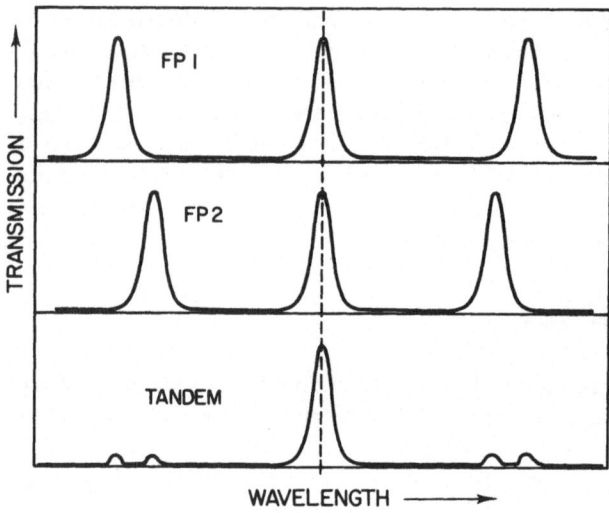

Fig. 6.3. Elimination of neighboring interference orders in a tandem arrangement of two unequal interferometers. Only ghosts of neighboring orders remain

Only if $\lambda_1 = \lambda_2$ will light be transmitted through the combination. In this way it is possible to increase the FSR by a factor of 10 to 20 over that of the single interferometer, although as shown in Fig. 6.3, small ghosts remain of the suppressed orders.

The practical limitation in the use of tandem interferometers has been due to the problem of scan synchronization. Thus, to scan the transmitted wavelength, it is necessary to increment the mirror spacings L_1 and L_2 by δL_1 and δL_2 such that

$$\frac{\delta L_1}{\delta L_2} = \frac{L_1}{L_2}. \qquad (6.30)$$

The only previously known method of achieving this involved pressure scanning [6.42] which gives only a limited scan range, inadequate for the type of experiments to be described below, and suffering from the stability problems discussed earlier.

Very recently, *Dil* et al. [6.43] successfully operated a tandem combination of two piezoelectrically scanned interferometers using electronic feedback to lock the scans together. The relative scan amplitudes are adjusted manually to satisfy the above synchronisation condition. The main disadvantages of such a system are distortions introduced by nonlinearities in the piezoelectric scan (which will generally be different for the two interferometers) and the difficulty in changing the FSR.

A simpler and more flexible technique is described below which achieves synchronisation of two or more interferometers based on a single translation stage common to all interferometers. However, first the construction of the basic interferometer is described since this shows significant changes with respect to the previous practice.

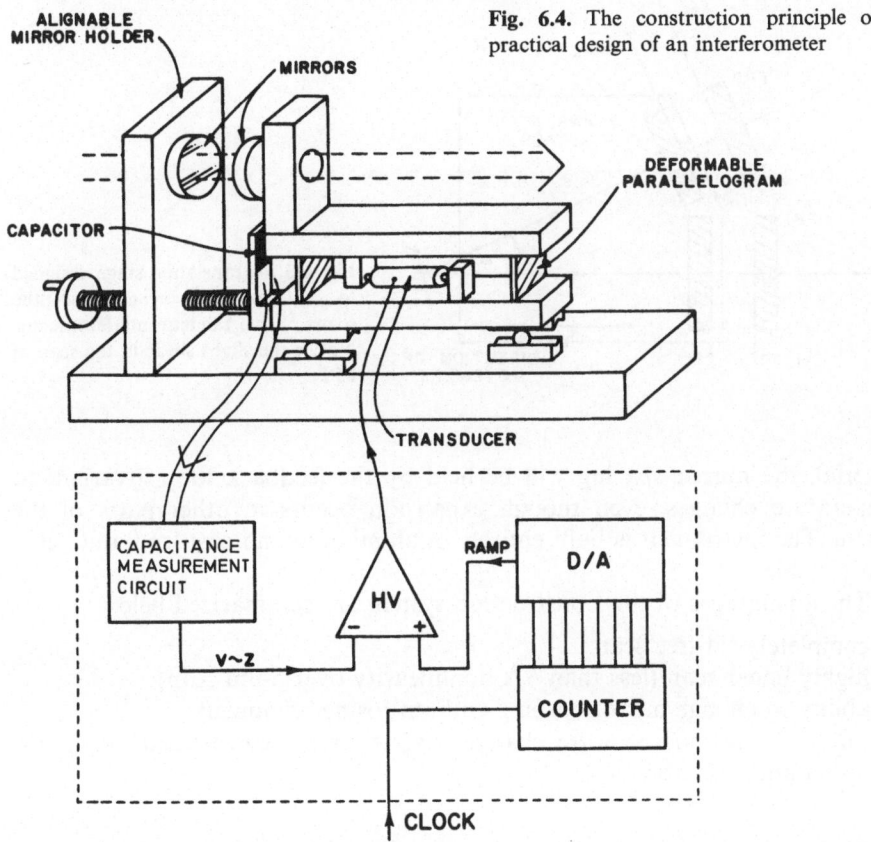

a) Basic Interferometer Construction

The principle of the construction of the new interferometer is illustrated in Fig. 6.4. A scanning stage consisting of a deformable parallelogram [6.44–46] rides on top of a roller translation stage. The former, actuated by a piezoelectric transducer, provides completely tilt-free movement of the interferometer mirror over scan lengths up to 10 μm or more. The latter enables the coarse mirror spacing to be set to the desired value in the range 0–50 mm. The roller translation stage is sufficiently collinear that a movement of several mm leaves the mirrors aligned parallel to better than 1 fringe across the total mirror diameter. A novelty of the construction is the use of a small parallel plate capacitor to measure the scan displacement. The associated electronics produces a voltage accurately proportional to the capacitor spacing, this voltage being used in a feedback loop in order to linearize the scan displacement with respect to the applied scan voltage. Furthermore, it is seen from Fig. 6.4 that the mirror spacing is related to the capacitor spacing *only* through the length of the translation stage screw. Thus, provided this screw is made of low-expansion

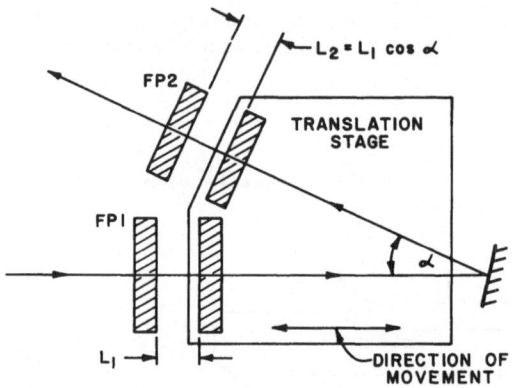

Fig. 6.5. A translation stage designed to automatically synchronise the scans of two tandem interferometers. Notice the slight shear in the scan of FP2

material, the mirror spacing will be held by the feedback loop invariant to temperature changes, even though expansion occurs in other parts of the system. The instrument is built entirely of aluminium and cast iron and yet is highly stable.

The advantages of this construction system are summarized below:

a) completely tilt-free scan
b) highly linear scan (less than 5 Å nonlinearity over 5-μm scan)
c) ability to change mirror spacing without losing alignment
d) stable against temperature change despite simple construction of aluminium and cast iron.

b) Tandem Interferometer

By a slight modification to the scanning stage of the instrument depicted in Fig. 6.4, it is possible to make a synchronously scanning tandem interferometer [6.47]. The new scanning stage viewed from above is shown in Fig. 6.5. The scanning mirrors of two interferometers are mounted on the same scanning stage, one with the mirror axis parallel to the scan direction, the other offset by an angle α. The scan of the second interferometer therefore has a light shear although this has no effect on its use as an interferometer. It is clear that the spacings of the two interferometers satisfy

$$L_2 = L_1 \cos\alpha \qquad\qquad (6.31)$$

and that the synchronization condition (6.30) is satisfied.

There is a further very important, but less obvious, benefit of this method of building a tandem interferometer and this is the fact that the two interferometers may be mounted in close proximity to each other. By sharing the same environment, this tandem system is thus much less sensitive to temperature fluctuations than a system built of two separated interferometers.

For the experiments reported here, an angle $\alpha \simeq 20°$ was used. It should be pointed out that many variations on this scheme are possible, for example, α may be chosen close to 90° (typically 85°) and the first interferometer may be replaced with a confocal interferometer for high resolution work. It is also apparent that more than two interferometers may be synchronously scanned by this technique.

Although Fig. 6.5 shows only a single pass through each interferometer, in practice each interferometer is used in a multipass mode. The alternative of multipassing the whole tandem combination is equally satisfactory.

Long term stabilisation of the instrument is achieved by feedback stabilisation as described above where successive corrections are applied to the alignment stacks to maintain the interferometers parallel and with the correct relative spacing to maintain synchronisation.

6.3 Light Scattering from Acoustic Phonons in Opaque Materials

In this section, the theory presented in Sect. 6.1 is extended to include scattering from opaque materials. There are several ways in which the opacity makes itself felt. Firstly, the high optical absorption limits the scattering to a volume close to the sample surface. This has an influence on the wave vector conservation rule. Secondly, the effect of the surface on the excitations cannot be ignored. This manifests itself in the reflection of bulk excitations at the surface and indeed in the appearance of new excitations associated with the surface. Thirdly, the ripples produced on the surface by these excitations scatter light and for highly opaque materials, may be the dominant scattering source.

These effects are discussed below and compared with experimental observations.

6.3.1 Effect of Optical Absorption

It was pointed out in the derivation of (6.8) that the delta function $\delta_{k^s, k^i + q}$ arises only if the scattering volume V is large in all dimensions compared to the wavelengths involved.

In a scattering experiment from an opaque medium the scattering volume will be restricted perpendicular to the sample surface due to the high optical absorption. The effect of this can be calculated by noting that the wave vector transfer $K = k^s - k^i$ is now complex. Denoting

$$K = K' - iK'', \tag{6.32}$$

the integral I_v of (6.8) now becomes

$$I_v = \int u_{ik} \exp[-i(K' - iK'') \cdot r] dV. \tag{6.33}$$

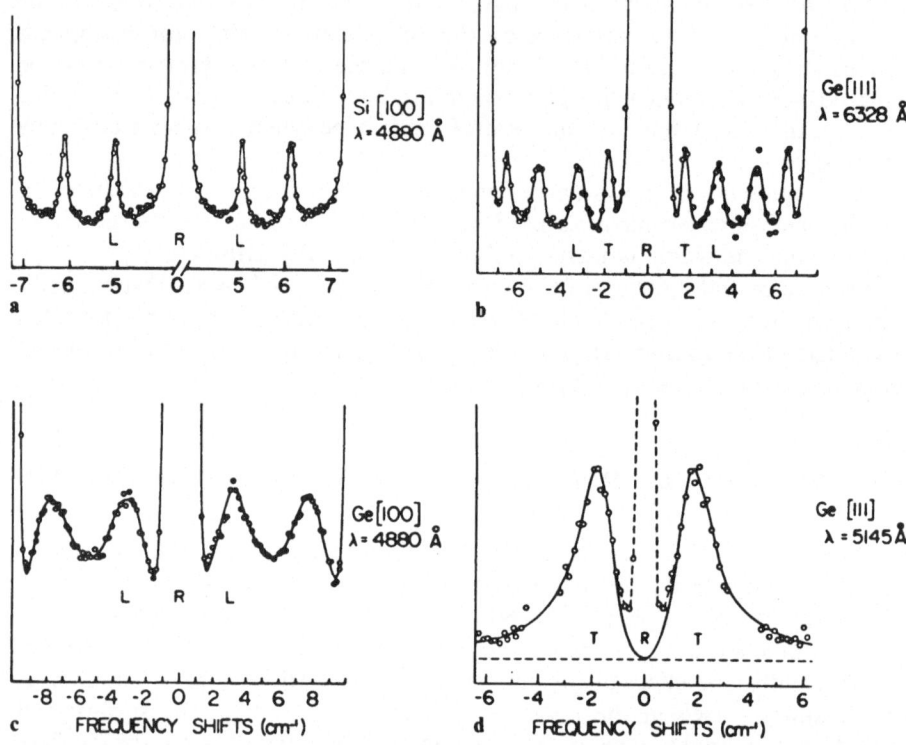

Fig. 6.6a–d. Brillouin backscattering spectra of Si and Ge demonstrating the increasing linewidth due to high optical absorption. The optical absorption coefficient increases from $10^4 \, \mathrm{cm^{-1}}$ in (**a**), through $2 \times 10^5 \, \mathrm{cm^{-1}}$ in (**b**), to about $6 \times 10^5 \, \mathrm{cm^{-1}}$ in (**c**) and (**d**). The spectrum (**d**) taken with the tandem interferometer shows clearly the asymmetric line shape

Under the simplifying assumption that K is perpendicular to the sample surface, we may write $K = K_z$ and

$$I_v = \frac{i}{2} e_i q \delta_{q_x, 0} \delta_{q_y, 0} L_x L_y \int_0^\infty \exp[i(q_z - K_z')z - K_z''z]\,dz$$

$$= \frac{i}{2} e_i q_z L_x L_y \delta_{q_x, 0} \delta_{q_y, 0} \frac{1}{i(q_z - K_z') - K_z''} \tag{6.34}$$

which leads to a q dependence of the scattered intensity:

$$\frac{dP}{d\Omega} \sim \frac{1}{(q - K')^2 + K''^2} \sim \frac{1}{(q/k_0 - 2\eta)^2 + (2\kappa)^2}, \tag{6.35}$$

where the complex refractive index has been defined by

$$n = \eta + i\kappa. \tag{6.36}$$

Equation (6.35) shows that in the presence of optical absorption, the phonon wave vector q is no longer uniquely defined, but rather that all phonons within a range $\Delta q \sim K''$ around the value $q = K'$ may scatter light into the direction k^s. For acoustic phonons, the phonon energy is proportional to the wave vector and so the Brillouin peaks in the presence of optical absorption become broadened with a full width at half maximum (FWHM) of

$$\Delta \omega = 4 v k_0 \kappa. \tag{6.37}$$

The peak position remains $\omega = 2 \eta v k_0$, as obtained from (6.3). The Lorentzian linewidth of (6.35) has been derived by several authors [6.48, 49].

Experimentally, a broadening effect of this nature was observed in CdS by *Pine* [6.50] and demonstrated more convincingly later by *Sandercock* [6.48] in measurements on Si and Ge and by *Benner* et al. [6.51] on the tungsten bronzes. The Si and Ge measurements are reproduced in Figs. 6.6a–c. Notice the marked increase in linewidth as the absorption coefficient increases from $10^4 \, \mathrm{cm}^{-1}$ to $6 \times 10^5 \, \mathrm{cm}^{-1}$. The spectra, having been taken with a multipass interferometer, show the overlapping interference orders typical of the Fabry–Perot. In Fig. 6.6d, a depolarised spectrum from transverse phonons in Ge is shown for comparison, taken with the tandem multipass interferometer. The lineshape is far better resolved in the absence of overlapping interference orders.

It is noticeable, however, that the lineshape of Fig. 6.6d is not a symmetrical Lorentzian predicted by (6.35). The origin of this asymmetry was pointed out by *Dresselhaus* and *Pine* [6.52] and is discussed in the following section.

6.3.2 Allowance for Phonon Reflection

Dresselhaus and *Pine* [6.52] showed that the above simple derivation of the linewidth in opaque materials ignores the fact that the phonon q is coherently reflected into $-q$ at the sample surface. These authors derive an asymmetric lineshape in place of the symmetric Lorentzian above. Later calculations by *Dervish* and *Loudon* [6.53], *Loudon* [6.54], *Subbaswamy* and *Maradudin* [6.55], and *Rowell* and *Stegeman* [6.56] consistently show a different, albeit still asymmetric, lineshape. It appears that the model chosen by *Dresselhaus* and *Pine* in their calculation is incorrect.

The derivation of the lineshape follows by noting that the phonon reflection may be included by writing the phonon displacement in the form [6.53]

$$e = e_0 [\exp(i q \cdot r) + \exp(-i q \cdot r)] \tag{6.38}$$

so that the strain $u_{ik}(q)$ in (6.33) becomes replaced by

$$u_{ik} = u_{ik}(q) - u_{ik}(-q). \qquad (6.39)$$

This strain is zero at the surface as required by the boundary condition. The integral $I_v(q)$ of (6.34) becomes

$$I_v = I_v(q) - I_v(-q) \qquad (6.40)$$

leading to a scattered intensity with a lineshape of the form

$$\frac{dP}{d\Omega} \sim \left| \frac{1}{i(q-K')+K''} - \frac{1}{i(-q-K')+K''} \right|^2$$

$$\sim \frac{4(q/k_0)^2}{[(q/k_0)^2 - 4\eta^2 + 4\kappa^2]^2 + (8\eta\kappa)^2}. \qquad (6.41)$$

This expression is equivalent to (2.106).

The q^2 term in the numerator makes this expression asymmetric. The FWHM is still the same as for the Lorentzian [Eq. (6.37)] but the peak position is somewhat shifted to the value

$$\omega = 2vk_0(\eta^2 + \kappa^2)^{1/2}. \qquad (6.42)$$

It is worth emphasizing that in a transparent material, the phonon q gives rise to the anti-Stokes scattering while $-q$ contributes only to the Stokes process. Only in an optically opaque material can both q and $-q$ contribute together to the same process. The fact that q and $-q$ are, furthermore, coherent is responsible for the particular form of the line shape derived in (6.41). This line shape is shown by the solid line fitted to the data of Fig. 6.6d.

6.3.3 Phonon Modes in the Presence of Surfaces and Interfaces

In Sect. 6.3.1 the sample surface appeared only as an electromagnetic boundary. In Sect. 6.3.2 its additional influence as a mechanical boundary affecting the excitation spectrum of the solid was seen. This latter influence must be considered in more detail, including the case of oblique incidence, i.e., wave vector transfer $K = K_z + K_x$.

The conclusion of Sect. 6.3.1 that K_z is more or less undefined in opaque materials still applies, however, the mechanical boundary now permits new excitations specific to the surface having wave vector $q_x = K_x$ parallel to the surface. These additional excitations, namely, Rayleigh, Lamb and Love modes, introduce new features into the spectrum of the scattered light. The scattering from the coherently reflected phonons discussed in Sect. 6.3.2 is, in

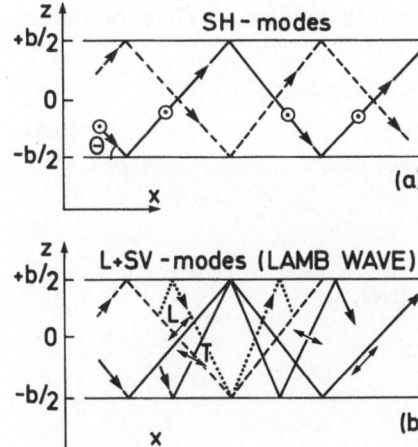

Fig. 6.7. (a) Partial wave pattern for shear horizontal (SH) wave propagation in a slab; (b) Partial wave pattern for mixed longitutinal (L) and shear vertical (SV) propagation in a slab of thickness b (Lamb waves)

fact, just the particularly simple case of normal incidence scattering from Lamb waves in a semi-infinite slab.

A good feeling for the nature of these modes can be obtained following the eloquent treatment of *Auld* [6.57] using the very simple transverse resonance technique. It is convenient to first derive the modes of a plate. The results for the plate may then be carried over to the case of a semi-infinite sample by allowing the plate thickness to approach infinity.

a) Modes of a Plate

The symmetry of the problem with respect to the plane through the middle of the plate ($z=0$) requires that the waves be either symmetric or antisymmetric with respect to this plane. We thus require at least two plane waves with wave vectors k and $(\bar{k})$, where $\bar{k}$ is obtained by reflecting k on the symmetry plane. The simplest case is that of shear horizontal waves (Fig. 6.7a). In this case, the polarization remains horizontal upon reflection and only two waves, k and $\bar{k}$, are required to fulfill the boundary conditions (no force in the z direction). This boundary condition requires that the displacement amplitude vanishes at $z = \pm b/2$, i.e.,

$$\frac{\omega}{v_t}\sin\theta = k_{zt} = \frac{m\pi}{b}, \quad m = 1, 2, 3 \ldots, \tag{6.43}$$

where the angle θ is defined in Fig. 6.7a. Equation (6.43) has a continuum of solutions with a lower cutoff for $\omega = v_t \pi/b$.

The situation is more complicated whenever longitudinal waves are involved upon reflection on the plate boundary; these waves are partly converted into shear vertical waves (SV) and four plane waves (L with k_1 and $\bar{k}_1$ and SV with k_t and $\bar{k}_t$) are required to fulfill the boundary conditions. These boundary conditions lead to [6.57]:

$$\frac{\tan(k_{zt}b/2)}{\tan(k_{zl}b/2)} = -\frac{4\beta^2 k_{zt}k_{zl}}{(k_{zt}^2 - k_{zl}^2)^2} \quad \text{(symmetric)} \tag{6.44}$$

and

$$\frac{\tan(k_{zt}b/2)}{\tan(k_{zl}b/2)} = -\frac{(k_{zt}^2 - k_{zl}^2)^2}{4\beta^2 k_{zt}k_{zl}} \quad \text{(antisymmetric)}.$$

The mode frequencies ω are given by

$$\begin{aligned}
k_{zl}^2 &= \left(\frac{\omega}{v_l}\right)^2 - \beta^2 \\
k_{zt}^2 &= \left(\frac{\omega}{v_t}\right)^2 - \beta^2.
\end{aligned} \tag{6.45}$$

There are three frequency regimes implicit in the (6.45):

a) $\omega/\beta > v_l$. In this case, both k_{zl} and k_{zt} are real and the solutions are normal bulk modes.

b) $v_l > \omega/\beta > v_t$. Here, k_{zt} is real but k_{zl} is purely imaginary.

The solutions represent bulk transverse modes combined with a longitudinal component which is strongly localised at the surface. These modes are referred to as *Lamb* [7.58] waves.

c) $\omega/\beta < v_t$. In this case, both transverse and longitudinal components are localised at the surfaces.

These modes are related to the *Rayleigh* [6.59] surface mode of a semi-infinite solid. In fact, the sum of the lowest order symmetric and antisymmetric solutions (6.4.4) in the limit $\beta b \to \infty$ is just the Rayleigh solution.

In the limit $\beta b \to \infty$, the discrete Lamb and shear horizontal (SH) solutions form continua in which the SH modes are indistinguishable from normal bulk modes. The Lamb modes, on the other hand, are bulk transverse modes with an additional longitudinal component localized at the surface.

The modes discussed above which have surface character are localised typically within a wavelength from the surface and would not, therefore, normally be seen in a light scattering experiment from a transparent material where bulk effects would dominate. In opaque materials, however, the scattering volume is near the surface and surface effects dominate. In fact, as described below, in the limit of high opacity the entire spectrum arises from ripples produced in the surface by both bulk and surface excitations.

6.3.4 Ripple Scattering

It has been known for many years that surface ripples can scatter light [6.60]. Recently, *Mishra* and *Bray* [6.61] have shown that the ripple mechanism can be the dominant mechanism for scattering from semiconductors. The effect has been calculated in detail by *Loudon* [6.62], *Subbaswamy* and *Maradudin* [6.55], *Rowell* and *Stegeman* [6.56], *Loudon* and *Sandercock* [6.63], *Velasco* and *Garcia-Moliner* [6.64], and in several papers by *Marvin* et al. [6.65, 66]. The essence of the theory is presented here.

Consider the scattering configuration of Fig. 6.8 for scattering from fluctuations $e_z(0)$ in the surface whose undisturbed location is the plane $z=0$. For incident light polarised in the plane of the surface, the scattered field E^s is given by [6.63]

$$E^s = -2i\ell_z^i \frac{(k_z^i - \ell_z^i)}{(k_z^s - \ell_z^s)} \cdot E^i \cdot e_z(0),$$
(6.46)

where the wave vectors are defined in Fig. 6.8, the scattered wave vector k^s inside the material being necessary in order to satisfy the boundary conditions.

The scattered power is given by

$$\frac{1}{P}\left(\frac{dP}{d\Omega}\right) = \frac{\omega_s^2}{4\pi^2 c^2} \cdot A\cos^2\theta_s \frac{\overline{|E^s|^2}}{|E^i|},$$
(6.47)

where A is the total area of the sample.

For the particularly simple case of backscattering, $\theta_i = \theta_s = \theta$ and $\ell_z^s = -\ell_z^i$ so that (6.46, 47) yield

$$\frac{1}{P}\left(\frac{dP}{d\Omega}\right) = \frac{\omega_s^4}{\pi^2 c^4} \cdot A\cos^4\theta \cdot R(\theta) \cdot \overline{e_z(0)^2},$$
(6.48)

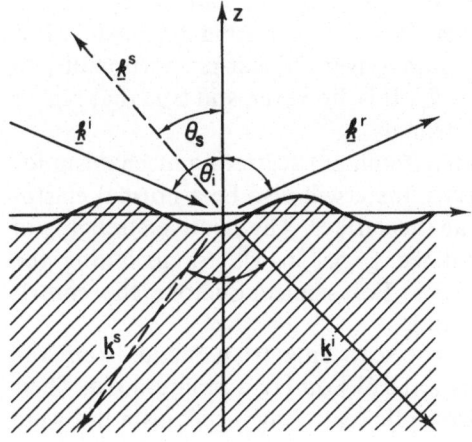

Fig. 6.8. The wave vectors involved in scattering from a surface ripple

where $R(\theta)$ is the reflection coefficient for light incident at an angle θ, polarised in the plane of the surface.

The surface ripple $e_z(0)$ arises from excitations of wave vector $\boldsymbol{q}$ satisfying

$$q_x = (\boldsymbol{k}^s - \boldsymbol{k}^i)_x. \tag{6.49}$$

The magnitude and spectrum of the ripple can be calculated [6.63] yielding a differential scattered power

$$\frac{1}{P}\left(\frac{d^2P}{d\Omega d\omega}\right) = \frac{\omega_s^4}{\pi^2 c^4} \cdot \cos^4\theta \cdot R(\theta) \cdot \frac{k_B T}{\pi \varrho v_t^3 q_x^2} \cdot \mathrm{Re}\{\ldots\}, \tag{6.50}$$

with

$$\mathrm{Re}\{\ldots\} = \mathrm{Re}\left\{\frac{\omega v_t^3 q_x^2 [(\omega/v_l)^2 - q_x^2]^{1/2}}{4v_t^4 q_x^2 [(\omega/v_l)^2 - q_x^2]^{1/2}[(\omega/v_t)^2 - q_x^2]^{1/2} + (\omega^2 - 2v_t^2 q_x^2)^2}\right\}. \tag{6.51}$$

Expression (6.51) contains the spectral information which may be subdivided into three spectral regions depending on whether

a) $\omega/q_x > v_l$,

b) $v_l > \omega/q_x > v_t$,

c) $v_l, v_t > \omega/q_x$.

As discussed in Sect. 6.3.3a these regions refer, respectively, to scattering from bulk phonons, Lamb waves and Rayleigh waves. The exact form of the spectrum depends on the ratio v_t/v_l. The theory does not give a width to the Rayleigh wave peak, but merely an integrated intensity.

Notice that (6.50) gives the scattered intensity proportional to the reflectivity $R(\theta)$. Metals and opaque materials will, therefore, generally scatter more strongly than transparent dielectrics. It should be stressed that the proportionality to $R(\theta)$ applies only for polarisation in the plane of the surface, and only for backscattering ($\theta_i = \theta_s$) or forward scattering ($\theta_i - = \theta_s$). For polarisation in the sagittal plane, there is no longer a linear proportionality to $R(\theta)$ except for forward scattering ($\theta_i = -\theta_s$). It is, however, still true that higher reflectivity materials will scatter more strongly.

The ripple scattering mechanism is the dominant scattering mechanism for opaque metals. For less opaque materials, there will also be a normal elasto-optic contribution to the scattering near the surface. The relative strengths of the two mechanisms have been calculated [6.55, 56, 64, 66], including the interference between the two mechanisms. For semiconductors, the contributions may be roughly equal in magnitude.

The first experimental results for light scattering from metals were the pioneering measurements of *Dil* and *Brody* [6.67] on Hg and liquid Ga. Their data could not be well fitted by an elasto-optic scattering mechanism but were

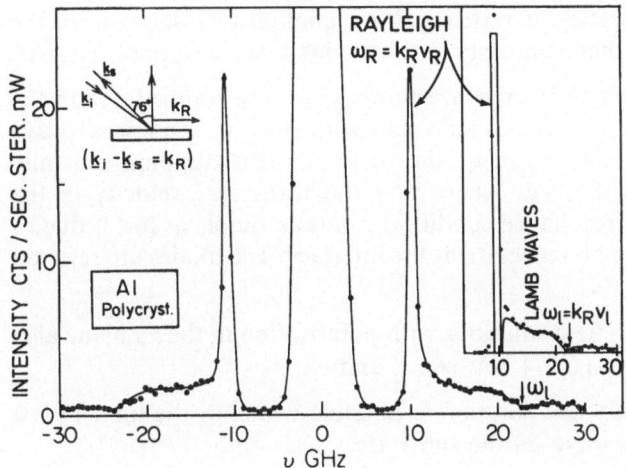

Fig. 6.9. Brillouin back-scattering spectrum of polycrystalline Al taken from [6.69]. Inset shows the theoretical fit to the data [6.62]

later shown to be well described by the ripple associated with the bulk longitudinal phonons [6.56, 68]. More recently, measurements from solid metal surfaces have been made [6.69] using the tandem multipass interferometer. The measurements of Al reproduced in Fig. 6.9 show the complete spectrum of scattering from the Rayleigh surface wave, Lamb waves and bulk waves in excellent accord with the calculated spectrum of (6.51). The shoulder in the spectrum due to scattering from Lamb waves can only be clearly observed using the extended free spectral range of the tandem interferometer. Measurements of the Rayleigh wave have just been reported by *Senn* et al. [6.70] in α-Si, by *Harley* and *Fleury* [6.71] in some layer compounds, and by *Chang* et al. [6.72] in metallic glasses. *Vacher* et al. [6.73] have reported lifetime measurements of the Rayleigh wave.

It should be noted that the lineshape derived in (6.51) assumes a thermal equilibrium phonon distribution. For nonequilibrium situations, for example, the external generation of an intense phonon beam q, the spectrum will be a sharp peak at energy vq, provided that (6.49) is satisfied. An example of non-equilibrium scattering can be seen in the experiments of *Mishra* and *Bray* [6.61] on intense phonon beams in GaAs.

Some attempts have been made to verify the reflectivity dependence of the ripple scattering by evaporation of a thin Al film onto less reflective substrates. Interpretation of such measurements is complicated by the fact that the film can seriously modify the excitation spectrum, as discussed below.

6.3.5 Modes of a Plate on a Half Space

Assuming the lower surface of the plate to be the interface with the half space, the reflection coefficient for phonons at the lower plate surface is now no longer unity. This fact seriously complicates the analysis since the transmitted waves must be included. This section is restricted to a qualitative discussion of the modes with emphasis on their similarity to the modes of the unsupported plate.

Love Modes. The solutions of interest are those which are trapped within the slab, corresponding to total internal reflection with only an evanescent wave extending into the substrate. It can be shown [6.57] that trapping can only occur for the SH waves if $v_t' > v_t$, where v_t' is the transverse velocity in the substrate. The transverse resonance condition is not as simple as (6.43) due to the phase change occuring on reflection at the interface. The modes are referred to as Love waves.

Generalised Lamb Waves. These solutions with polarisation in the xz plane also depend on the relationship [6.57] between v_t' and v_t.

a) $v_t' \ll v_t$. In this case, only one solution is possible which in the limit $\beta b \to 0$ becomes the Rayleigh wave on the substrate.

b) $v_t' \gg v_t$. Here there is an infinite number of solutions. As before there are symmetric and antisymmetric solutions which are here labeled M_{1i} and M_{2i}, respectively. Of special interest are the lowest order modes M_{11} and M_{21}. In the limit $\beta b \to 0$, only M_{11} is trapped and this corresponds to the Rayleigh wave on the bare substrate. As βb increases from zero, M_{21} is the next mode to be trapped. This is called the *Sezawa* [6.74] mode. In the limit $\beta b \to \infty$, the Sezawa mode becomes the Rayleigh mode on the top surface of the slab.

Under certain conditions for $v_t' \simeq v_t$ the mode M_{21} becomes a bound interface mode known as the *Stoneley* [6.75] wave. The Stoneley wave velocity v_S must satisfy the condition

$$v_R' < v_S < v_t',$$

where v_R' is the Rayleigh wave velocity on the bare substrate.

The extent to which these modes can scatter light is discussed in the next section.

6.3.6 Scattering from Films on Substrates

For scattering from supported metallic films, the important scattering mechanism will be from the surface ripple. For more transparent films, there will be, in addition, an elasto-optic contribution from the film and substrate, and a ripple contribution from the interface. Appropriate calculations have been carried out by *Bortolani* et al. [6.76, 77] for films of Al and Au on Si. The spectra show the expected structure due to scattering from Rayleigh waves and discrete Sezawa or Lamb modes, together with a continuum from the bulk modes. The reported experimental results [6.76, 77] show good agreement with the theory. The Al/Si measurements can be explained entirely in terms of the surface ripple; the

Au/Si measurements indicate an additional elasto-optic contribution. Some similar measurements on Al films on SiO_2 have recently been reported [6.78].

In an interesting and analogous experiment, *Rowell* and *Stegeman* [6.79] used the light guiding property of a transparent medium to observe scattering from thin film excitations.

The Love modes can also scatter light but not by the ripple mechanism since the Love waves are polarised in the plane of the surface and therefore do not have a ripple component. *Albuquerque* et al. [6.80] have calculated the scattered intensity from Love waves in SiO_2 on MgO as a function of film thickness. The scattered light is depolarised.

6.4 Scattering from Magnons in Opaque Materials

Measurements of light scattering from magnons in yttrium iron garnet (YIG) [6.81] demonstrated the usefulness of the light scattering technique and essentially confirmed the theory presented in Sect. 6.1.2a. In particular, the dependence of magnon frequency on wave vector and magnetic field was measured and the unusual asymmetry in the Stokes – anti-Stokes intensities predicted by (6.24) was observed. Although YIG is rather opaque in the visible, the effects of opacity only appear in the scattered intensity through the dissipative contributions [6.20] to K, M (6.24). Further measurements have been reported on $CoCO_3$, $CrBr_3$, $FeBO_3$ and $CoCO_3$. All these materials are, to some degree, transparent and as a result, the light probes only bulk excitations. A brief review of these earlier measurements has been given by *Borovik-Romanov* and *Kreines* [6.82].

More recently, some measurements on highly opaque ferromagnets have been reported in which both bulk and surface features have been observed. The high degree of opacity alters the form of the spectrum, as discussed below. It should be borne in mind, however, that the scattering only occurs via fluctuations in the dielectric constant. There is no surface ripple associated with magnons.

6.4.1 Light Scattering in Highly Opaque Magnets

The calculation of the scattered intensity proceeds as for phonons, whereby the high optical absorption α leads to a q dependent cross-section as in expressions (6.35, 41). Again, the coherent reflection of the spin wave at the surface must be taken into account with the difference from the phonon case, however, that there may be a phase change on reflection, depending on the degree of pinning of the spins at the surface. Thus, the spin wave reflection coefficient is $+1$ in the absence of pinning ranging to -1 for complete pinning.

Using the dispersion relation 17, *Cottam* [6.83] has calculated the spectral line shape as a function of optical absorption, exchange constant and degree of

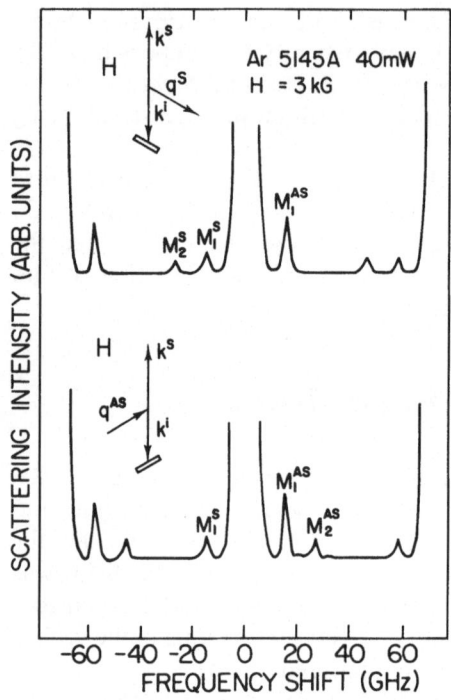

Fig. 6.10. Brillouin spectra of EuO showing surface (M_2) and bulk magnons (M_1), [6.85]

pinning. *Camley* and *Mills* [6.84] have performed numerical calculations including the scattering from the Damon–Eshbach surface mode. Remembering that for backscattering the Stokes scattering involves emission of a spin wave in the same direction as the incident light, while for antistokes the spin-wave direction is reversed, it is apparent from the discussion in Sect. 6.1.2 and Fig. 6.1 that the Damon-Eshbach mode will only appear on one side of the spectrum depending on the orientation of M_s.

This strange asymmetry in the spectrum for scattering from the Damon–Eshbach mode was first observed by *Grunberg* and *Metawe* [6.85] in measurements on EuO. A spectrum is reproduced in Fig. 6.10 where the Damon–Eshbach mode is seen only in the Stokes spectrum. A puzzling feature of these results is that the observed spin-wave energies lie rather far from the expected positions. This discrepancy has not yet been explained. The linewidths are narrow as expected for the low value of the exchange constant D.

Measurements in Fe and Ni [6.86] show an asymmetry similar to that in EuO, but due to the higher exchange constant, the bulk wave peaks become appreciably broadened, as shown in Fig. 6.11, which gives a series of spectra for different propagation directions relative to the applied field. As the propagation direction θ decreases towards the critical angle for surface mode propagation, the surface mode is seen to decrease in energy. The increasing linewidth arises due to the increase in the exchange induced decay into bulk

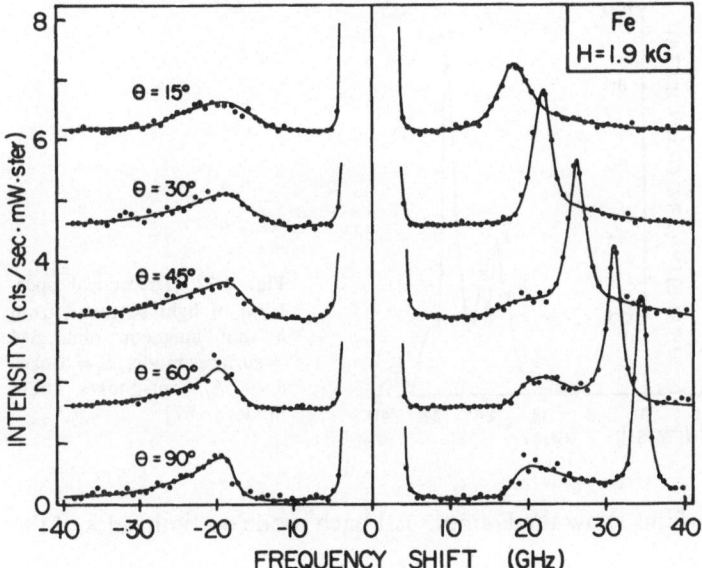

Fig. 6.11. Brillouin spectra of Fe showing variation of bulk and surface magnon peaks as a function of propagation direction. θ is the angle between M_s and q [6.86]

magnons. Notice that the bulk mode peaks for Stokes and anti-Stokes are dissimilar. Equation (6.24) predicts such an effect due to lack of time-reversal symmetry. However, even in the limit that $K \gg G$ or $K \ll G$ when (6.24) predicts no assymmetry, the calculations of *Camley* et al. [6.87] show that an asymmetry nevertheless persists for scattering from opaque materials. This asymmetry arises from the lack of reflection symmetry of a magnon at the surface and is the same asymmetry responsible for the nonreciprocity in the Damon–Eshbach mode propagation.

A series of measurements on sputtered films of magnetic alloys have been reported by *Malozemoff* et al. [6.88] in which the usefulness of the light scattering method for determining g-factors and magnetisation is demonstrated. On the thinnest films, an interesting structure has been observed in the spectra, as discussed below.

6.4.2 Light Scattering from Thin Magnetic Films

For thin magnetic films, the bulk spin-wave modes become strongly quantized in the z direction. *Camley* et al. [6.87] have performed calculations for the situation where the film thickness and optical penetration are comparable. From these calculations they draw several conclusions:

a) light scattered intensity depends strongly on the degree of surface pinning – upper and lower surface pinning have different effects on the spectra;

b) modes seen on the same side of the spectrum as the Damon–Eshbach mode have intensities which fluctuate strongly as a function of q_z;

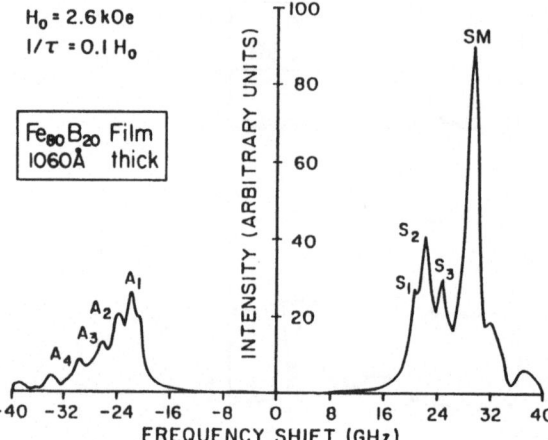

Fig. 6.12. Theoretical spectrum of light scattered from a thin magnetic film. SM = surface mode, S_i = Stokes and A_i anti-Stokes bulk modes [6.87]

c) sufficiently thin films show the Damon–Eshbach mode on both sides of the spectrum;

d) pinning has the largest effect on the lowest order (low q_z) modes.

A typical spectrum calculated by *Camley* et al. [6.87] is reproduced in Fig. 6.12. This spectrum is in good agreement with the measurements reported by *Grimsditch* et al. [6.89] on ~ 1000 Å films of $Fe_{80}B_{20}$.

6.5 Scattering from Diffusive Excitations

The derivation of the *Brillouin* scattered intensity in Sect. 6.1.1 assumed scattering only from strain-dependent fluctuations in the dielectric constant. In fact, temperature or entropy fluctuations may also modulate the dielectric constant, as was pointed out by *Einstein* [6.90] in his discussion of critical opalescence.

Consider a simple fluid described by the independent thermodynamic variables ϱ and T. A fluctuation $\delta\varepsilon$ may be expressed as

$$\delta\varepsilon = \left(\frac{\partial\varepsilon}{\partial\varrho}\right)_T \delta\varrho + \left(\frac{\partial\varepsilon}{\partial T}\right)_\varrho \delta T. \tag{6.52}$$

In most liquids, the term $\left(\frac{\partial\varepsilon}{\partial T}\right)_\varrho$ is rather small and so the fluctuation $\delta\varepsilon$ arises predominantly through the density dependence of ε. In terms of the variables entropy S and pressure P, the fluctuation $\delta\varepsilon$ may then be approximated as [2.34]:

$$\delta\varepsilon \simeq \left(\frac{\partial\varepsilon}{\partial\varrho}\right)_T \left[\left(\frac{\partial\varrho}{\partial P}\right)_S \delta P + \left(\frac{\partial\varrho}{\partial S}\right)_P \delta S\right]. \tag{6.53}$$

The first term in the bracket describes the scattering from adiabatic sound waves as discussed above (Brillouin scattering from acoustic phonons), while the second term describes scattering from entropy fluctuations. Since entropy fluctuations are normally overdamped, this latter contribution usually appears as a peak centered around $\omega = 0$ with a width Γ determined by the thermal diffusivity

$$\Gamma(\text{FWHM}) = 2Dq^2. \tag{6.54}$$

Landau and *Placzek* [6.91] calculated the ratio R_{LP} of the scattered intensity in the central peak I_c to that in the Brillouin peak I_B:

$$R_{LP} = \frac{I_c}{I_B} = \frac{C_p}{C_v} - 1. \tag{6.55}$$

This is a good approximation for most liquids. In a solid, $C_p \simeq C_v$ and so the above derivation gives $R_{LP} \sim 0$. It was shown by *Wehner* and *Klein* [6.92], however, that the direct term $(\partial\varepsilon/\partial T)_\varrho$ in a solid is by no means negligible (for a discussion of this term in diamond-like semiconductors, see [6.93]). Inclusion of this term yields a new ratio R_{WK} for the central to Brillouin intensity where

$$R_{WK} = R_{LP}\left[1 + \frac{C_p}{\alpha C_v} \cdot \frac{(\partial\varepsilon/\partial T)_\varrho}{(\partial\varepsilon/\partial\varrho)_S}\right]^2 \tag{6.56}$$

with the conclusion that R_{WK} for solids may be comparable to the value observed in liquids (α is the volume coefficient of thermal expansion).

On a microscopic scale, the indirect scattering process discussed by *Landau* and *Placzek* [6.91] describes the interaction of light with a single phonon q which, via anharmonicity, couples to pairs of phonons q_1 and $q_1 - q$.

The entropy fluctuations appear as fluctuations in the phonon density of the phonons q_1. In the microscopic picture, the Landau–Placzek scattering therefore describes the indirect coupling of light to phonon-density fluctuations. The direct mechanism of *Wehner* and *Klein* [6.92] involving the term $(\partial\varepsilon/\partial T)_\varrho$ involves the direct coupling of light to pairs of phonons q_1 and $q_1 - q$ and so describes the direct coupling of light to phonon-density fluctuations. Both direct and indirect mechanisms in this approximation, which assumes at all times local thermodynamic equilibrium, lead to a single central peak with a width given by expression (6.54).

That this is not the whole story was pointed out by *Klein* [6.93]. The phonon-density fluctuations cannot be described solely by fluctuating regions in local thermodynamic equilibrium but rather the additional "dielectric" fluctuations of *Cowley* and *Coombs* [6.95] must be included. These dielectric fluctuations are fluctuations in phonon density away from local thermodynamic equilibrium. This higher frequency regime cannot be described by

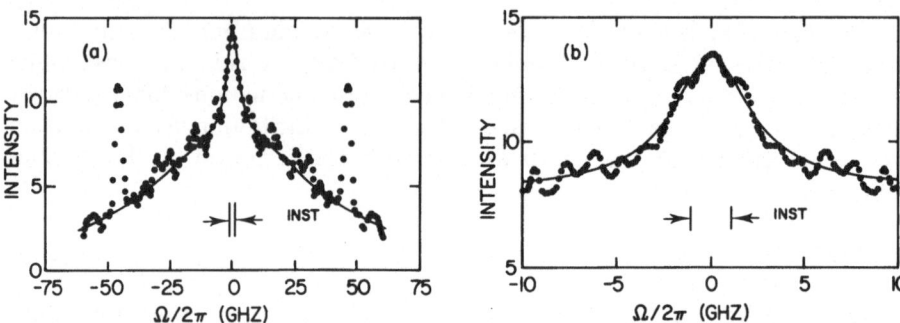

Fig. 6.13a, b. Measurements of the double central peak structure in KaTO$_3$ [6.34]. The narrow mode is shown at higher resolution in (**b**)

phonon transport theory, but the scattering process can rather be regarded as a two-phonon difference process involving the emission of a phonon q_1 with simultaneous absorption of a phonon $q_1 - q$ from the same phonon branch. Such a process is dominated by phonons q_1 near the zone boundary where the density of states is high and leads to a broad central peak extending out to the Brillouin peaks. If the lifetime of the zone boundary phonons is short ($< \omega^{-1}$ where ω is the Brillouin frequency), then the central peak may be broadened to beyond the Brillouin peaks. The linewidth of this peak arising from the dielectric fluctuations should be essentially independent of q.

Fleury and *Lyons* [6.34, 96] have performed measurements on a variety of materials and have observed this predicted double central peak. Their results most relevant to the present discussion are those on SrTiO$_3$ and KTaO$_3$ [6.34]. The latter measurements are reproduced in Fig. 6.13 and show the central mode structure as observed far away from the phase transition. Considerable structure is still visible in the spectra arising from the use of the iodine filter to remove the elastic scattered light. The central peaks show no singular behaviour on approaching the incipient ferroelectric phase transition (at $\simeq 0\,\mathrm{K}$), indicating no dependence of the central peak on the ferroelectric ordering. The narrow peak has a width in good agreement with that calculated from (6.54). The q dependence of the linewidth is not inconsistent with a q^2 behaviour, although this has not been conclusively demonstrated. The broad peak is found to be q independent as expected but the temperature dependence of the width and intensity is not that expected for a difference process involving two zone-boundary phonons. It is interesting to note that *Field* et al. [6.97] have reported a similar double central peak in the superionic conductor RbAg$_4$I$_5$. They explain their results in terms of a model specific to a superionic conductor. It is interesting to conjecture that a more general mechanism may be involved, particularly since the double peak structure has been observed in a variety of different materials.

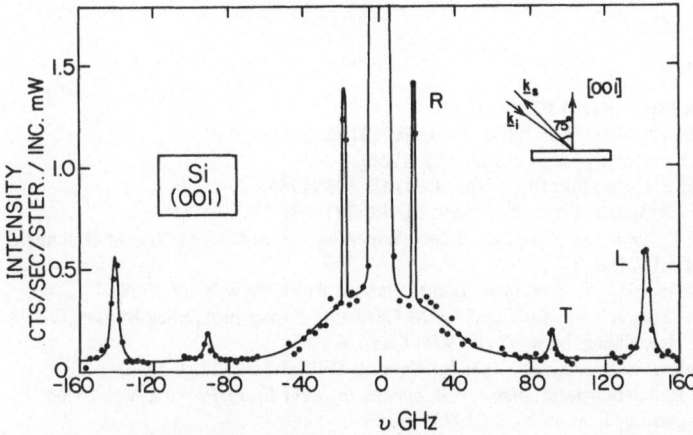

Fig. 6.14. Backscattering spectrum of Si [6.69] showing bulk (T, L) and Rayleigh surface phonons (R) together with a central peak

A final example is presented in Fig. 6.14 for scattering from Si [6.69]. This shows the spectrum of Fig. 6.6a as measured with the tandem multipass interferometer. A central peak is clearly seen. *Wehner* and *Klein* [6.92] predicted a strong diffusion mode in this material, however, the width observed ($\sim 60\,\text{GHz}$) is considerably smaller than that predicted by expression (6.53) ($\sim 160\,\text{GHz}$). The observed width is close to what would be expected for a difference process involving two transverse phonons, in which case the absence of the diffusive peak must be explained.

6.6 Conclusions

It has been the intention of this chapter to present theoretical and experimental results to demonstrate the wide range of application of Brillouin scattering in opaque materials. The high optical absorption and inherent heating due to the incident laser beam implies that light scattering is not a good tool to study phase transitions in these materials. As a result, the whole topic of phase transitions has not been touched upon although this would have been particularly relevant to the central modes [6.96] discussed in the previous section. Despite the local heating, light scattering may, nonetheless, offer a means of studying the onset of phase transitions. In this respect, it is interesting to note that a soft bulk mode implies the existence of a soft surface mode [6.98]. *Kragler* [6.99] has proposed studying the transition in Nb_3Sn by measuring the softening of the appropriate Rayleigh mode.

Acknowledgement. It is a pleasure to thank Dr. D. Baeriswyl for many helpful discussions.

References

6.1   L. Brillouin: Ann. Phys. (Paris) **17**, 88 (1922)
6.2   L. I. Mandelshtam: Zh. Russ. Fiz. Khim. Ova. **58**, 381 (1926)
6.3   G. B. Benedek, K. Fritsch: Phys. Rev. **149**, 647 (1966)
6.4   R. W. Gammon, H. Z. Cummins: Phys. Rev. Lett. **17**, 193 (1966)
6.5   E. M. Brody, H. Z. Cummins: Phys. Rev. Lett. **21**, 1263 (1968)
6.6   J. R. Sandercock: In *2nd Int. Conf. on Light Scattering in Solids*, ed. by M. Balkanski (Flammarion, Paris 1971) p. 9
6.7   I. L. Fabelinskii: *Molecular Scattering of Light* (Plenum Press, New York 1968)
6.8   A. S. Pine: In *Light Scattering in Solids*, ed. by M. Cardona, Topics in Applied Physics, Vol. 8 (Springer, Berlin, Heidelberg, New York 1975) Chap. 6
6.9   W. Hayes, R. Loudon: *Scattering of Light by Crystals* (Wiley, New York 1978)
6.10   J. R. Sandercock: In *Festkörperprobleme – Advances in Solid State Physics*, Vol. 15 ed. by H. J. Queisser (Pergamon, Braunschweig 1975) p. 183
6.11   L. Landau, G. Placzek: Phys. Z. Sowjetunion **5**, 172 (1934)
6.12   L. D. Landau, E. M. Lifshitz: *Electrodynamics of Continuous Media* (Pergamon Press, New York 1958) Chap. 14
6.13   R. Vacher, L. Boyer: Phys. Rev. B**6**, 639 (1972)
6.14   R. W. Dixon: IEEE J. QE-3, 85 (1967)
6.15   H. Küppers: Phys. Status Solidi. **37**, K 59 (1970)
6.16   O. Keller, C. Sondergaard: Jpn. J. Appl. Phys. **13**, 1765 (1974)
6.17a D. F. Nelson, M. Lax: Phys. Rev. Lett. **24**, 379 (1970)
6.17b M. Cardona, G. Güntherodt (eds.): *Light Scattering in Solids II*, Topics in Applied Physics, Vol. 50 (Springer, Berlin, Heidelberg, New York 1982)
6.17c C. Kittel: *Quantum Theory of Solids* (Wiley, New York 1966) p. 67
6.18   R. W. Damon, J. R. Eshbach: J. Phys. Chem. Solids **19**, 308 (1961)
6.19   P. A. Fleury, R. Loudon: Phys. Rev. **166**, 514 (1968)
6.20   W. Wettling, M. G. Cottam, J. R. Sandercock: J. Phys. C**8**, 211 (1975)
6.21   H. Le Gall, J. P. Jamet, B. Desormiere: In *2nd Int. Conf. on Light Scattering in Solids*, ed. by M. Balkanski (Flammarion, Paris 1971) p. 170
6.22   P. Jacquinot: Rep. Prog. Phys. **23**, 268 (1960)
6.23   J. R. Sandercock: In *2nd Int. Conf. on Light Scattering in Solids*, ed. by M. Balkanski (Flammarion, Paris 1971) p. 9
6.24   S. M. Lindsay, I. W. Shephard: J. Phys. E**10**, 150 (1977)
6.25   C. Roychoudhuri, M. Hercher: Appl. Opt. **16**, 2514 (1977)
6.26   M. Hercher: Appl. Opt. **7**, 951 (1968)
6.27   D. S. Cannell, J. H. Lunacek: Rev. Sci. Instrum. **44**, 1651 (1973)
6.28   J. R. Sandercock: U.S. Patent 4,014,614 (1977)
6.29   R. Vacher, H. Sussner, M. v. Schickfuss: Rev. Sci. Instrum. **51**, 288 (1980)
6.30   W. Proffitt, L. M. Fraas, P. Cervenka, S. P. S. Porto: Appl. Opt. **10**, 531 (1971)
6.31   G. E. Devlin, J. L. Davis, L. Chase, S. Geschwind: Appl. Phys. Lett. **19**, 138 (1971)
6.32   P. E. Schoen, D. A. Jackson: J. Phys. E**5**, 519 (1972)
6.33   K. Lyons, P. A. Fleury: J. Appl. Phys. **42**, 4898 (1976)
6.34   K. Lyons, P. A. Fleury: Phys. Rev. Lett. **37**, 161 (1976)
6.35   R. Y. Chiao, B. P. Stoicheff: J. Opt. Soc. Am. **54**, 1286 (1964)
6.36   D. A. Jackson, E. R. Pike: J. Phys. E**1**, 394 (1968)
6.37   S. Fray, F. A. Johnson, R. Jones, S. Kay, C. J. Oliver, E. R. Pike, J. P. Russell, C. Senett, J. O'Shaughnessy, C. Smith: In *Light Scattering Spectra of Solids*, ed. by G. B. Wright (Springer, Berlin, Heidelberg, New York 1969) p. 139
6.38   D. Bechtle: Rev. Sci. Instrum. **47**, 1377 (1976)
6.39   J. R. Sandercock: J. Phys. E**9**, 566 (1976)

6.40   A. Asenbaum: Appl. Opt. **18**, 540 (1979)
6.41   M. Yamada, H. Ikeshima, Y. Takahashi: Rev. Sci. Instrum. **51**, 431 (1980)
6.42   D. S. Cannell, G. B. Benedek: Phys. Rev. Lett. **25**, 1157 (1970)
6.43   J. G. Dil, N. C. J. A. van Hijningen, F. van Dorst, R. M. Aarts: Appl. Opt. **20**, 1374 (1981)
6.44   R. V. Jones, I. R. Young: J. Sci. Instrum. **33**, 11 (1956)
6.45   D. J. Bradley: J. Sci. Instrum. **39**, 41 (1962)
6.46   S. E. Gustafsson, M. N. Khan, J. van Eijk: J. Phys. E **12**, 1100 (1979)
6.47   J. R. Sandercock: US Patent 4,225,236 (1980)
6.48   J. R. Sandercock: Phys. Rev. Lett. **28**, 237 (1972)
6.49   R. K. Wehner: Opt. Commun. **6**, 174 (1972)
6.50   A. S. Pine: Phys. Rev. B**5**, 3003 (1972)
6.51   R. E. Benner, E. M. Brody, H. R. Shanks: J. Solid. State Chem. **22**, 361 (1977)
6.52   G. Dresselhaus, A. S. Pine: Solid State Commun. **16**, 1001 (1975)
6.53   A. Dervisch, R. Loudon: J. Phys. C**9**, L669 (1976)
6.54   R. Loudon: J. Phys. C**11**, 2623 (1978)
6.55   K. R. Subbaswami, A. A. Maradudin: Phys. Rev. B**18**, 4181 (1978)
6.56   N. L. Rowell, G. I. Stegeman: Phys. Rev. B**18**, 2598 (1978)
6.57   B. A. Auld: *Acoustic Fields and Waves in Solids*, Vol. 2 (Wiley, New York 1973)
6.58   H. Lamb: Proc. R. Soc. London A**93**, 114 (1917)
6.59   Lord Rayleigh: London Math. Soc. Proc. **17**, 4 (1887)
6.60   L. Bergman: *Ultrasonics* (Bell, London 1938) p. 164
6.61   S. Mishra, R. Bray: Phys. Rev. Lett. **39**, 222 (1977)
6.62   R. Loudon: Phys. Rev. Lett. **40**, 581 (1978)
6.63   R. Loudon, J. R. Sandercock: J. Phys. C**13**, 2609 (1980)
6.64   V. R. Velasco, F. Garcia-Moliner: Solid State Commun. **33**, 1 (1980)
6.65   A. M. Marvin, V. Bortolani, F. Nizzoli: J. Phys. C**13**, 299 (1980)
6.66   A. M. Marvin, V. Bortolani, F. Nizzoli: J. Phys. C**13**, 1607 (1980)
6.67   J. G. Dil, E. M. Brody: Phys. Rev. B**14**, 5218 (1976)
6.68   A. Dervisch, R. Loudon: J. Phys. C**11**, L291 (1978)
6.69   J. R. Sandercock: Solid State Commun. **26**, 547 (1978)
6.70   W. Senn, G. Winterling, M. Grimsditch, M. Brodsky: Physics of Semiconductors 1978, Inst.
       Phys. Conf. Ser. No. 43, p. 709
6.71   R. T. Harley, P. A. Fleury: J. Phys. C**12**, L863 (1979)
6.72   P. H. Chang, A. P. Malozemoff, M. Grimsditch, W. Senn, G. Winterling: Solid State Commun.
       27, 617 (1978)
6.73   R. Vacher, H. Sussner, M. Schmidt: Solid State Commun. **34**, 279 (1980)
6.74   K. Sezawa: Bull. Earthquake Res. Inst. **3**, 1 (1927)
6.75   R. Stoneley: Proc. R. Soc. London A**232**, 447 (1955)
6.76   V. Bortolani, F. Nizzoli, G. Santoro, J. R. Sandercock, A. M. Marvin: In *Proc. VIIth Intern.
       Conf. on Raman Spectroscopy*, ed. by W. F. Murphy (North-Holland, New York 1980) p. 442
6.77   V. Bortolani, F. Nizzoli, G. Santoro, A. Marvin, J. R. Sandercock: Phys. Rev. Lett. **43**, 224
       (1979)
6.78   H. Sussner, J. Pelous, M. Schmidt, R. Vacher: Solid State Commun. **36**, 123 (1980)
6.79   N. L. Rowell, G. I. Stegeman: Phys. Rev. Lett. **41**, 970 (1978)
6.80   E. L. Albuquerque, R. Loudon, D. R. Tilley: J. Phys. C**13**, 1775 (1980)
6.81   J. R. Sandercock, W. Wettling: Solid State Commun. **13**, 1729 (1973)
6.82   A. S. Borovik-Romanov, N. M. Kreines: J. Magn. Magn. Mater. **15–18**, 760 (1980)
6.83   M. G. Cottam: J. Phys. C**11**, 165 (1978)
6.84   R. E. Camley, D. L. Mills: Phys. Rev. B**18**, 4821 (1978)
6.85   P. Grunberg, F. Metawe: Phys. Rev. Lett. **39**, 1561 (1977)
6.86   J. R. Sandercock, W. Wettling: J. Appl. Phys. **50**, 7784 (1979)
6.87   R. E. Camley, T. S. Rahman, D. L. Mills: Phys. Rev. B**23**, 1226 (1981)
6.88   A. P. Malozemoff, M. Grimsditch, J. Aboaf, A. Brunsch: J. Appl. Phys. **50**, 5885 (1979)

6.89   M. Grimsditch, A. Malozemoff, A. Brunsch: Phys. Rev. Lett. **43**, 711 (1979)
6.90   A. Einstein: Ann. Phys. **33**, 1275 (1910)
6.91   L. D. Landau, G. Placzek: Phys. Z. Sowjetunion **5**, 172 (1934)
6.92   R. K. Wehner, R. Klein: Physica **62**, 161 (1972)
6.93   M. Cardona: In *Atomic Structure and Properties of Solids*, ed. by E. Burstein (Academic Press, New York 1972) p. 514
6.94   R. Klein: Proc. NATO Study Inst. *Anharmonic Lattices, Structural Transitions and Melting*, ed. by T. Riste (Nordhoff, Leiden 1974) p. 161
6.95   R. A. Cowley, G. J. Coombs: J. Phys. C**6**, 121 (1973)
6.96   P. A. Fleury, K. B. Lyons: Solid State Commun. **32**, 103 (1979)
6.97   R. A. Field, D. A. Gallagher, M. V. Klein: Phys. Rev. B**18**, 2995 (1978)
6.98   L. Bjerkan, K. Fossheim: Solid State Commun. **21**, 1147 (1977)
6.99   R. Kragler: Solid State Commun. **35**, 429 (1980)

7. Resonant Light Scattering Mediated by Excitonic Polaritons in Semiconductors

By C. Weisbuch and R. G. Ulbrich

With 32 Figures

Excitonic polaritons are coupled-mode excitations made up from excitons interacting with photons (for an overview of the field of polaritons in 1972 and reprints of excellent selected key papers, see [7.1]). They represent the elementary excitations propagating in insulators or semiconductors with frequencies in the vicinity of the fundamental absorption edge. Although many aspects of these excitations such as absorption [7.2–6], luminescence [7.7–11], nonlinear optics [7.12, 13], and reflectivity [7.14–16] have been studied in the past, they still appeared to many semiconductor physicists much more as speculative constructions, rather than being the true excitations of the crystal which must be used for adequate descriptions of optical experiments. The recent advent of tunable lasers stimulated a very thorough investigation of these excitations and lead to the elucidation of several key features of excitonic polaritons. In particular, resonant Brillouin scattering (RBS) [7.17, 18] and hyper-Raman scattering [7.19] experiments allowed the measurement of the polariton dispersion curve in a large number of semiconductors and demonstrated the quasi-ubiquity of exciton polariton phenomenona in reasonably pure, direct-gap semiconductors at sufficiently low temperatures.

Besides being a unique tool for the study of excitonic polaritons themselves, experiments on light scattering mediated by excitonic polaritons have a number of features which set them quite apart from other light scattering methods: the isolated, extremely sharp resonance behavior facilitates the clear-cut distinction of different polariton-crystal coupling mechanisms. The enormous cross sections render the usually stringent laser light-rejection requirements much easier to fulfill. Commercially available double or even single [7.20] grating monochromators can be used for RBS measurements, which can thus be pursued in almost any laboratory. More sophisticated multiple-pass Fabry-Perot interferometers have to be used in the case of nonresonant BS [7.21] because of its inherent low signal levels.

Theoretical considerations on exciton-polariton mediated light scattering came after a very natural development of the studies of resonant light scattering. Uncorrelated electron-hole pairs were first considered by *Loudon* [7.22, 23]: he showed that the *indirect* mediation of phonon light scattering by electron-hole pairs in a *three-step process* is more efficient than the *direct two-step* iterated photon–phonon interaction. The influence of sample opacity was considered in a simplified way, and we shall see below that great care must actually be exercised when absorption corrections are to be made. *Ganguly* and

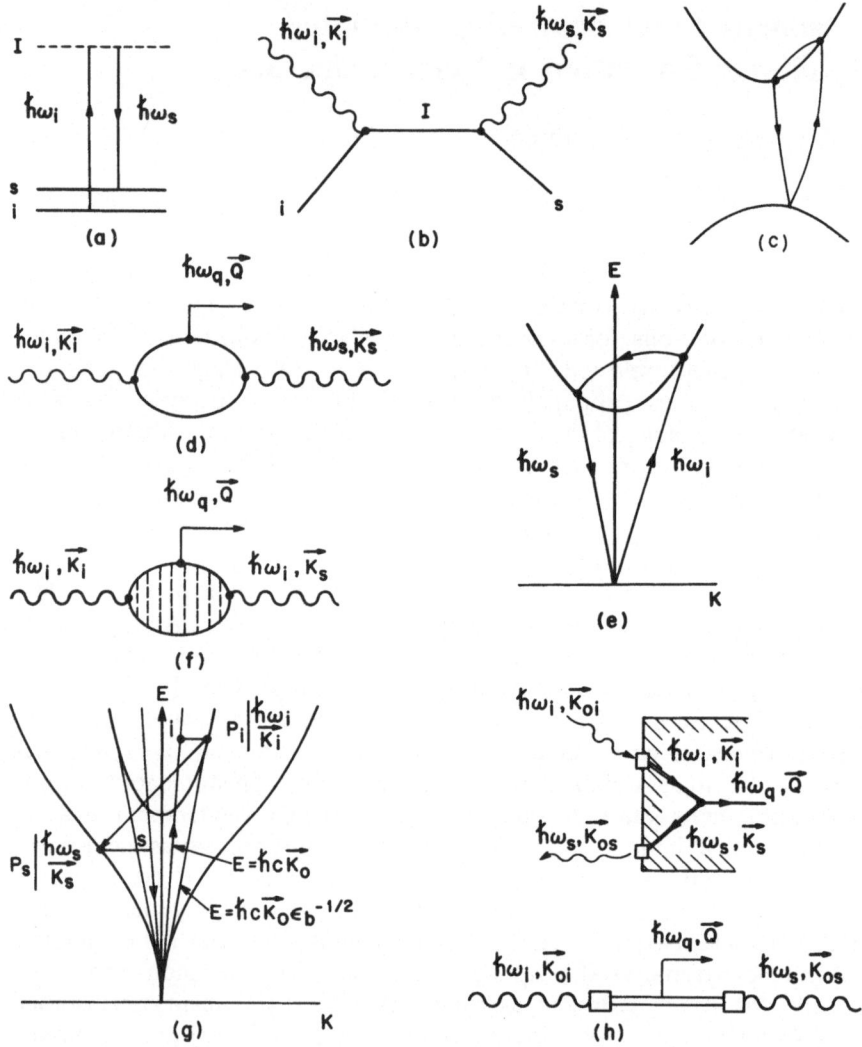

Fig. 7.1a–h. Various descriptions of light scattering by phonons in insulators or semiconductors: (a) energy levels of initial (i), intermediate (I) and final (s) state and energies $\hbar\omega_i$ and $\hbar\omega_s$ of incident and scattered photons; (b) schematic diagram with two vertices which represent *direct* iterated *photon–phonon* interactions; (c), (d) *indirect* photon-phonon interaction, mediated by electron-hole pairs: the incoming photon creates an electron-hole pair (real or virtual), and one of the two particles interacts with the lattice-phonon via the electron-phonon interaction (only the e-phonon vertex is shown). The process is now treated in a *third*-order perturbation scheme, as compared to the second-order perturbations depicted in (a), (b); (e), (f) more refined scheme based on a pair-description of the excited electronic state. The dashed lines indicate e–h-pair correlation due to Coulomb interaction (excitonic effects); (g), (h) diagonalization of the exciton-photon coupling and consideration of the exciton-phonon interaction as a perturbation lead to the polariton description of exciton-mediated light scattering. Excitonic polariton, and vacuum (resp. background ε_b "dressed") photon dispersion curves are shown together with diagrams stressing the first-order character of light scattering in this description. ($\sim\!\!\sim\!\!\sim$ photon in vacuum, □ conversion of photon into polariton state, ══ excitonic polariton, ⟶ phonon)

Birman [7.24] included the electron–hole correlation in the intermediate state. This led to the exciton description of the intermediate crystalline elementary excitation. Finally, the exact diagonalization of the exciton–photon interaction leading to polariton states was carried out [7.25–28]. This development can be schematized in a diagrammatic representation as in Fig. 7.1a–h.

7.1 Outline

The recent upsurge in activity is due to the appearance of convenient tunable sources (mainly cw dye lasers) in the spectral range of the high-purity and best characterized directgap semiconductors, GaAs and CdS. This has led (i) to a recent development of the literature in a field which was already widely documented, and (ii) to the appearance of numerous excellent recent review articles [7.29–31]. We will, therefore, not attempt to review the whole subject of polaritons or try to give a complete bibliography of the subject. We shall rather present our own perception of the field of light scattering mediated by excitonic polaritons. It will be the point of view of experimentalists, and the corresponding limited set of references (although amounting to ~200), should be considered in that context. Very useful and more detailed reviews one some aspects presented here are available in the various chapters of a treatise on excitons to appear soon [7.29]. The recent review by *Bendow* [7.30] of resonant Raman scattering (RRS) mediated by excitonic polaritons provides a comprehensive theoretical background to the descriptive approach adopted here. A detailed account of exciton–phonon coupling was recently given by *Yu* [7.31]. General information on light-scattering phenomena in solids can be found in [7.32, 33]. Useful, more elementary treatments were given by *Cummins* [7.34], *Walker* [7.35], and *Mooradian* [7.36].

The discussion of polariton-mediated light scattering will be made along the following lines: Sect. 7.2 will expose the basics of exciton–polariton phenomena: excitonic energy levels and quantum states, exciton interactions, and diagonalization of the exciton–photon interaction into exciton–polariton states. As the specific *optical* properties due to the exciton–polariton resonance are of paramount importance to light scattering phenomena, we also describe some recent results on polariton reflection, propagation and transmission. We then proceed in Sect. 7.3 to give a simple description of light scattering in the polariton framework. The concepts are then illustrated in the following sections on Brillouin scattering (Sect. 7.4), one-phonon Raman scattering (Sect. 7.5), multiphonon scattering and polarization effects (Sect. 7.6), nonlinear phenomena (Sect. 7.7), and electronic Raman scattering (Sect. 7.8). We conclude in Sect. 7.9 by summarizing the results obtained in these resonant light scattering experiments and by stressing remaining interpretational problems and future developments.

7.2 Basic Properties of Excitonic Polaritons

7.2.1 Static Properties of Excitons

Excitons are electronic excitations of pure (i.e., intrinsic) semiconductors and insulators [7.37, 38] consisting of Coulomb-correlated electron–hole pairs. For most semiconductors, the lowest of these excitations occur with their minimum energy at low wave vectors, typically near the Brillouin zone center. The energy levels are readily calculated in the *isotropic, parabolic, nondegenerate* band case to give (Fig. 7.2)

$$E_n(K) = E_G - \frac{R^*}{n^2} + \frac{\hbar^2 K^2}{2M_X} = E_{n,\,T} + \frac{\hbar^2 K^2}{2M_X}. \tag{7.1}$$

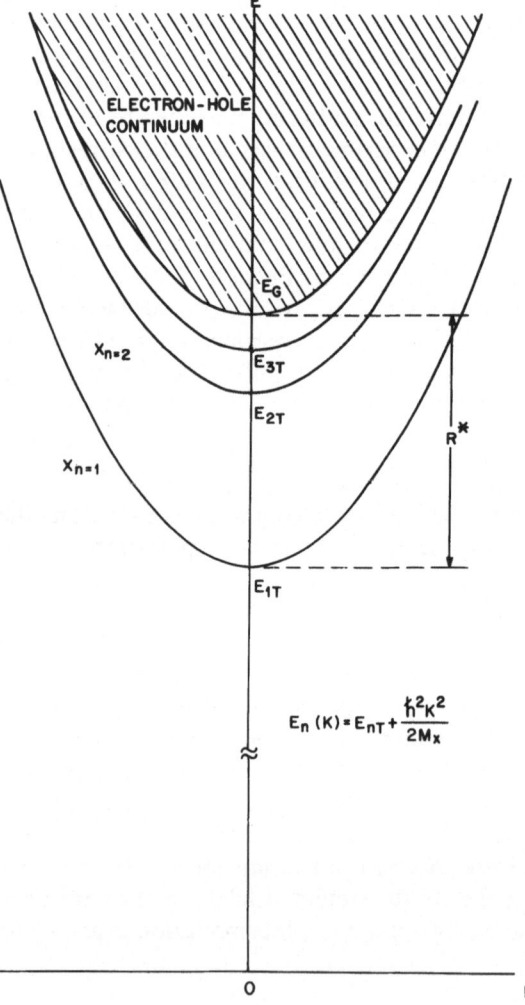

$$E_n(K) = E_{nT} + \frac{\hbar^2 K^2}{2M_X}$$

Fig. 7.2. Exciton energy levels E_n as a function of exciton total wave vector K. n is the principal quantum number for discrete states resulting from the (hydrogen-like) *relative* motion of electron and hole, R^* is the corresponding Rydberg energy. The electron-hole continuum represents unbound (but still correlated) e–h-pairs with total energy above the energy gap E_G

Here n is the hydrogenic exciton principal quantum number and $E_{n,\text{T}}$ the nth transverse exciton energy. It is assumed that the effective-mass approximation hold, i.e., that the Rydberg energy R^* is small compared to the bandgap energy E_G. R^* is given by (SI units)

$$R^* = \frac{\mu_0 e^4}{2(4\pi\varepsilon_0)^2 \hbar^2 \varepsilon^2} = \frac{e^2}{2(4\pi\varepsilon_0)\varepsilon a_\text{B}}, \tag{7.2}$$

where μ_0 is the reduced exciton mass ($\mu_0^{-1} = m_\text{e}^{-1} + m_\text{h}^{-1}$, with m_e and m_h the electron and hole effective masses), ε is the dielectric constant of the medium *at the Rydberg energy* and a_B the exciton Bohr radius. Depending on the relative sizes of R^* and the LO-phonon energy, one uses as ε either the static dielectric constant ε_0 or the high-frequency dielectric constant ε_∞. The last term of (7.1) represents the translational kinetic energy of the exciton as a function of its wave vector K and translational mass $M_X = m_\text{e} + m_\text{h}$. Typical values for R^* range from a few meV's to tens of meV's, while the exciton Bohr radius ranges from 20 to 300 Å (see below Columns 5 and 6 of Table 7.1).

This two-band exciton model is a very convenient and useful approximation of exciton phenomena, but it does not describe in totality the complexity of the rich fine structure of excitonic levels:

The case of *single* bands is mainly encountered in anisotropic wurtzite crystals such as CdS, but there the *isotropic* approximation is usually poor. Due to the direction-dependent exciton parameters, the energy levels do not, in general, exactly form a hydrogenic series [7.39].

Even for single bands, *k-linear terms* (which are present in crystals without inversion symmetry) can eventually remove the spin degeneracy away from the zone center [7.40].

For *degenerate valence bands* (zinc-blende cubic semiconductors such as GaAs and CdTe), two types of holes, heavy and light, have to be included in the dynamical exciton problem. This leads to certain consequences: existence of "heavy" and "light" excitons, anisotropy of the exciton dispersion, nonparabolic dispersion of excitons (even though conduction and valence bands are parabolic), and nonhydrogenic series of excitonic levels [7.41–43] (Fig. 7.3).

Other effects also play a role in the correct description of exciton energy levels: corrections in the dielectric approximation have to be made, mainly in the exciton ground state [7.44]. The exciton–phonon interaction modifies the effective electron–hole interaction strength through polaron effects [7.44–46]. Finally, various exchange-type interactions between electron and holes must be taken into account [7.47–50].

The fine structure of excitons is therefore a rather wide subject which has given rise to a vast amount of literature. We want to address the reader to some recent reviews in the field ([Ref. 7.38, Chaps. 1 and 2], [7.50–52]) and point out that experimental studies were essentially limited until recently to reflectivity

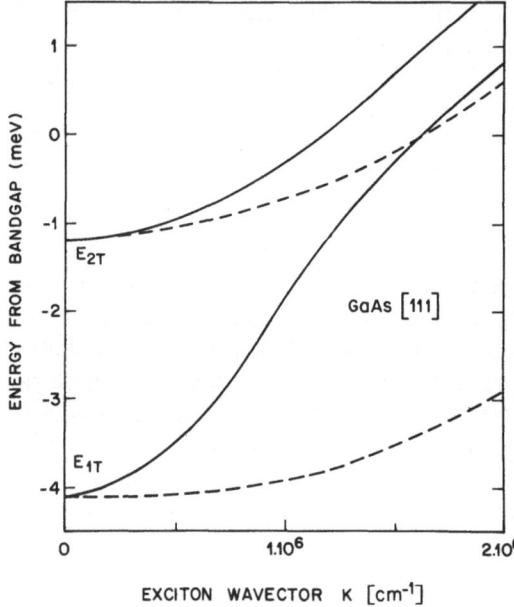

Fig. 7.3. Energy-wave vector dispersion curves for $n=1$ and $n=2$ excitons in a typical zinc-blende semiconductor with degenerate valence band edge (GaAs; $\Gamma_8 \times \Gamma_6$ exciton; after [7.43]). Nonparabolicity and nonhydrogenic level spacing are due to hole degeneracy. Full lines: "light" excitons; dashed lines: "heavy" excitons

and luminescence measurements under externally applied magnetic fields or mechanical stress ([7.53], for GaAs [7.10a], for CdS [7.54a], for ZnTe [7.54b], for ZnSe [7.54c], and for a review on magnetic field measurements [7.55]).

7.2.2 Dynamical Properties of Excitons

The analysis of light-scattering experiments requires some knowledge about the interactions of excitons with various excitations. Some of these interactions are by now well known, such as the exciton–photon [7.37, 56, 57], exciton–phonon [7.58, 59] and exciton-isolated shallow-impurity interactions (see, in general, [7.60]). Others are still unresolved, in particular those leading to the Urbach tail [7.61, 62] of absorption below the exciton resonance. Although some recent progress has been made in that field [7.63–65], a complete description of exciton dissipation mechanisms is still lacking. This fact sets severe limitations on *quantitative* evaluations of light scattering phenomena, as will be discussed below.

The dynamic properties of excitons depend crucially on the *hierarchy of couplings* [7.59]: if all the couplings of single-mode excitations with excitons are weak compared to incoherent perturbations of the excitonic states, then perturbation theory can describe the time evolution of excitonic states. If one coupling is much stronger than all others and than the incoherent perturbations, one has to consider the coupled-mode excitation resulting from this strong coupling. This leads to excitonic polaritons in the case of dominant exciton–photon coupling, and to excitonic polarons for dominant exciton–

phonon coupling. Several couplings can also be strong simultaneously. The situation which is the most common and will be dealt with here assumes the following hierarchy: (i) electron–hole interaction leading to exciton states, (ii) exciton–photon interaction leading to excitonic polaritons. All other interactions are supposed to be weak.

a) Exciton–Photon Interaction

The exciton–photon interaction has been theoretically treated by *Elliott* [7.56, 57]. Due to wave vector conservation, the interaction occurs (Fig. 7.4) in the ω, K plane at the crossing point (near E_T) between the exciton dispersion curve (the parabola $E_{n,T} + \hbar^2 K^2 / 2 M_X$) and the photon dispersion curve (straight line $\hbar c K / \sqrt{\varepsilon_b}$, where ε_b is the *"background"* *dielectric constant* at the exciton energy, taking all crystalline excitations into account but the exciton level under consideration). The strength of the exciton–photon interaction can be described equivalently by a number of interrelated quantities [7.37] such as the oscillator strength per unit cell f_{exc}, the exciton polarizability α or the longitudinal-transverse splitting E_{LT}. This last quantity is directly connected with measurable polariton properties. The exciton–photon interaction strength *for the ls-exciton ground state* [7.37, 57] can be expressed by

$$f_{exc} \approx \frac{1}{E_G} P^2 \frac{\Omega}{\pi a_B^3}, \tag{7.3}$$

$$4\pi\alpha \approx \frac{4\pi e^2}{m_0} \frac{1}{\Omega} \frac{f_{exc}}{E_G^2} = \frac{4\pi e^2 \hbar^2}{m_0 E_G^3} \frac{P^2}{\pi a_B^3}, \tag{7.4}$$

$$E_{LT} = \frac{2\pi\alpha}{\varepsilon_b} E_G \approx \frac{2\pi e^2 \hbar^2}{\varepsilon_b m_0 E_G^2} \frac{P^2}{\pi a_B^3}. \tag{7.5}$$

P^2 is the usual reduced interband matrix element ($P^2 = 2|\langle s|P|x\rangle|^2/m_0$) between s conduction band states and x valence band states, Ω the volume of the elementary cell, m_0 the free electron mass. The $\approx$ sign stands for the isotropic spherical hydrogenic approximation made in the evaluation of f_{exc} and $4\pi\alpha$.

Two limiting cases occur:

(i) weak exciton–photon coupling
Through the absorption process, a photon of energy ω and momentum K is transformed into an exciton ω, K. Before the *reverse* process occurs, namely, the conversion of the exciton ω, K into a photon ω, K, the exciton is scattered in a time Γ^{-1} to other states by its coupling to other crystal excitations like, e.g., impurities, phonons and deep centers. The absorption of light is then mainly determined by the exciton–photon interaction, with a lineshape due to the exciton-crystal interactions [7.60, 66]. The concept of weak exciton–photon

coupling is a *relative* one, depending on the size of Γ, i.e., on temperature, crystal purity and perfection. The light absorption *rate* is determined by the exciton–photon coupling.

(ii) strong exciton–photon coupling
Before any scattering event, the exciton ω, K converts back into a photon ω, K, which again converts into an exciton and so forth. One cannot consider the propagation of excitons and photons separately, but only that of the coupled mode, the excitonic polariton. It can be described in a purely *classical* manner as in the early description of phonon plaritons by *Huang* [7.67, 68]. Quantum mechanically, it corresponds to the diagonalization of the total Hamiltonian of the coupled system "exciton–photon" with its eigenstates, the polariton states. The time oscillation of the system from one component to the other is described by the standard Rabi formula [7.69, 70] with frequency ω_{LT}. This description fails when the transition time between exciton states due to external perturbations is shorter than the time required for the polariton state to oscillate from one state to the other, i.e., when $\Gamma > \omega_{LT}$ [7.69, 70].

b) Exciton–Phonon Coupling

The exciton–phonon coupling strength is usually weak compared to the exciton binding energy R^*. Therefore, it can be treated as a perturbation to the exciton states, as first described by *Ansel'm* and *Firsov* [7.58]. In this case, the exciton–phonon coupling is merely the sum of the electron–phonon and hole–phonon interactions acting separately on the electron and hole bound in the exciton complex. The corresponding matrix elements for the standard interaction mechanisms can easily be evaluated for exciton intraband or interband transitions (i.e., between exciton states with same or different n's). We give here only results for the intraband case.

(i) acoustic phonon scattering by deformation potential (dp) interaction

$$\langle n, K, n_q | \mathscr{H}_{\text{exc}-\text{ph}} | n, K', n_q \pm 1 \rangle_{\text{dp}} = \sqrt{\frac{\hbar}{2\varrho u_s}} \, q^{1/2} (E_c q_e - E_v q_h) \sqrt{n_q + \tfrac{1}{2} \pm \tfrac{1}{2}} \,. \tag{7.6}$$

$|n, K, n_q\rangle$ represents the crystal quantum state with an exciton in band n and momentum K and n_q phonons with momentum q. The phonon momentum q is related to K and K' through the energy and momentum conservation equations. In (7.6), ϱ is the crystal density, u_s the phonon velocity, E_c and E_v the conduction and valence band single-electron deformation potentials, q_e and q_h the Fourier transform of the electron or hole charge distribution functions [7.58]. For instance, for the ls state of an hydrogenic exciton (see also [Ref. 7.71, Sect. 2.39])

$$q_e = \left[1 + \left(\frac{m_h}{m_e + m_h} \frac{q a_B}{2}\right)^2\right]^{-2}, \quad q_h = \left[1 + \left(\frac{m_e}{m_e + m_h} \frac{q a_B}{2}\right)^2\right]^{-2}. \tag{7.7a}$$

Usually $q < 2a_B^{-1}$, so that $q_e \approx q_h \approx 1$.

(ii) acoustic phonon scattering by piezoelectric (pe) interaction

In noncentrosymmetric crystals, acoustic phonons can be piezoelectric-active in certain directions. Thus, a longitudinal piezoelectric macroscopic field is created [7.72, 73]. This is in particular the case for TA phonons in the [110] direction of cubic zinc-blende structure or perpendicular to the c-axis of hexagonal wurtzite structure compounds. TA phonons usually lead to a very small deformation potential due to the near isotropy of the deformation potential (see Chap. 6; however dp-mediated TA scattering has been observed in CuBr [7.74]) and the fact that in *isotropic* materials, scattering by TA phonons is forbidden. Therefore, TA phonon scattering is dominated by the piezoelectric mechanism and will occur only in those directions for which the TA phonon is indeed piezoelectric active. For zinc-blende semiconductors with $K, K', q \parallel$ [110] direction, the matrix element is

$$\langle n, Kn_q | \mathcal{H}_{\text{exc}-\text{ph}} | n', K', n_q \pm 1 \rangle_{\text{pe}} = \sqrt{\frac{\hbar}{2\varrho u_s}} \frac{4\pi e\, e_{14}}{\varepsilon_0 q^{1/2}} (q_e - q_h). \tag{7.7b}$$

The angular variation of the matrix element is given in [7.31, 72, 73].

This interaction energy is the potential energy of the electron and hole in the electric field created by the phonon. Therefore, it vanishes in several limiting cases for which $q_e - q_h \approx 0$:

— Large q ($qa_B \gg 1$), which average to zero the effect of the electric field over the exciton wave function ($q_e \approx 0$, $q_h \approx 0$), due to its rapid spatial oscillations.
— Small q ($qa_B \ll 1$), in this case a uniform pe field over the exciton wave function exists, giving cancellation of the electrostatic interactions with the electron and the hole.
— $m_e = m_h$, which implies the same spatial extension of electron and hole relative wave functions and again exactly cancelling interactions with the pe field.

In the perfectly hydrogenic approximation, for $qa_B/2 \ll 1$,

$$q_e - q_h \approx (qa_B)^2 (m_e - m_h)/2M_X. \tag{7.8}$$

(iii) LO phonon scattering mediated by the Fröhlich (Fr) interaction

As was already very widely studied ([Ref. 7.32, Chap. 3], [7.75, 76], see also [Ref. 7.71, Sect. 2.3.8]) the dominant contribution to RRS by LO phonons is the "forbidden" scattering mediated by the Fröhlich interaction. Its matrix elements due to the longitudinal electric field created by LO phonons is given by

$$\langle n, K, n_q | \mathcal{H}_{\text{exc}-\text{ph}} | n, K', n_q \pm 1 \rangle_{\text{Fr}} = \frac{1}{q} \left[\frac{2\pi\hbar\omega_{\text{LO}} e^2}{V} \left(\frac{1}{\varepsilon_\infty} \frac{1}{\varepsilon_0} \right) \right]^{1/2} (q_e - q_h) \sqrt{n_q + \frac{1}{2} \pm \frac{1}{2}}, \tag{7.9}$$

where $\hbar\omega_{\text{LO}}$ is the LO phonon energy, V the volume of the crystal.

c) Exciton–Carrier and Exciton–Impurity Interaction

By treating excitons as two-particle complexes, one can calculate their interactions with other charges using Coulomb interaction in lowest order (with or without screening). The problem bears a strong resemblance to molecular scattering in atomic physics. Calculations have been done up to now for *free* particle scattering (i.e., electron–exciton and exciton–exciton scattering) [7.77–80], the only exception being the calculation by *Zinets* and *Sugakov* [7.81] of *elastic* scattering of excitons on ionized donors. Enormous cross

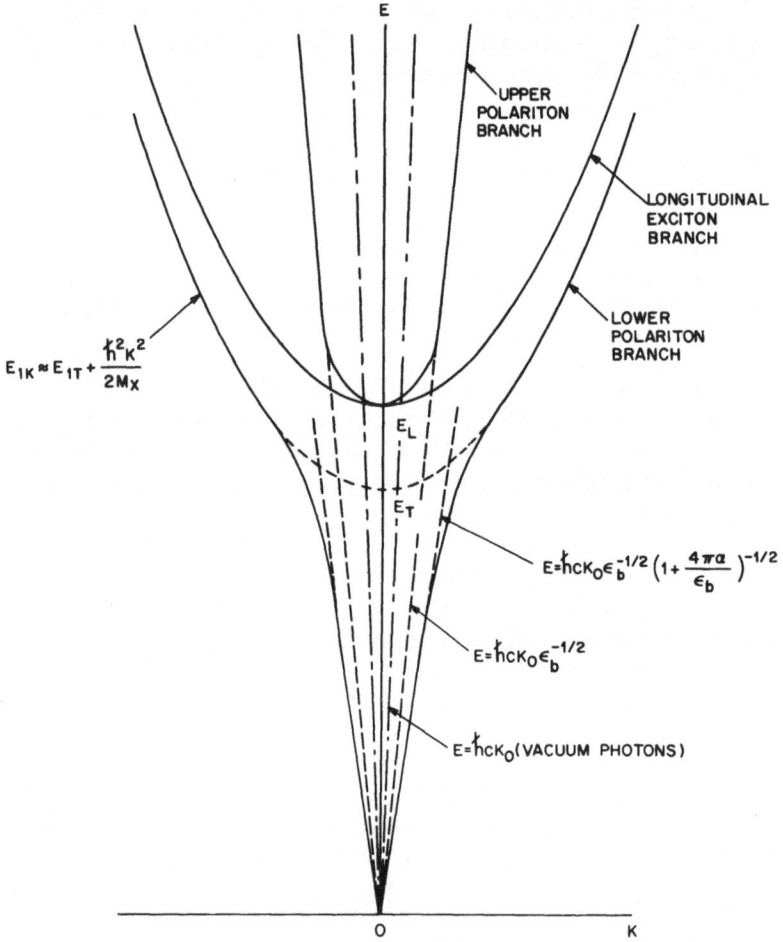

Fig. 7.4. Dispersion curves for excitonic polaritons (i.e., the exciton-photon coupled mode excitations) in the vicinity of a single exciton resonance E_T. Note that the asymptotic photon-like polariton states above E_L are renormalized by all excitations of the crystal above E_L (photon "dressed" according to the background dielectric constant ε_b) and below E_T by the same plus the excitonic level (total dielectric constant $\varepsilon_b + 4\pi\alpha$). Vacuum photon and longitudinal exciton (uncoupled to light) dispersion curves are also shown

sections (up to $\approx 300\, a_B^2$) are found in that case at low kinetic energies. *Inelastic* scattering of polaritons on donors will be considered in Sect. 7.7. Cross sections of the order of a_B^2 are observed. These results point to the strong resemblance with atomic physics calculations scaled to semiconductor parameters, i.e., using a dielectric constant and effective masses wherever needed.

7.2.3 Excitonic Polaritons–Static Properties

As mentioned earlier, excitonic polaritons are classically the coupled-mode excitations (quantum mechanically the eigenmodes) of the coupled system "exciton + photons". The dispersion relation, expressed by the dielectric constant $\varepsilon(K, \omega)$, can be found classically [7.68] or quantum mechanically [7.2, 3] in the single exciton-band approximation, as given by

$$\varepsilon(K, \omega) = \frac{c^2 K^2}{\omega^2(K)} = \varepsilon_b + \frac{4\pi\alpha\omega_n^2(K)}{\omega_n^2(K) - \omega^2(K)}, \tag{7.10}$$

where $\omega_n(K)$ is the *unperturbed* exciton dispersion given by (7.1). The polariton dispersion curves are then obtained by solving (7.10) for $\omega(K)$. The results are shown in Fig. 7.4, along with the longitudinal exciton branch $E_L(K)$, which is unperturbed as being not coupled to light, but shifted upwards by $E_{LT} = E_L - E_T$ due to the long-range Coulomb exchange interaction between the electron and hole [7.37]. There are four parameters characterizing the $n = 1$ exciton-polariton dispersion curves in this simple case: $E_T = E_G - R^*$, ε_b, E_{LT} [(or α or f_{exc})] and M_X. Table 7.1 lists exciton parameters for usual semiconductors as well as calculated and measured values for E_{LT}. The remarkable agreement between theory and experiment for E_{LT} (well within a factor of 2) comes from the good experimental knowledge of exciton parameters and the insensitivity of E_{LT} with respect to the fine structure of the excitons; only fractional changes are expected from fine structure corrections.

We mainly deal here and in the following with 1s exciton-polaritons. As f_{exc}, α and E_{LT} depend on n as n^{-3}, one expects polariton effects to be less important for excited exciton levels $n = 2, 3, \ldots$. However, the quantity to be compared to E_{LT} is Γ, and it might well happen that polariton-like states describe light propagation phenomena even at the $n = 2$ exciton resonance of pure semiconductors with small values of E_{LT}. This was strikingly evidenced in the time-of-flight measurements (see Sect. 7.2.4 and [7.110]) of *Ulbrich* and *Fehrenbach* in GaAs (Fig. 7.9) where the slowing down of light due to the polariton effect was indeed well-observed in the region of the $n = 2$ exciton, although its LT splitting should only be ≈ 0.01 meV. In the case of small radii (i.e., large R^*) excitons, the smallness of a_B still provides a value of E_{LT} in the meV range for excited states.

The influence of damping (interaction with phonons, impurities and other defects) is not accounted for in (7.10). As mentioned above, polariton effects disappear when $\omega_{LT} \ll \Gamma$ [7.3]. More quantitative descriptions of damping

Table 7.1. Exciton parameters in various compounds. (Values in *italics* are measured values)

	E_G [eV] [a]	ε [$\hbar\omega = R^*$] [b]	m_e [m_0] [b]	m_h [m_0] [b]	a_B [Å] [c]	R^* [meV]	E_T [eV]	P^2 [eV] [d]	E_{LT} [meV]
GaAs	1.5192 [1]	*12.55* [2]	*0.0667* [3]	[100] $m_{lh}=0.082$ [4] $m_{hh}=0.45$	136	4.2 [1]	*1.5150* [5]	28.9	0.086 / *0.08* [5]
InP	1.4234 [6]	*12.35* [7]	*0.0803* [3]	[100] $m_{lh}=0.12$ [8] $m_{hh}=0.56$	119	4.9 [6]	1.4185 [6]	20.7	0.108 / *0.14* [6]
GaSb	0.8137 [9]	*15.69* [10]	*0.041* [11]	[100] $m_{lh}=0.046$ [12] $m_{hh}=0.26$	327	1.4	0.8123	22.4	0.0135 / *0.035* [13]
CdS	2.581 [14]	$\varepsilon_\parallel=8.64$ [15] $\varepsilon_\perp=8.28$	0.208 [16]		30	28 [14]	2.5529 [14]	21	2.99 / *1.88* [17] / *1.86* [14]
CdSe	1.8402 [15]	$\varepsilon_\parallel=9.25$ [15] $\varepsilon_\perp=8.75$	$0.120\parallel$ [18] $0.115\perp$ [18]		53.6	*16* [15]	*1.8242* [15]	20	1.00 / *0.5* [19] / *0.95* [20]
CdTe	1.606 [15]	9.65 [15] / 9.0 [21]	0.096 [21,22]	[110] $m_{lh}=0.12$ [23] $m_{hh}=0.81$	72	*10.4* [24]	*1.5954* [25]	22.8	0.546 / *0.4* [25] / *0.5* [26]
ZnSe	2.818 [15]	8.66 [15]	0.16 [27]		43.8	19 [15]	2.8023 [28]	23	0.874 / *1.45* [8]
ZnTe	2.394 [32]	9.67 [15]	0.122 [29]	[100] $m_{lh}=0.154$ [30] $m_{hh}=0.64$	58	12.8 [31]	2.377 [32]	24.5	0.67 / *0.6* [31]
CuCl	3.392 [33]	3.6 [34]			10.5	190 [33]	3.2025 [33,35]	20	91 / *5.7* [35]
CuB$_R$	3.07 [36]	5.4 [37]			13.3	105 [36]	2.9644 [37]	20	39 / *12.2* [37]
GaSe	2.1294 [38]	$\varepsilon_\parallel=6$ [38] $\varepsilon_\perp=10.4$			46.7	19.5 [38]	2.1099 [38]	25	1.5
PbI$_2$	2.527 [39]	$\varepsilon_\parallel=8.4$ [39] $\varepsilon_\perp=12.6$			23	30 [40]	2.497 [40]	25	6.55 / *8* [39]
HgI$_2$	2.3694 [39]	$\varepsilon_\parallel=5.8$ [41] $\varepsilon_\perp=6.8$			34	34 [41]	2.3354 [42]	25	3.94 / *5.33* [42]

[a] Usually deduced (at low temperatures) from exciton measurements by $E_G = E_T + R^*$.

[b] Only cyclotron resonance measurements are listed.

[c] Calculated in the spherical approximation: $a_B = e^2/4\pi\varepsilon_0\varepsilon R^*$.

[d] Deduced from effective mass and g-factors, as described in [43], otherwise estimated.

Table 7.1. References

[1] D.D.Sell: Phys. Rev. B6, 3750 (1972)

[2] G.E.Stillman et al.: Solid State Commun. 9, 2245 (1971)

[3] J.M.Chamberlain et al.: Proc. Int. Conf. on the Physics of Semiconductors, Warsaw, 1972, ed. by M.Miasek (PWN, Warsaw 1972) p. 1012

[4] M.S.Skolnick et al.: J. Phys. C9, 2809 (1976)

[5] R.G.Ulbrich, C.Weisbuch: Phys. Rev. Lett. 38, 865 (1977)

[6] F.Evangelisti et al.: Phys. Rev. B9, 1516 (1974)

[7] C.Hilsum et al.: Solid State Commun. 7, 1057 (1969)

[8] J.Leotin et al.: Solid State Commun. 15, 693 (1974)

[9] A.Filion, E.Fortin: Phys. Rev. B12, 5803 (1975)

[10] M.Hass, B.W.Henvis: J. Phys. Chem. Sol. 23, 1099 (1962)

[11] D.A.Hill, C.F.Schwerdtfeger: J. Phys. Chem. Solids 35, 1533 (1974)

[12] R.A.Stradling: Proc. of the Int. Conf. on the Applications of High-Magnetic Fields in Semiconductor Physics, Würzburg, 1974

[13] W.Rühle et al.: Phys. Status Solidi B73, 255 (1976)

[14] I.V.Makarenko et al.: Phys. Status Solidi B98, 773 (1980)

[15] B.Segall, D.Marple: In Physics and Chemistry of the II–VI Compounds, ed. by M.Aven, J.S.Prener (North-Holland, Amsterdam 1967) p. 317

[16] F.Pollak: In II–VI Semiconducting Compounds, Proc. of the 1967 Int. Conf., Providence, ed. by D.G.Thomas (Benjamin, New York 1967) p. 552

[17] F.Evangelisti et al.: Phys. Rev. B10, 4253 (1974)

[18] L.Eaves et al.: J. Phys. C5, L19 (1972)

[19] C.Hermann, P.Y.Yu.: Phys. Rev. B21, 3675 (1980)

[20] V.A.Kiselev et al.: Phys. Status Solidi. B72, 161 (1975)

[21] C.W.Litton et al.: Phys. Rev. B13, 5392 (1976)

[22] A.Mears, R.A.Stradling: Solid State Commun. 7, 1267 (1967)

[23] R.Romestain, C.Weisbuch: Phys. Rev. Lett. 45, 2067 (1980)

[24] R.G.Ulbrich, C.Weisbuch: Unpublished (1979)

[25] R.G.Ulbrich, C.Weisbuch: In Festkörperprobleme, Vol. 18, ed. by J.Treusch (Vieweg, Braunschweig 1978) p. 217

[26] W.Dreybrodt et al.: Phys. Rev. B21, 4692 (1980)

[27] J.L.Merz et al.: Phys. Rev. B6, 545 (1972)

[28] B.Sermage, G.Fishman: Phys. Rev. (to be published)

[29] B.Clerjaud et al.: Phys. Rev. B19, 2056 (1979)

[30] R.Stradling et al.: Solid State Commun. 6, 665 (1968)

[31] H.Venghaus, B.Jusserand: Phys. Rev. B22, 932 (1980)

[32] Y.Oka, M.Cardona: Solid State Commun. 30, 447 (1979)

[33] S.Suga et al.: Phys. Status Solidi B55, 355 (1973)

[34] A.Feldman, D.Horowitz: J. Opt. Soc. Am. 59, 1406 (1969)

[35] B.Hönerlage et al.: Phys. Rev. Lett. 41, 49 (1979)

[36] H.J.Mattausch, Ch.Uihlein: Phys. Status Solidi B96, 189 (1979)

[37] B.Hönerlage et al.: Phys. Rev. B22, 797 (1980)

[38] R.LeToullec et al.: Phys. Rev. B22, 6162 (1980)

[39] J.Biellmann et al.: In Polaritons, ed. by E.Burstein, F.de Martini (Pergamon Press, London 1974) p. 183

[40] D.Fröhlich, R.Kenklies: Nuovo Cim. B38 433 (1977)

[41] F.Sakuma et al.: J. Phys. Soc. Jpn. 45, 1349 (1978)

[42] T.Goto, Y.Nishima: Solid State Commun. 31, 751 (1979)

[43] C.Hermann, C.Weisbuch: Phys. Rev. B15, 823 (1977)

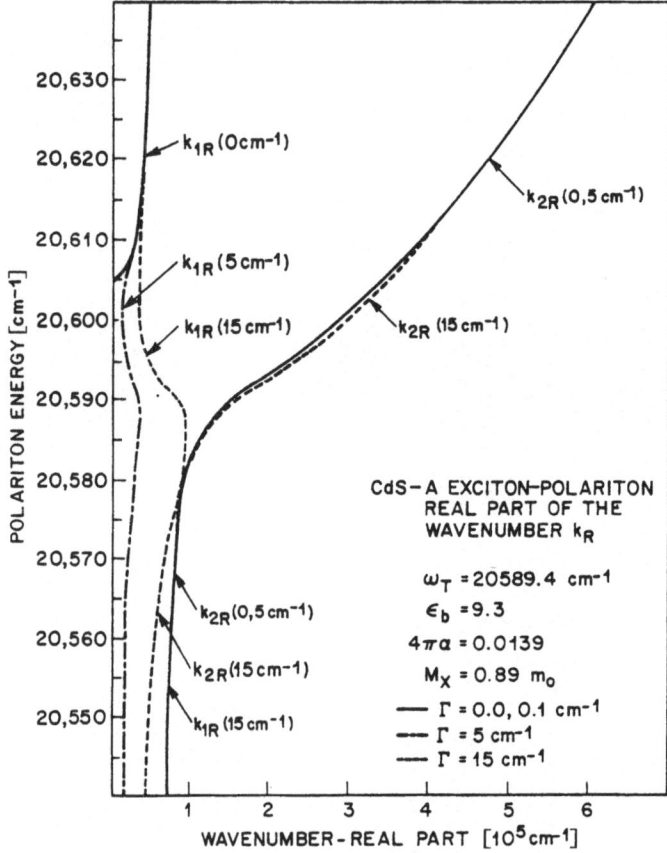

Fig. 7.5. Polariton dispersion curves of the A-exciton in CdS in the presence of damping. The plots show the real part of the wave vector K according to the formula

$$\varepsilon(K, \omega) = \frac{C^2 K^2}{\omega^2(K)} = \varepsilon_b + \frac{4\pi\alpha\omega^2(K)}{\omega_T^2 + \dfrac{\hbar^2 K^2}{m^*}\omega_T - \omega^2(K) - i\omega(K)\Gamma}.$$

The value of $4\pi\alpha$ corresponds to an LT splitting $E_{LT} \sim 15\,\text{cm}^{-1}$. (After M. Matsushita, unpublished)

include a *phenomenological* constant factor Γ in the denominator of (7.10). Although this procedure is questionable both theoretically and to describe experimental results, as will be discussed below, it helps describe the transition from undamped polaritons to strongly damped ones [7.82, 83]. Only rather small modifications to the polariton dispersion curve appear at small values of Γ (below $0.1\,\omega_{LT}$), but drastic changes of $\omega(K)$ (see Fig. 7.5) occur around a critical value Γ_c given by [7.82]

$$\Gamma_c = 2^{3/2}(\hbar\omega_T^2\omega_{LT}/M_X c^2)^{1/2} = 2^{3/2}\left(\frac{\omega_T}{\omega_{LT}}\right)^{1/2}\left(\frac{\hbar\omega_T}{M_X c^2}\right)^{1/2}\omega_{LT}^{-1}. \tag{7.11}$$

For standard values of $M_X \approx m_0$, $\omega_{LT} \approx 1\,\text{meV}$, $\hbar\omega_T \approx 2\,\text{eV}$, one finds $\Gamma_c \approx 0.3\,\omega_{LT}$, equivalent to Hopfield's estimate $\Gamma_c \lesssim \omega_{LT}$.

The two coupled modes obtained from (7.10) correspond quantum mechanically to the eigenstates resulting from the diagonalization of the total Hamiltonian of the exciton–photon system. Several properties of this diagonalization need to be emphasized, as they will prove to be fundamental in the analysis of polariton-mediated light scattering [7.28].

i) Far from resonance, the modes are either exciton-like or photon-like (see Fig. 7.4): the lower polariton branch (LPB) is photon-like at small ω, K's and exciton-like at large ω, K's. The reverse is true for the upper polariton branch (UPB). In the resonance region, the amount of state admixture is determined by the transformation coefficients from unmixed states to coupled states. *Hopfield* [7.28] showed that the exciton admixture is almost total in a wide range around the resonance, *not determined by E_{LT}, but by $\hbar\omega_c = \sqrt{E_{LT} \cdot E_T}$. The polariton wave function has thus a mixed character in the region $|\hbar\omega - E_T| \lesssim \hbar\omega_c$, although its dispersion seems almost unperturbed beyond $|\hbar\omega - E_T| \gtrsim E_{LT}$.*

ii) The indices of refraction of the two branches, defined by $n_{1,2} = cK_{1,2}/\omega$, change significantly only in the region $|\hbar\omega - E_T| \lesssim E_{LT}$. Accordingly, this is also the spectral region where the reflection and transmission coefficients may vary.

iii) The polariton group velocity $v_g = d\omega/dK$ is strongly influenced by the polariton effect in the wider range $|\hbar\omega - E_T| < \hbar\omega_c$. It can be approximated outside the exact resonance by [7.28]

$$v_g \approx c\varepsilon_b^{-1/2}[1 + \omega_c^2/2(\omega_T - \omega)^2]^{-1}. \tag{7.12}$$

In the resonance region, it approaches the exciton velocity $\hbar K/M_X$.

This simple picture of polaritons in the case of a single isotropic exciton band has been extended to more complicated exciton levels: k-linear-term-split excitons (as for the B exciton of CdS [7.40]), multiple exciton branches of zinc-blende structure semiconductors [7.84, 85] and interacting exciton resonance [7.86, 87]. Equation (7.10) is generalized into

$$\varepsilon(K, \omega) = \frac{c^2 K^2}{\omega^2} = \varepsilon_b + \sum_i \frac{4\pi\alpha_i\omega_i^2(K)}{\omega_i^2(K) - \omega^2}. \tag{7.13}$$

The index i characterizes the various exciton resonances.

7.2.4 Optical Properties of Polaritons

Coupled-mode propagation introduces specific optical properties of those crystals in which excitonic polaritons exist. These optical properties have been very widely studied as they provide the experimental access to polariton phenomena [7.14, 88–90]. Besides important modifications of reflectivity transmission and absorption phenomena, the polariton induces an additional optical property of the medium, namely, spatial dispersion: the dependence of the dielectric function both on ω and K is equivalent to a position-dependent dielectric constant $\varepsilon(r - r')$ in the medium. The presence of interfaces in all experimental situations destroys the translational symmetry of $\varepsilon(r - r')$. The optics of such media clearly represent very interesting theoretical problems [7.91] of a rather complicated nature. We will only outline here some of the properties and problems relevant to light scattering, deferring the reader to recent reviews of the subject for a more complete treatment [7.88 91].

(i) Additional boundary conditions (ABC) and spatial dispersion

Any analysis of an experimental situation requires the knowledge of the connection between outside and inside microscopic electromagnetic fields. In usual optics, this connection is determined by the boundary conditions for the electric and magnetic fields E and H derived from Maxwell's equations [7.92], which allow the knowledge of the polarization vector P. In the case of spatial dispersion, several polarization waves P_i (each corresponding to a different polariton branch) may propagate at a given energy (for example above E_L, because of the upwards bowing of the LPB) and therefore, *additional boundary conditions* (ABC's) are required to determine the various field amplitudes in the crystal, and from those the reflection and transmission coefficients at the interface. While it has been shown that the general type of ABC can be written [7.88]

$$P + \lambda \frac{dP}{dz}\bigg|_{z=0} = 0 \qquad (7.14)$$

(where P is the total polarization vector and z the normal direction to surface), conservation laws (i.e., for energy flow [7.93]) and microscopic models impose some restrictions on the choice of λ.

(ii) Interface effects: the dead-layer model

In the preceding paragraph (i), the properties of the medium were considered as space-homogeneous. Actually, the presence of an interface modifies exciton eigenstates and the dielectric properties of the medium near the interface. *Hopfield* and *Thomas* [7.14] showed that in the case of Wannier (i.e., large radii) excitons, this effect could be approximated by considering an "exciton dead layer" near the surface, of thickness $d \approx$ the exciton diameter $2a_B$, where the dielectric constant is the background dielectric constant ε_b. Beyond this layer, the medium should retain the dielectric properties of the infinite medium, determined by (7.10). The interface problem is thus solved with a *three-media system*, using normal BC's at the vacuum–crystal interface and Maxwell's BC's plus Pekar's ABC $P = 0$ at the dead-layer interface in the crystal. This set of *two* rules constitutes the Hopfield and Thomas ABC.

(iii) Recent developments of ABC theory

Before discussing the analysis of experimental data any further, which has been made up to now essentially with this ABC model, let us describe more recent developments of the ABC theory. Around 1972, a number of analyses [7.94] claimed to have been able to solve completely Maxwell's equations (i.e., find all fields inside the crystal) and *deduce* the ABC by assuming a priori a dielectric function of the type $\theta(z) \, \varepsilon(K, \omega)$, where $\theta(z)$ is the step function at the interface ($\theta = 0$ outside the crystal, $\theta = 1$ inside) and $\varepsilon(K, \omega)$ is the dielectric function given by (7.10). It was soon recognized that this assumption is too strong, not taking into account the perturbation to exciton states near the interface. It is actually equivalent to *assuming* a priori an ABC. It became clear that the knowledge of the dielectric function $\varepsilon(r, r', \omega)$ everywhere is necessary in order to determine all the field in the crystal (i.e., the ABC) [7.88, 95, 96], but no self-consistent calculation of a realistic $\varepsilon(r, r', \omega)$ has been made up to now for Wannier excitons. A recent evaluation [7.96] tends to favor Pekar's ABC ($P = 0$) through a variational approach of the exciton–surface problem.

(iv) Experimental results concerning ABC's: reflectivity

From points (i) and (ii) it appears that the minimum number of parameters defining exciton polaritons in the Hopfield–Thomas model is rather high: E_T, ε_b, M_X, E_{LT}, d and one ABC. Eventually one can introduce damping of polaritons in a phenomenological way in (7.10) and adjust it to experiments. How well does such a procedure perform? The original paper of Thomas and Hopfield produces a very convincing fit for the CdS reflectivity at low temperatures. In GaAs, the reflectivity was also well accounted for by this model [7.10a]. Schottky barrier measurements in

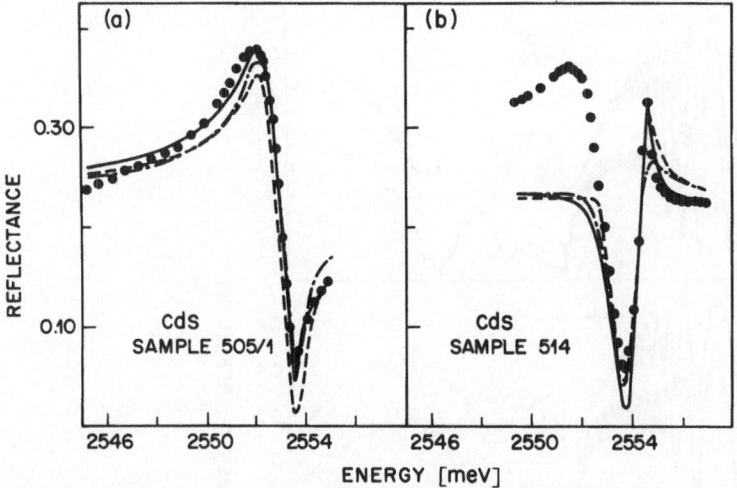

Fig. 7.6a, b. Comparison of theoretical reflectivity spectra calculated according to *different* ABC's using *different parameters* to fit the experimental spectrum (dotted line) for two samples with a thin (**a**) or thick (**b**) dead layer. Theoretical spectra are calculated with the Pekar-Thomas-Hopfield [PTH] (full line), Ting et al. [T] (dashed-dotted line) and Aggarwal et al. [A] (dashed line) ABC. The values for the dead layer thickness L and damping factor Γ are for the two samples and respective ABC's: Sample 505/1: $\Gamma(\text{meV}) = 0.14$ [PTH], 0.66 [A], 0.37 [T]; $L(\text{Å}) = 98$ [PTH], 133 [A], 154 [T]. Sample 514: $\Gamma(\text{meV}) = 0.20$ [PTH], 0.45 [A], 0.42 [T]; $L(\text{Å}) = 181$ [PTH], 210 [A], 217 [T]. Note the remaining discrepancy on the low-energy side of the reflectivity curve of sample 514. From [7.16]

GaAs and InP provided voltage-varying spectra which could be well interpreted in a varying dead layer model [7.15, 97]. As more experiments were done, however, it became clear that very different reflectivity curves could be produced experimentally, requiring different sets of parameters to fit the results [7.97, 98]. The very documented case of CdS can be taken as an example where varied, grossly different reflectivity spectra can be produced by changing surface conditions [7.98]. Also, a recent study of reflectivity as a function of exciton impact ionization in GaAs requires an analysis beyond the simple dead-layer model to explain the results [7.99]. The limits of the model are best exemplified in the detailed analysis of *Patella* et al. for the reflectivity of CdS [7.16]. They show (Fig. 7.6) that by trying to fit several ABC's proposed in the literature, one can obtain a satisfactory agreement with the experimental data, provided *different, adjusted values of Γ and d are used for different* ABC's. This means in short that dead-layer and damping effects overwhelm any specific influence of ABC's in these reflectivity measurements.

(v) Transmission measurements
Absorption (i.e., transmission) measurements are also very dependent on ABC's (see the lucid discussion in [7.100]). Absorption (i.e., dissipation of radiant energy into crystalline degrees of freedom) is controlled by the interactions of excitonic polaritons with other crystal excitations. The scattering probability depends on the ABC through the polariton-branch dependence of the scattering probability. However, due to the difficulty of absorption (i.e., transmission) measurements with available crystals (well exemplified by the controversy about GaSe [7.101]), it was not possible to obtain a useful application of transmission measurements to ABC determination, despite the extensive documentation gathered on CdS by *Voigt* and his co-workers [7.102].

(vi) Additional waves interference
A major breakthrough in polariton optics appeared in 1973 when the so-called additional-wave interference (AW) due to the multiple polariton branches above E_{L} was reported by *Kiselev* et al.

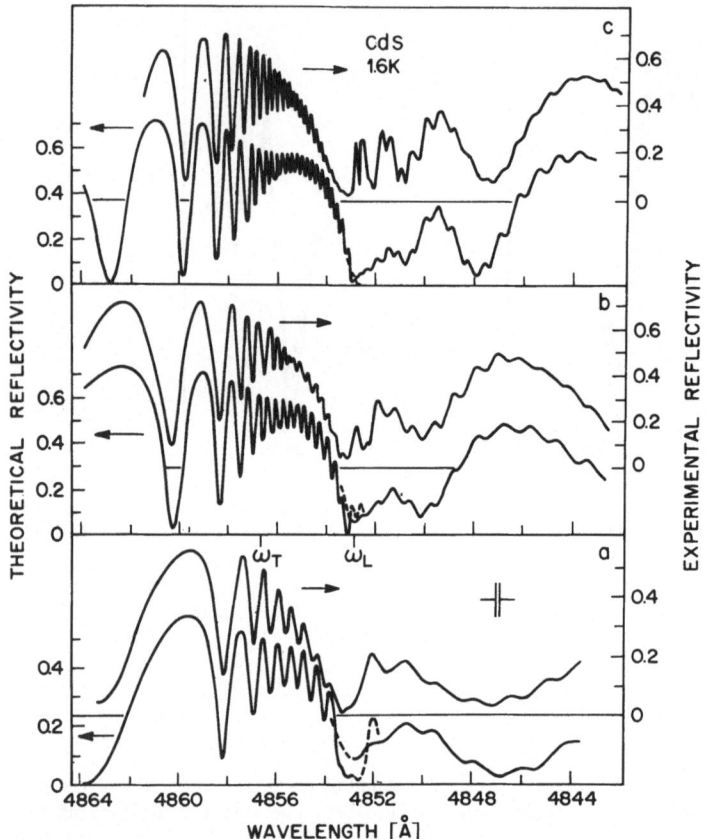

Fig. 7.7a–c. Calculated and measured transmission and reflectivity spectra of ultra-thin (≦1 µm) CdS platelets around the A-exciton polariton resonance and the observation of so-called "additional wave interferences" due to degenerate polariton branches above E_L caused by spatial dispersion. Thicknesses: 0.51 µm (**a**), 0,34 µm (**b**), 0.20 µm (**c**). From [7.103]

[7.5, 103] on ultrathin CdS crystals (<1 µm) both in reflection and transmission measurements (Fig. 7.7). Although the observed structure is very sensitive to the relative intensities of polariton modes, thus to ABC's, it was not possible to obtain a straightforward determination of the ABC. As pointed out recently in the most recent study by *Makarenko* et al. [7.103], some remaining discrepancies between theory and experiment are unexplainable in a simple dead-layer model.

In our opinion, the accounts by *Patella* et al. [7.16] and *Makarenko* et al. [7.103] give an excellent picture of the state of the art of reflectivity and transmission measurements and their failure to predict ABC's. A very recent experiment on CdS *prism-shaped samples* with thicknesses of 16 µm and 80 µm enabled *Broser* et al. [7.104] to directly measure the polariton index of refraction from the deflection of a transmitted beam. The spectral dependence of the transmitted beams and their interference should also allow for a determination of ABC's [7.105], but this has not been attempted up to now.

(vii) Polariton propagation velocity

Spectacular evidence of the polariton propagation mode is provided by time-of-flight measurements [7.91, 106]. As in any coupled-mode propagation, one expects the group velocity of the fast

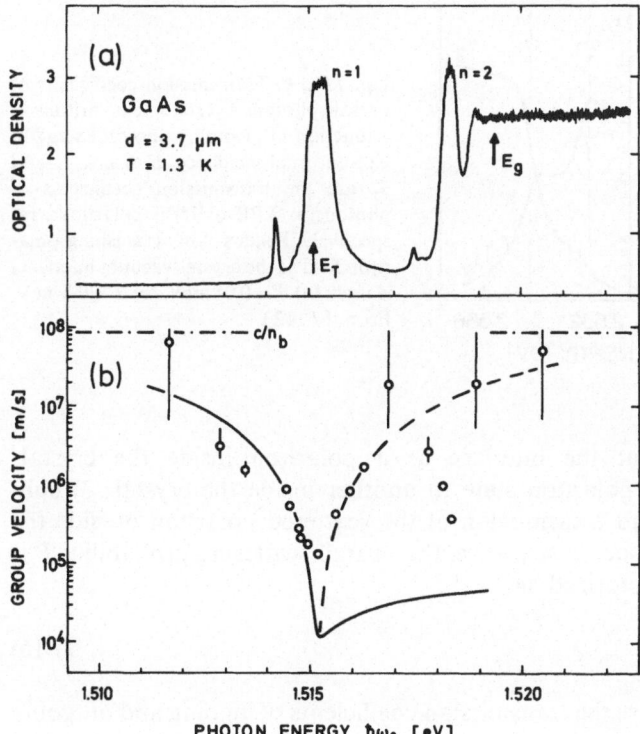

Fig. 7.8a, b. Measured group velocity of excitonic polariton wave packets of 12 psec duration in a thin (3.7 μm) GaAs crystal. The full and dashed lines indicate theoretical group velocities for lower and upper branch polaritons (including only the $n=1$ exciton resonance). Note the clear experimental observation also of the $n=2$ exciton resonance. From [7.110]

excitations (here the photons) to be drastically diminished due to the bending of the dispersion curves. In addition, in the case of optics of damped, dispersive media, one has to distinguish between energy propagation velocity v_E and group velocity $v_g = d\omega/dK$ [7.30, 107]. Very recent experiments in CuCl and GaAs [7.108–110] have shown both the important slowing down of polariton wave packets at resonance and the equality of energy propagation velocity (measured) to the (calculated) group velocity, because of the smallness of polariton damping (Fig. 7.8).

7.3 Light Scattering in the Polariton Framework

The importance of polariton propagation for light scattering phenomena was recognized very early, mainly in the pioneering work of *Ovander* [7.12, 111]. As discussed in greater detail by *Bendow* [7.30], two main approaches to describe light scattering by excitonic polaritons have been pursued. The first (called approach A in the following) is to consider a scattering event as the *succession of three steps* (Fig. 7.1g, h): 1) transmission of

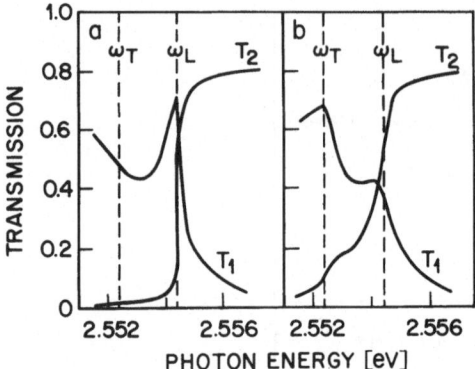

Fig. 7.9a, b. Transmission coefficients of outside photons as excitonic polaritons as a function of incoming photon energy in CdS (normal incidence, E, $K \perp c$). T_1 and T_2 are the transmission coefficients of photons as LPB or UPB polaritons, respectively. Pekar's ABC (vanishing polariton field at the crystal-vacuum interface) is used. (**a**) $\Gamma = 0.05$ meV, (**b**) $\Gamma = 0.5$ meV. From [7.112]

an incoming photon at the interface as a polariton inside the crystal; 2) scattering from one polariton state to another inside the crystal; 3) subsequent propagation and transmission of the scattered polariton outside the crystal as a photon. In such a sequence, the overall scattering probability P_{out} (or efficiency) can be factorized as

$$P_{out} = P_{in} T(\omega_i) T(\omega_s), \tag{7.15}$$

where $T(\omega_i)$ and $T(\omega_s)$ are the transmission coefficients of ingoing and outgoing polaritons at the crystal interface (Fig. 7.9) and P_{in} is the scattering probability of polaritons *inside* the crystal. When the outside probability is expressed per unit angle, care must be exerted to relate outside angles $d\Omega_{out}$ to the corresponding inside angle $d\Omega_{in}$. A very common approximation for near-normal angles is $d\Omega_{out} = n^2 d\Omega_{in}$, where n^2 is the relative index of refraction (for a thorough discussion, see the work by *Lax* and *Nelson* [7.113]). The second approach (B) considers quantum states extending over the whole space, asymptotically behaving as photon states or polariton states far away on one side or the other of the crystal–vacuum interface [7.114]. A scattering event then consists of the transition from an incident photon state to another (scattered) photon state. In this B approach, the whole resonance behavior is taken into account in the branching coefficients between photons and polariton states, while only photon-like densities of states and velocities are involved. Approach A, on the contrary, has almost constant coupling coefficients (such as transmission, exciton–phonon coupling) but has a resonant behavior through polariton densities of sates and velocities. Whereas approach B is more rigorous in principle, it has been little studied and we shall therefore use the first approach which has the advantage of pointing out more clearly the basic physical assumptions and consequences of the polariton description of light scattering. It also permits a rather direct comparison of experimental results with theory. Working out both theories down to their ultimate consequences should, in principle, yield equivalent results. The second approach accounts in a natural way for interference effects between multibranch polaritons. The influence of multiple scattering events on the occurence of such interferences has not, however, been tested up to now. From here on, only the three-step model of scattering (Fig. 7.1h) will be used and discussed.

7.3.1 Kinematic Properties

In the polariton formalism used here, scattering events are ascribed to transitions between polariton states within the crystal. The *kinematic* properties of polaritons (i.e., direct consequences from the dispersion curve) will therefore play a major role. From the examination of dispersion curves such as shown on Fig. 7.10, a number of distinctive properties of polariton-mediated scattering can be expected.

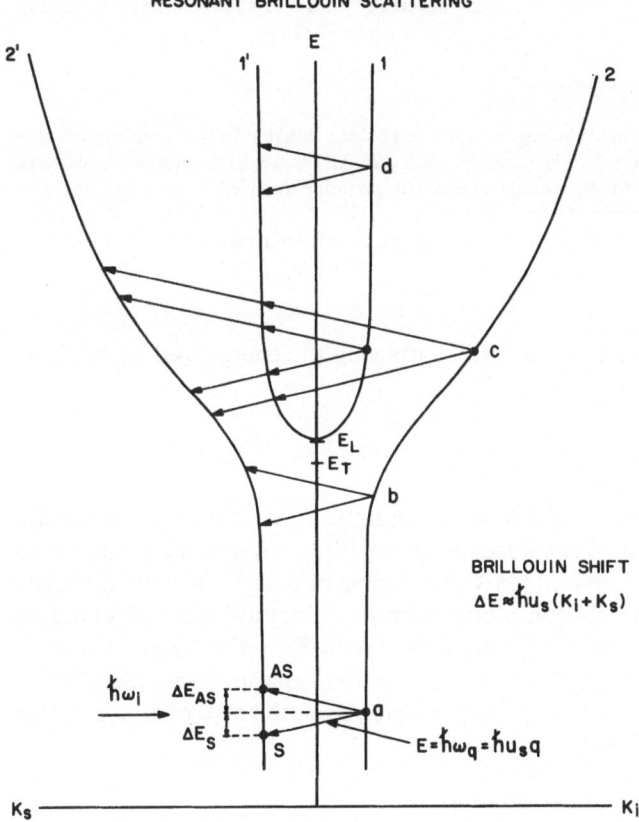

Fig. 7.10. Excitonic polariton dispersion curve and possible Brillouin-scattering events for four different incident photon energies. Note the occurrence of Brillouin multiplets at and above the exciton energy E_L (schematic)

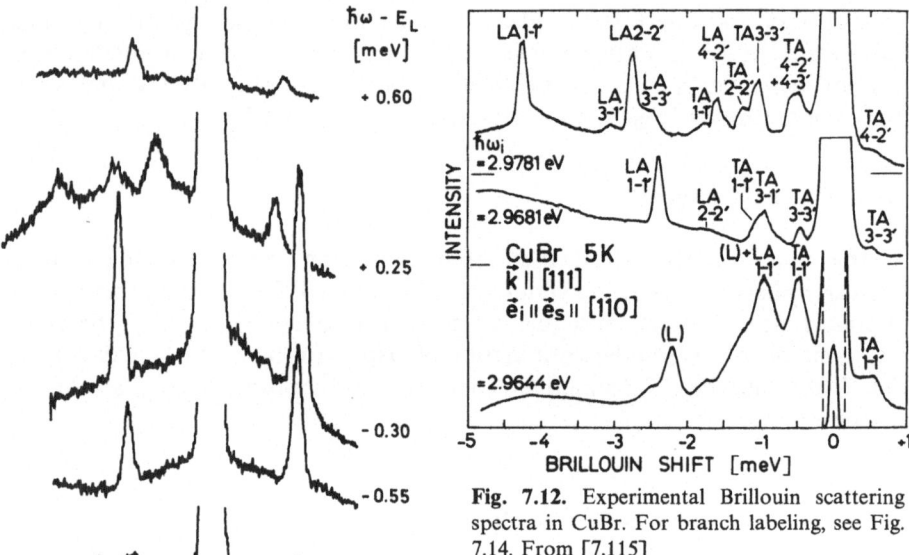

Fig. 7.12. Experimental Brillouin scattering spectra in CuBr. For branch labeling, see Fig. 7.14. From [7.115]

Fig. 7.11. Experimental Brillouin scattering in GaAs at 12 K for different incident photon energies $\hbar\omega$ around the exciton resonance E_L. The observed peak shift ΔE relative to the elastically scattered laser light is a direct measure for the energy of acoustic phonons involved in the scattering processes (Stokes to the left, anti-Stokes to the right). From [7.139]. Note the asymmetry of the shifts, larger AS efficiencies than S ones for $\hbar\omega$ below E_L, multiple broadened peaks at $\hbar\omega = E_L + 0.25$ meV

a) Dispersion Effects

In acoustic phonon scattering, transitions obeying the energy and momentum conservation relations

$$K_i + K_s = q$$

$$\hbar\omega_s = \hbar\omega_i \pm \hbar u_s q = \hbar\omega_i \pm \Delta E$$

can be represented, in the ω, K space, by straight lines connecting polariton branches with upwards or downwards slope u_s. Phonons with large momenta and energies can be generated. In GaAs, scattering of polaritons with momenta two times larger than the photon momentum in the crystal were observed by RBS (Fig. 7.11). In ZnSe and CuBr, polaritons with K's up to 10 times photon K's were reported [7.20, 115]. Besides such direct kinematic effects, the effect of dispersion shows up in the scattering efficiencies, as in the case of K-dependent scattering ("forbidden" Raman scattering).

b) Stokes Versus Anti-Stokes Scattering

The usual relationships between Stokes (S) and anti-Stokes (AS) scattering are not preserved [7.116]. For example, in usual Brillouin scattering, the slight energy asymmetry between S and AS peaks, of the order of u_s/c $= \Delta E/\hbar\omega_i \approx 10^{-5}$, is never detected in standard (nonresonant) measurements.

In the case of polariton scattering, however, due to the *strong dispersion*, the q's of S and AS phonons (downwards or upwards lines in the ω, K space) can differ markedly and lead to asymmetric multiplets (see, e.g., Fig. 7.11). The connection between S and AS intensities was examined in detail by *Loudon* [7.116] on the basis of the time-reversal symmetry of light scattering experiments. The usual relationship is $I_s/I_{AS} = [n(\omega_q) + 1]/n(\omega_q)$, where $n(\omega_q)$ is the number of phonons per mode with frequency $\omega_q = u_s|q|$ at temperature T, given by $n(\omega_q) = [\exp(\hbar\omega_q/kT_B) - 1]^{-1}$. However, this assumes that scattering probabilities vary slowly over an energy range of $\approx \hbar\omega_q$. This is *not* the case in RBS of polaritons, where the density of final states ($4\pi k^2/v_g$ in the spherical band approximation) which enters the transition probability can vary as much as a factor of ten over the energy interval separating S and AS peaks. In fact, one can observe AS peaks larger than S ones (see, e.g., Fig. 7.11), an extraordinary situation in light scattering experiments.

c) Multimode Scattering

Above E_L, several polariton modes can propagate in the crystal, leading to multichannel scattering. In RBS, the finite dispersion of acoustic phonons allows the various channels to be energy-resolved, as shown in Fig. 7.10. This was the major prediction of the BZB paper [7.17] which forecasted the observation of an octuplet of Brillouin peaks in CdS (4S, 4AS) instead of the usual doublet. However, due to the more or less strong coupling of the modes to external photons (through transmission coefficients) and strongly different damping coefficients, not all modes are to be observed simultaneously. The first multiple RBS peaks were observed in GaAs [7.117]. There, due to the existence of heavy and light excitons, up to 9 peaks should be observed on either side of the unshifted Rayleigh line. Actually, the maximum number of simultaneously observed lines was only 3, but interbranch as well as intrabranch scattering was observed in the first experiment [7.18,117]. Again, better cases seem to be ZnSe and CuBr in which many more peaks were observed (Fig. 7.12). In CdS and CdSe, $2 - 2'$ intrabranch and $2 - 1'$, $1 - 2'$ interbranch scatterings were observed early [7.118, 119], but it is only very recently that the UPB $1 - 1'$ intrabranch scattering was observed, thanks to the use of multipass Fabry-Perot interferometers [7.120, 121].

7.3.2 Scattering Efficiencies

We shall use a very straightforward derivation of the scattering cross section exposed in greater detail in various places [e.g., 7.30]. From first-order perturbation theory, the scattering probability per unit time from a polariton state $|K_i\rangle$ into the polariton state $|K_s\rangle$ is

$$\left(\frac{dP}{dt}\right)_{K_i \to K_s} = T_i T_s \frac{2\pi}{\hbar} |\langle K_i| V |K_s\rangle|^2 \delta(\hbar\omega_i - \hbar\omega_s + \hbar\omega_q). \tag{7.16}$$

The scattering probability into a solid angle $d\Omega_{in}$ inside the crystal is obtained by summing over the $|K_s\rangle$-states contained in $d\Omega_{in}$, i.e.,

$$\sum_{K_s} \rightarrow \frac{K_s^2}{(2\pi)^3} d\Omega_{in} \left(\frac{\partial E_s}{\partial K}\right)^{-1}_{K=K_s} = \frac{K_s^2 d\Omega_{in}}{(2\pi)^3 \hbar v_{gs}}. \tag{7.17}$$

The scattering probability per unit time and per unit angle is then

$$\frac{dP}{dt d\Omega_{in}} = T_i T_s \frac{K_s^2}{4\pi^2 \hbar^2 v_{gs}} |\langle K_i| V |K_s\rangle|^2. \tag{7.18}$$

This can be transformed in the scattering probability (efficiency) for a finite crystal of length L, *assuming negligible damping*, taking into account the time used by polaritons to pass through the crystal $dt = L/v_{Ei}$, where v_{Ei} is the energy transport velocity of incoming polaritons. For small damping, v_{Ei} can be taken as equal to the polariton group velocity v_{gi}. Inserting in (7.18), we obtain

$$\frac{dP}{d\Omega_{in}} = T_i T_s \frac{K_s^2 L}{4\pi^2 \hbar^2 v_{gi} v_{gs}} |\langle K_i| V |K_s\rangle|^2. \tag{7.19}$$

To connect this probability *inside* the crystal with the measured probability *outside*, we use the near-normal incidence relation $d\Omega_{out} \approx n_s^2 d\Omega_{in}$, where n_s is the relative index of outgoing polaritons.

A number of additional approximations are usually made in order to compute the outside efficiency

$$\eta = \frac{dP}{d\Omega_{out}} = T_i T_s \frac{K_s^2 L}{4\pi^2 \hbar^2 n_s^2 v_{gi} v_{gs}} |\langle K_i| V |K_s\rangle|^2. \tag{7.20}$$

The matrix element $\langle K_i| V |K_s\rangle$ is that of polaritons. As polaritons interact through their exciton content with phonons or defects, the matrix element $\langle K_i| V |K_s\rangle$ is only approximately equal to that of the excitons with the same wave vector. *Hopfield* has however shown [7.28] that the approximation is good over a wide range around the exact resonance energy such as $|\hbar\omega - E_T| \lesssim E_C = \sqrt{E_T E_{LT}}$.

In the usual case of an absorbing medium with a thickness much larger than the absorption depth (i.e., appearing semiinfinite), one considers L to be an effective scattering length L_{eff} of polaritons before damping. In a simplified model of Raman scattering [7.23], *Loudon* showed that L takes then the form $L_{eff} = (\alpha_i + \alpha_s)^{-1}$, where α_i and α_s are the absorption coefficients for the incoming and outgoing photons, respectively. It is obvious that conceptual difficulties arise here, as L_{eff} depends on the type of scattering being studied. In all cases, inelastic scattering should be considered in L_{eff}. Whether elastic processes have to be included in L_{eff} (or not) *depends on the scattering process:*

they have no influence on the events which are only sensitive to the *energy* of incoming polaritons (such as allowed Raman scattering). On the contrary, they must be considered for scattering processes such as RBS or forbidden Raman scattering which depend on the *momentum* of polaritons either through the energy position of the scattering peak (LA phonon RBS) or through the k-dependence of the matrix element (forbidden RRS), or through both (piezoelectric RBS). The (rarely) available information on L_{eff} comes from transmission experiments which incorporate both elastic and inelastic processes as a cause for the disappearance of transmitted polaritons.

In the case of multiple polariton branches, the separation of the scattering probability into expressions of the form (7.15) with different T's is in principle not valid, as interference effects between scattering channels might occur. One should then resort to the more general description (B) of polariton scattering. However, it is not clear at the moment that this is compulsory, as the influence of multiple elastic scattering events will destroy the phase coherence between polariton amplitudes and their interference. This effect has, however, not been examined up to now and we shall use the simple form (7.15) even in the case of multiple branches.

A major question which arises when considering the mechanism of polariton-mediated light scattering is whether it can be distinguished from the standard third-order pertubative approaches to light scattering. As pointed out by *Bendow* in his review of RRS in solids [7.30], no experiment was available at this time (1977) to clearly show the role of excitonic polaritons in RRS. It will be the purpose of Sects. 7.4–8 to produce such evidence, relying mostly on the *kinematic* (i.e., linked to the dispersion) properties of polaritons and their dynamics. The scattering *efficiency* alone does not clearly prove the mediation through excitonic polaritons as neither the polariton-mediated nor the perturbation efficiencies have been calculated in realistic enough models, taking into account the preceding remarks.

7.4 Resonant Brillouin Scattering (RBS)

Resonant Brillouin scattering of excitonic polaritons is the most direct evidence of the existence of these coupled-mode excitations, as predicted by BZB in 1972 and verified in GaAs in 1977 [7.18, 117]. It is thus a recent, active field, and RBS proves to be a very useful technique for the study of these excitations. Earlier experiments dealing with the *bandgap* resonance of BS have been performed [7.122–124] in a situation where exciton effects are not pronounced (room-temperature measurements usually). As they are not relevant to polariton-mediated phenomena, they will not be discussed here. Reviews of nonpolariton BS in semiconductors were given by *Pine* [7.123] and *Sandercock* [7.125] (see also the most recent contribution by Sandercock in Chap. 6). The description of RBS in the polariton framework has been discussed before by the present authors [7.126]. More detailed accounts of RBS work done since then can be found in recent review articles [7.31, 127].

7.4.1 Polariton Dispersion Curves

The most direct application of RBS is the determination of polariton dispersion curves. From a measurement of the Brillouin shift ΔE
$= \hbar u_s |K_i - K_s| \approx \hbar u_s (K_i + K_s)$ in the backscattering configuration as a function of incident photon energy $\hbar \omega_i$, one can deduce the ω_i, K_i relationship. Actually, the simplest way of performing the analysis of experimental data is to generate polariton dispersion curves from a set of parameters $(E_T, E_{LT}, M_X \ldots)$, calculate numerically the resulting Brillouin shifts and to adjust the parameters so as to fit the data. The task is greatly simplified by noting that the various parameters affect different features of the dispersion curves (Fig. 7.10): E_T sets the energy at which the dispersion curve is significantly bent; E_{LT} mainly represents the distance between the points of strongest curvatures of the UPB and LPB; the exciton mass is determined by the dispersion of the LPB at large wave vectors (see Fig. 7.14 below).

One often encountered problem lies in the poor knowledge of u_s and ε_b, the former because of the lack of reliable sound velocitys measurements at low temperatures, the latter because of the difficulty in isolating the dielectric background constant in the vicinity of an exciton resonance: usually ε_b is larger than ε_∞ due to the nearby bandgap and other exciton resonances (see, e.g., [7.87]). Measuring ΔE in the photon-like polariton region below the exact resonance where dispersion is negligible, one obtains $\Delta E \approx 2 \hbar u_s K_i \approx 2 \sqrt{\varepsilon_b} u_s \hbar \omega_i / c$; therefore, one determines ε_b (of which it is a very good measurement) or u_s, knowing the other quantity. The real problem arises when neither of them is known.

As discussed in Sect. 7.2.4, the strong dependence of reflectivity measurements on uncontrolled surface parameters (surface quality, purity) renders the determination of polariton parameters quite hazardous from such measurements. RBS, on the contrary, enables one to test excitations well inside the crystal up to $\approx L_{eff}$ and is thus much less dependent on surface properties. This explains the success of the RBS technique to determine exciton parameters together with the ease of its implementation: to our knowledge, RBS has been observed in every semiconductor tried, with the single exception of InP [7.128]. Thus, the exciton parameters in GaAs [7.18], CdS [7.118, 119, 129], CdTe [7.126], CdSe [7.130], ZnSe [7.20, 131], ZnTe [7.132], PbI$_2$ [7.133], CuBr [7.115, 134] were determined, most often for the first time. Table 7.2 gives a résumé of the results obtained. As an example, Fig. 7.13 shows the fit obtained in GaAs using a *two-band* model of polariton (i.e., *one* exciton band). A two-band model is, in this case, sufficient to account for the peaks, although one would require in principle a three-band model because multibranch scattering occurs only in a very restricted energy range (~ 0.1 meV around E_L) where opacity broadening prevents the observation of more than three Stokes peaks (see Fig. 7.11, for example). Therefore, only the $2 - 2'$ and $1 - 1'$ scatterings were identified unambiguously*. As pointed out in [7.18], the $2 - 2'$ scattering allows

---

*   *Note added in proof:* In a recent study R. Sooryakumar and P. E. Simmonds report having observed the intermediate branch scattering [Solid State Commun. **42**, 287 (1982)].

Table 7.2. Exciton parameters measured by resonant Brillouin scattering (RBS) (see references in text)

	E_T [eV]	E_{LT} [meV]	$u_s[10^5\,\mathrm{cm\,s^{-1}}]$	ε_b	M_X [m_0]	Additional remarks
GaAs [100]	1.515	0.08	4.8 LA 3.36 TA	12.56	$M_{hX}=0.7$	Due to the smallness of E_{LT} the intermediate polariton branch could not be resolved
CdTe [110]	1.595_4	0.4	3.41 LA 1.87 TA	10.1	$M_{hX}=2.4$	Preliminary analysis in a two-branch polariton model
ZnSe [100]	2.8023	1.45	4.21 LA 2.67 TA	8.18	$M_{hX}=1.11$ [100] $M_{lX}=0.3$	M_{hX} [110] $=1.95$; M_{lX} [110] $=0.37$ exchange splitting $\delta=-0.1\pm0.1$ meV
ZnTe	2.3804	0.8		7.1	$M_{hX}=0.70$ [100] $M_{lX}=0.27$	M_{hX} [110] $=0.78$; M_{lX} [110] $=0.26$
CuBr	2.9633	11.6	3.33 LA	5.4	$M_{hX}=1.81$ [100] $M_{lX}=1.33$	Exchange splitting $\delta=0.1\pm0.05$ meV K-linear term $C_k=5\times10^{-10}$ eV cm M_{hX} [110] $=1.76$; M_{lX} [110] $=1.35$ M_{hX} [110] $=1.74$; M_{lX} [111] $=1.36$
CdS A exciton	2.5528	1.9	4.25 LA	$\varepsilon=9.3$	$M_\perp=0.89$ $M_\parallel=2.85$	
CdS B exciton	2.3923	$\omega_{LT\perp}=1.25$ $\omega_{LT\parallel}=1.1$	4.25 A	$\varepsilon\perp=7.2$ $\varepsilon\parallel=8.9$	$M_\perp=1.2$ $M_\parallel=0.74$	k-linear term coefficient $\phi=5.6\times10^{-10}$ eV cm
CdSe	1.824	0.50	3.57 LA	8.4	$M_\perp=0.40$ $M_\parallel=1.3$	
HgI$_2$ [001]	2.3355	5.3	1.6	6.8	$M=1.2$ [001]	ε_b should be larger than 6.8 to fit data below E_T M_X (100) $=0.68$ M_X (201) $=0.75$

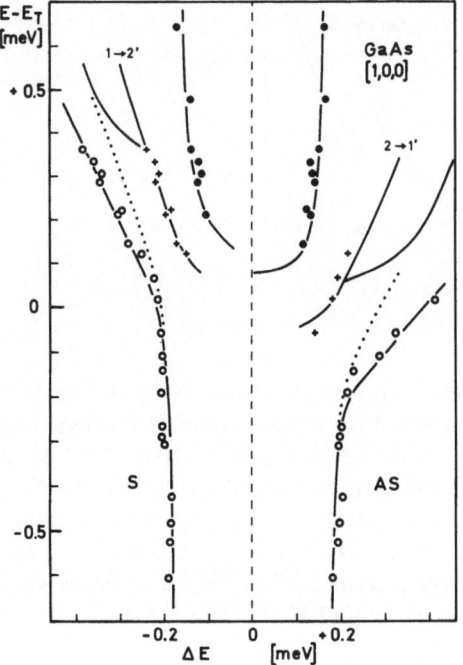

Fig. 7.13. Measured Brillouin shifts as a function of incident photon energy $\hbar\omega$ in GaAs. The full lines correspond to theoretical values calculated within the two-band, isolated-resonance approximation with parameters of Table 7.1. The dotted line shows the sensitivity of the fit to a change in the M_X value, from $0.7\,m_0$ to $0.3\,m_0$. From [7.126]

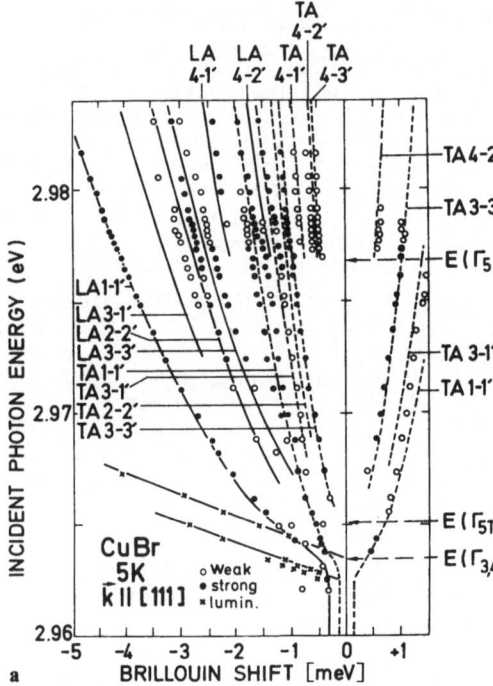

a

Fig. 7.14a, b. Measured and calculated Brillouin shifts in CuBr (**a**). The various polariton branches in this case of degenerate valence band structure are shown in (**b**), as calculated from (7.13). Exchange and k-linear terms give rise to up to five polariton branches in the [110] direction. From [7.115]

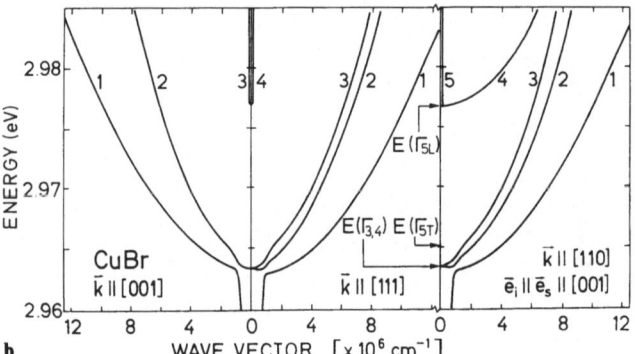

b

the precise determination of the heavy-exciton mass, of which it was the first measurement in a semiconductor. (Compare, for instance, with the insensitivity of reflectivity measurements to exciton mass as reported in [7.10a].) The good agreement of the mass $M_X = 0.7 \, m_0$ with the theoretical evaluations [7.42, 43] is a strong support for the successive perturbation description of polariton energy levels, even in the degenerate valence band case. Since then, RBS has been measured in other cubic semiconductors and the full three-band dispersion curve [7.84, 85] has been determined in ZnSe and CuBr (Fig. 7.14).

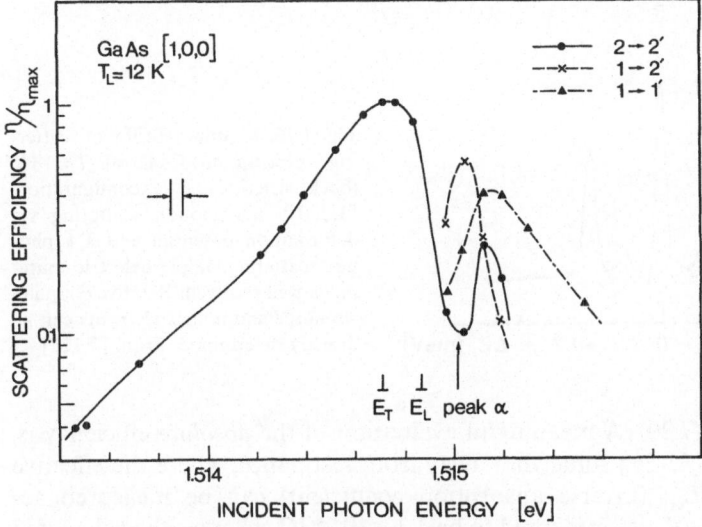

Fig. 7.15. Measured absolute scattering efficiency per steradian for LA phonon scattering in [1,0,0] GaAs (Stokes process, backward scattering configuration) at $T = 12$ K as a function of incident photon energy. E_T denotes the transverse exciton energy, peak α the spectral position of the maximum absorption in the $n = 1$ exciton resonance. From [7.126]

7.4.2 Phonon Coupling

The various exciton-phonon couplings were discussed in Sect. 7.2.2. Depending on conditions, one or another coupling will be allowed and/or dominant. LA coupling is always allowed through the deformation potential mechanism. On the other hand, piezoelectric scattering is usually observable only for those modes which have a longitudinal electric field associated with the phonon field. This fact helps to assign some of the observed Brillouin shifts, by studying the orientational dependence of scattering intensities in these modes.

a) LA Phonon Scattering

This is the usually observed mode, as the coupling by deformation potential is in general allowed [7.125]. Figure 7.15 shows the variation of LA scattering efficiency with the incident photon energy in GaAs in a 2 meV interval around the $n = 1$ exciton resonance E_T. The *efficiency* changes only by about an order of magnitude in this interval, whereas the factor $(v_{gi} v_{gs})^{-1}$ in (7.20) should change ≈ 2–3 orders of magnitude. This indicates that the resonant increase in the scattering cross section is partially offset by an increase in competing scattering rates [a decrease of L_{eff} in (7.20)] as evidenced by absorption-transmission experiments. The problem of performing precise transmission measurements on the same (optically thick) samples on which RBS is measured, prevents at the moment a detailed comparison of the results of Fig. 7.15 with

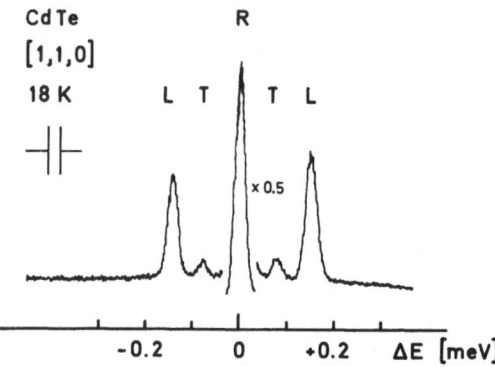

Fig. 7.16. Resonant Brillouin scattering spectrum in CdTe at $T = 18$ K (backscattering configuration, [1, 1, 0]). LA phonon scattering via deformation potential and TA phonon scattering via piezoelectric coupling is well resolved. Relative coupling strengths and matrix elements can be directly determined. From [7.126]

the predictions of (7.20). A meaningful evaluation of the *absolute* efficiency is, however, possible [7.126] somewhat away from resonance, where the effective scattering length L_{eff} (inverse absorption coefficient) can be measured accurately. At 1.5145 eV, η is measured to be $4.2 \times 10^{-4} d\Omega$, whereas the calculated value is $\approx 10^{-4} d\Omega$. This is a reasonable agreement, considering the approximations made.

b) TA Piezoelectric Scattering

The piezoelectric scattering mechanism can be nicely studied on the TA modes (for which the deformation potential mechanism is usually negligible). It would be superimposed on the deformation potential mechanism for the LA modes for which it is allowed ([111] LA in the zinc-blende symmetry [7.72]). The TA scattering was observed first in CdS (wurtzite structure) for $K, q \perp c$ [7.135], then in GaAs and CdTe [7.128] (zinc-blende structure) for $K, q \parallel$ [110], and in CuBr for $K, q \parallel$ [100], [110], and [111] [7.115, 134]. A typical spectra are shown in Fig. 7.16. A very interesting feature of such spectra is that they allow LA and TA scattering mechanisms to be tested *relative* to each other. If we use (7.6, 7, 20) to calculate the relative scattering efficiencies, we find for zincblende compounds

$$\frac{\eta_{TA}}{\eta_{LA}} \approx \frac{q^2 a_B^4}{4} \left(\frac{m_e - m_h}{m_e + m_h}\right)^2 \frac{16\pi^2 e^2 e_{14}^2}{\varepsilon_b^2 E_{dp}} \frac{u_{dp}}{u_{pe}}. \tag{7.21}$$

We have assumed that the final densities of states are equal for LA and TA events, and also the same effective scattering length L_{eff} and transmission coefficients. Away from exact resonance, these assumptions should hold well. The last factor in (7.21) represents the ratio of phonon occupation numbers at finite temperatures. Table 7.3 shows the calculated values of (7.21), along with the parameters used in the calculation and measured values away from resonance. The agreement is quite good for CdS, but worse for GaAs and CdTe if one considers that the parameters are well known (as well as the scattering

Table 7.3. Comparison of theoretical and experimental ratio of TA to LA phonon scattering efficiencies (7.21) and parameters used in the calculations

	GaAs [110]	CdTe [110]	CdS $(K, E\perp c)$
$\left(\dfrac{\eta_{TA}}{\eta_{LA}}\right)_{theor}$	0.095 at 1.515 eV	0.026 at 1.5947 eV	0.83 at 2.553 eV
$\left(\dfrac{\eta_{TA}}{\eta_{LA}}\right)_{exp}$	0.44	0.2	0.4
$m_e\,[m_0]$	0.0665 [a]	0.096 [a]	0.208 [b]
$m_h\,[m_0]$	0.138 [c]	0.34 [d]	0.98 [b]
$E_c - E_v\,[\text{eV}]$	9.6 [e]	4.8 [f]	2.2 [g]
$e_{14}\,[\text{Cb m}^{-2}]$	0.16 [h]	0.0335 [i]	0.210 [g]
ε_0	12.56 [k]	9,10.1 [o]	9.2 [b]
$a_B\,[\text{Å}]$	136 [m]	93 [n]	30 [b]
$q\,[10^5\,\text{cm}^{-1}]$	5.1	6	18.4
$u_{LA}\,[10^5\,\text{cm s}^{-1}]$	5.31	3.41	4.25
$u_{TA}\,[10^5\,\text{cm s}^{-1}]$	3.36	1.87	1.76
$(q_e - q_h)^2$	1.1×10^{-2}	6.6×10^{-3}	1.9×10^{-3}

[a] J.M. Chamberlain, P.E. Simmonds, R.A. Stradling, C.C. Bradley: In: Proc. 11th Conf. on the Physics of Semiconductors, Warsaw, 1972, ed. by M. Miasek (PWN, Warsaw 1972) p. 1016.
[b] J.J. Hopfield, D.G. Thomas: Phys. Rev. **122**, 35 (1961).
[c] M.S. Skolnick, A.K. Jain, R.A. Stradling, T. Leotin, J.C. Ousset, S. Askenazy: J. Phys. C9, 2809 (1976).
[d] V. Capek, K. Zimmermann, C. Konak, M. Popova, P. Polivka: Phys. Status Solidi B56, 739 (1973).
[e] F.H. Pollak, M. Cardona: Phys. Rev. **172**, 816 (1968).
[f] D.G. Thomas: J. Appl. Phys. Suppl. **32**, 2298 (1961).
[g] J.E. Rowe, M. Cardona, F.H. Pollack: In *II–VI Semiconducting Compounds*, ed. by D.G. Thomas (Benjamin, New York 1967) p. 112.
[h] G. Arlt, P. Quadflieg: Phys. Status Solidi **25**, 323 (1968).
[i] D. Berlincourt, H. Jaffe, L.R. Shiozawa: Phys. Rev. **129**, 1009 (1963).
[j] A.R. Hutson: J. Appl. Phys. Suppl. **32**, 2287 (1961).
[k] G.E. Stillman, D.M. Larsen, C.M. Wolfe, R.C. Brandt: Solid State Commun. **9**, 2245 (1971).
[l] C.W. Litton, K.J. Button, J. Waldman, D.R. Cohn, B. Lax: Phys. Rev. B13, 5392 (1976).
[m] D.D. Sell, S.E. Stokowski, R. Dingle, J.V. DiLorenzo: Phys. Rev. B7, 4568 (1973).
[n] P. Hiesinger, S. Suga, F. Willmann, W. Dreybrodt: Phys. Status Solidi B67, 641 (1975).
[o] R.G. Ulbrich, C. Weisbuch: In *Festkörperprobleme – Advances in Solid State Physics*, Vol. 18, ed. by J. Treusch, (Pergamon-Vieweg, Braunschweig 1978) p. 217.

matrix elements). It appears that the cancellation factor $q_e - q_h$, calculated in the hydrogenic approximation, is significantly underestimated: due to the valence band degeneracy, the exciton envelope wave function has not a purely $1s$ character, but is mixed with $2s$, $3s$, etc. excited states. The strong dependence of (7.21) on the exact "Bohr radius" (to fourth power!) indicates a very strong influence in this expression of the "tails" of the otherwise $1s$-like exciton envelope function. The exact calculation of the cancellation factor $(q_e - q_h)$ was not made due to the lack of precise exciton wave functions. One can, however, estimate from the perturbation admixture coefficients given by *Kane* [7.42] that the cancellation factor could be increased by approximately one order of magnitude.

A very convincing verfication of the preceding discussion is provided by the observation of $n=2$ exciton RBS in CdTe [7.136]. Only TA piezoelectric scattering is detected, yielding $(\eta_{TA}/\eta_{LA})_{x_{n=2}} > 10$. Considering that the LA scattering cross section should not have changed significantly between $n=1$ and $n=2$ excitons, this represents an increase of $\eta_{TA} \approx 50$ over $n=1$ exciton TA scattering. Such an increase in $q_e - q_h$ is well explained by a direct computation of q_e and q_h for $1s$ and $2s$ states. The inobservability of $n=2$ LA scattering peaks may be due to the stronger damping of $n=2$ excitons due to their larger radius, i.e., a smaller L_{eff} than for $n=1$ excitons.

7.4.3 Damping and Opacity Broadening

Up to now, only undamped polariton modes have been considered in our discussion of *kinematic* properties of RBS. However, besides determining the absolute scattering efficiency through L_{eff}, polariton damping has two additional important effects: it modifies the dispersion curve and broadens Brillouin peaks. Due to finite experimental resolution in the RBS measurements (up to now only grating spectrometers were used, with the exception of [7.118, 120, 121]), the dispersion curve modification could not be observed, but the broadening of the Brillouin peaks was. First observed by *Sandercock* in Si and Ge [7.21] under nonresonant conditions, it can be understood in the following way: polaritons conserve a well-defined K vector on a length $L_{eff} \approx (Im\{K\})^{-1} \approx \alpha^{-1}$. Correspondingly, the phonon wave vector q is uncertainty broadened by a quantity $\Delta q \approx L_{eff}^{-1} \approx Im\{K\}$. This results in a HWHM energy broadening $\Delta(\hbar\omega_q) = \hbar u_s Im\{K\} = \hbar u_s L_{eff}^{-1}$ and a relative broadening $\Delta(\hbar\omega_q)/\hbar\omega_q = Im\{K\}/Re\{K\}$. In a classical analysis, *Sandercock* [7.21] derived this result and also determined that the lineshape should be Lorentzian. More recent analyses [7.137] of nonresonant BS, taking into account phonon reflection at the interface, yield that the lineshape should actually be a skewed Lorentzian. In their original work on RBS, BZB [7.17] took damping into account in an elementary way, finding a Brillouin shift $\Delta E = 2\hbar u_s Re\{|K_i| + |K_f|\}$ and a Lorentzian lineshape with a relative FWHM line width $\Delta = 2 Im\{K_i + K_f\}/Re\{K_i + K_f\}$. A more detailed theory of RBS by *Tilley* [7.138], taking into account phonon reflection and damping, predicts skewed Lorentzian lineshapes of the form calculated by *Dervisch* and *Loudon* for the nonresonant case [7.137]. The Brillouin peak should occur at

$$q = [(Re\{K_i + K_f\})^2 + (Im\{K_i + K_f\})^2]^{1/2} \qquad (7.22)$$

and the FWHM should be

$$\Delta q = 2 Im\{K_i + K_f\}. \qquad (7.23)$$

Detailed formulas were also derived for Brillouin *efficiencies* in [7.138].

Lineshape studies of RBS have been quite limited, as the detailed experiments reported up to now were only performed with grating spectrometers which lack the resolution needed to verify (7.22, 23) in detail. *Weisbuch* and *Ulbrich* ([7.139], see also [Ref. 7.126, Fig. 7.13]) observed a significant broadening $\Delta(\hbar\omega_q)/\hbar\omega_q \approx 10\%$ at resonance in GaAs (see curve at $\hbar\omega - E_L$ $= 0.25$ meV of Fig. 7.11) which is in good agreement with the measured absorption coefficient. Opacity broadening was also reported in CdSe by *Hermann* and *Yu* [7.130, 140] who were able to account for their line width data by using (7.23). No lineshape analysis, however, was made.

7.4.4 RBS and the Problem of the Determination of ABC's

Although reflectivity and transmissivity experiments should provide tests of ABC's proposals, they prove very disappointing in that respect as discussed in Sect. 7.2.4. The main reason is that they do not sort out the various polariton waves propagating at a given energy above E_L. From the start [7.17], RBS was considered to be a promising experimental method to test ABC's: because of the dispersion of acoustic phonons, an energy discrimination between propagating polaritons belonging to different branches of the dispersion diagram is possible. Already in their original paper [7.17], BZB displayed theoretical evaluations of the spectral variation of the RBS cross section for various ABC's (Fig. 7.17): the various polariton branches are coupled to external photons

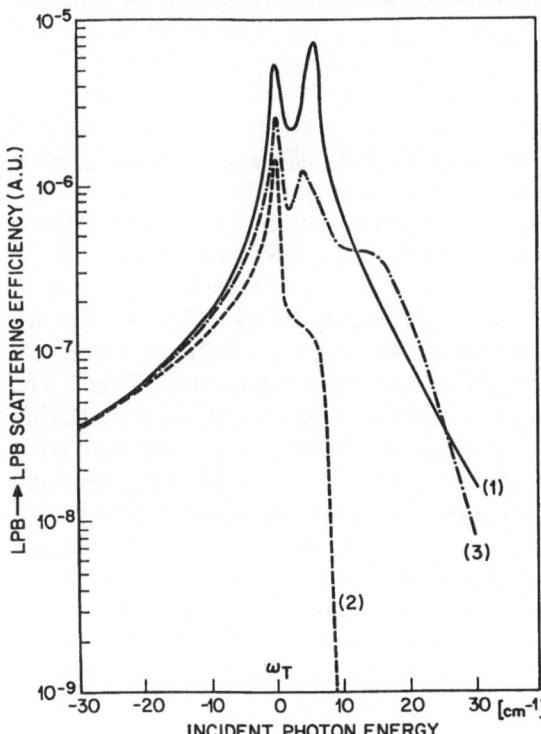

Fig. 7.17. Theoretical RBS efficiency spectra for three different additional boundary conditions (parameters for CdS, backward scattering, process $2 \rightarrow 2'$). From [7.17]

through different transmission coefficients, hence the dependence of the efficiency on ABC's (7.20). In this manner, RBS appears to be quite a powerful technique to solve the ABC problem. There are, however, a number of implicit assumptions which render the actual comparison of experimentally determined efficiencies with expected ones very risky:

(i) All interference effects between polaritons states are being neglected, although *Zeyher* et al. [7.114] showed that they produce a sizeable effect in RRS. Unfortunately, no explicit evaluation of these effects for a realistic case exists.

(ii) The evaluation of (7.20) requires the knowledge of L_{eff} for the various branches under consideration. It should *not* be the same for the various branches, and its evaluation from transmission or reflectivity data requires extensive, careful experiments on the same samples as those on which RBS is performed.

(iii) Multiple scattering events, such as those due to elastic scatterings, must be taken into account. They can strongly modify the n_s^2 factor relating inside and outside solid angles, i.e., efficiencies [7.113, 141].

(iv) One may wonder about the influence of extrinsic factors such as space-charge layers, surface and bulk impurities, surface roughness, etc., on the ABC's. We already know from reflectivity measurements that dead-layer effects strongly affect the branching of outside photon states to inside polariton states. Such a description is only a phenomenological description of the extrinsic interface effects which should be properly taken into account for a realistic description of the interface problem (see the fascinating account [7.142] of how similar reflectivity curves can be generated by vastly different exciton surface potentials).

With these remarks in mind, the reported RBS efficiencies do not provide enough information in order to solve the ABC problem. The authors concluded from data for GaAs that the observed spectral variation of η supports Pekar's choice of ABC [7.18]. More detailed experiments [7.126] showed later that the data cannot be simply reconciled with theory: instead of displaying a peak at resonance (minimum v_g), as predicted by (7.20) or by BZB [7.17], the efficiency curve (Fig. 7.15) displays a dip. This was analyzed as being due to the resonance in the *elastic* polariton scattering (EPS) probability by impurities [7.126]. The direct observation of EPS indeed shows that it overcomes all other scattering mechanisms at resonance [7.143]. However, no simple comparative analysis of RBS and EPS can reconcile the RBS data with a simple model of P_{out} resulting from the competition between only these two scattering processes. In such a model, the integrated RBS efficiency should be unity, as no other energy dissipative mechanism would exist. Since the actual *total* radiative efficiency is never larger than a few per cent, one must admit that other dissipative mechanisms are present. A very likely candidate is polariton dissociation by impurities [7.64, 65].

For CdS, *Yu* and *Evangelisti* [7.144] concluded from their analysis of the 2 − 2′ scattering efficiency in CdS that Pekar's ABC could not explain their results and only *Pattanayak* et al's ABC [7.94] could. The present authors believe that the evidence presented in [7.144] depends crucially on further implicit assumptions made, as discussed in the beginning of this paragraph. Some adjustments in the model of the interface potential and/or damping could change the theoretical evaluation of *the only observed 2 − 2′ scattering efficiency in* [7.144] quite drastically. More convincing evidence would at least require the *simultaneous observation and fitting* of 2→2′ and 1→1′ scattering in order to observe the switching of polariton propagation from one mode to another.

From the preceding discussion, it is clear that the *theoretical* evaluation of a specific ABC influence on RBS is extremely difficult due to the damping (with several possible mechanisms) and the delicate interface transition. Turning to the experiments, one might wonder what would be the minimum ingredients of an *experimental* determination of ABC's. It would clearly require a converging, redundant set of measurements such as reflectivity, transmission and RBS efficiency. A very useful additional piece of information would be provided by the lineshape analysis of RBS peaks which would give a measure of the damping directly.

7.5 LO Phonon Raman Scattering (RRS)

The theoretical aspects of polariton-mediated Raman scattering were thoroughly reviewed by *Bendow* [7.30] and compared to other theoretical treatments of RRS. Although numerous treatments of *exciton*-mediated RRS exist and despite vigorous activity in that field [7.145, 146], it must be emphasized that theoretical treatments of *polariton*-mediated RRS stayed much in the state of *Bendow*'s studies up to 1971 [7.27] with the later re-examination by *Zeyher* et al. [7.114] along approach B. Therefore, *quantitative* evaluations of experiments are not yet possible. As will become clear in the discussion of hot luminescence (HL) phenomena below, the analysis of polariton-mediated RRS requires a rather detailed investigation, taking into account elastic and inelastic multiple scattering in addition to the very specific propagation properties of polaritons. We shall, therefore, limit the discussion to recent experimental results which pertain to the polariton behavior of RRS, those being only a very small portion of existing data. Earlier experiments were reviewed by *Mooradian* [7.36], *Permogorov* [7.147], *Pinczuk* and *Burstein* in [7.32], *Martin* and *Falicov* [7.32], *Richter* [7.76], and *Yu* [7.31]. A very useful situation is encountered when both RBS and RRS are measured in the same crystal as in CdS [7.148] and CdTe [7.149]. One then obtains rather complete information on polariton–phonon interactions in the material which should be helpful in calculating the thermalization properties of polaritons.

7.5.1 Polariton-Mediated RRS Cross Sections

Formulas (7.9, 20) can be used to deduce a cross section for polariton-mediated RRS using approach *A*. Besides its fundamental approximate nature, the three-step description of RRS raises additional questions such as: What is the influence of the *multiple* intermediate states possible in the scattering process? From Fig. 7.18 which shows the schematics of a two-LO phonon scattering event, various exciton ground and excited states and band continua can participate as intermediate states. The weighing of the various states is complicated by the *K*-dependent nature of the Fröhlich coupling. It is an open question which mechanism mediates the RRS above the bandgap. Several authors have claimed that the contribution of *continuum* exciton states (i.e., electron–hole pairs) should predominate in that case (see, e.g., [7.150]). An additional complication is the necessity to actually deal with a *multiple* resonance situation, as the intermediate states can be real discrete or continuum exciton states.

It is remarkable that up to a recent date the in-going and out-going resonances predicted by (7.20) at the minima of v_{gi} and v_{gs} were not observed. In

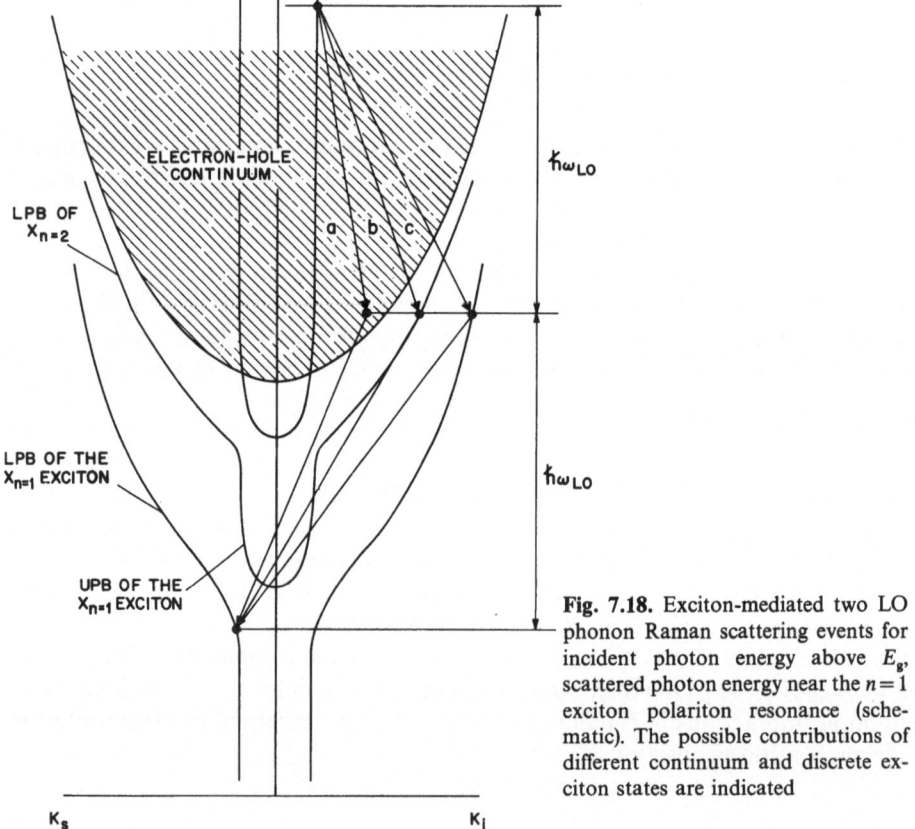

Fig. 7.18. Exciton-mediated two LO phonon Raman scattering events for incident photon energy above E_g, scattered photon energy near the $n = 1$ exciton polariton resonance (schematic). The possible contributions of different continuum and discrete exciton states are indicated

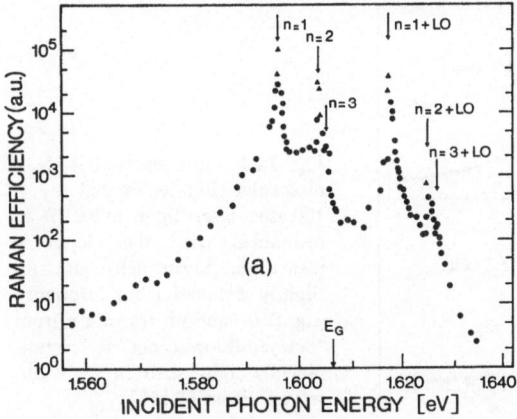

Fig. 7.19. Resonant Raman scattering efficiency for incident photon energies around and above the exciton resonances ($n = 1, 2, 3$) in CdTe. Ingoing *and* outgoing resonances [i.e., incident (resp. scattered) photon energy close to exciton level] were clearly observed. Triangles indicate events where no sharp RRS peak can be distinguished from broader luminescence bands (see Fig. 7.22). From [7.149]

the case of GaSe [7.151] and CdSe [7.152], they were mainly observed on the *absorption corrected cross section*. The directly measured *efficiency* was rather constant. In a recent study of RRS in CdTe [7.149], very large (≈ 5 orders of magnitude), sharp in-going and out-going resonances were observed near various exciton levels (Fig. 7.19) for the *uncorrected* efficiency. Although the data were not analyzed in detail along (7.20), a number of conclusions could be reached about the mediation of RRS by polaritons.

(i) *Separate* resonances are observed for different exciton levels which point out the correct use of the *isolated resonance* concept in the description of RRS.

(ii) The RRS cross section *above* the bandgap is still dominated by exciton effects, as is shown by the out-going resonance of Fig. 7.19 at $\hbar\omega_i \approx 1.62$ eV. The e-h pair mediated RRS therefore appears to be much smaller than that due to excitonic polaritons.

(iii) Due to the same difficulties as in the analysis of RBS data, no detailed conclusion can be drawn from the analysis of the spectral variation of the efficiency η. In particular, we cannot point out here whether the variation of η is typical of polariton-mediated or exciton-mediated RRS (see, e.g., the discussion in [7.30, 153]). It is only recently that the role of elastic multiple scattering in relaxing the K-conservation rule has been elucidated. This was at the origin of the discrepancies reported in several unfruitful attempts to correlate forward and backward RRS efficiencies mediated by the K-dependent Fröhlich mechanism [7.154]. The importance of impurity scattering on RRS and RBS simultaneously was demonstrated in CdSe [7.140].

7.5.2 Resonant Raman Scattering and Hot Luminescence (HL)

It has been a matter of controversy to decide whether RRS could be distinguished from HL [7.155]. Briefly stated, the problem is the following: in a scattering event, intermediate states are populated during only the very short

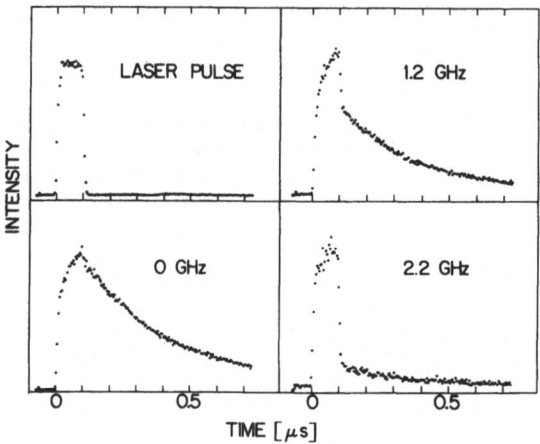

Fig. 7.20. Time-resolved RRS of molecular iodine excited by a 100 nsec laser light pulse (i) at resonance with the electronic transition (lower left) and (ii) slightly detuned from resonance (right). A smooth transition from "delayed fluorescence" to "instantaneous light scattering" is observed. From [7.156]

time $\hbar/\delta E$ given by the uncertainty relation (δE is the energy difference between the incident photons and the nearest excited energy level). Therefore, the scattering process is (i) instantaneous, (ii) does not introduce any broadening as the intermediate state is too short-lived to experience interactions with the surroundings. A large amount of literature has been devoted to the detailed study of how far these assertions still hold in the case of RRS where long-lived intermediate states can exist, and whether the scattering process could be decomposed in a three-step process, namely, absorption followed by relaxation and then emission. Long-lived states have been directly evidenced in the case of resonant excitation of I_2 molecules vapor (Fig. 7.20): *Williams* et al. [7.156] observed the transformation from a fast scattering process away from resonance to a slow process consisting of a real molecular-level excitation followed by spontaneous emission with the ordinary lifetime [7.156, 157]. This possibility of describing the resonant scattering process in successive steps, excitation-relaxation-emission, has been studied in detail for standard electronic systems using stochastic forces to describe intermediate state interactions [7.158]. A fundamental parameter is the *damping* of the intermediate state, which determines the degree of correlation (i.e., the ratio of RRS to HL) of scattered photons with the incident photons. The situation is somewhat more complicated in the polariton case and has not been studied up to now in detail. It must first be emphasized that the scattering process is no more instantaneous: the finite polariton propagation time in and out of the crystal introduces delays between the incident light pulse and the scattered light pulse. The transient spectroscopy of RRS was not performed, but a closely related experiment was performed by *Planel* et al. [7.159] in CdSe: the time of flight of polaritons was measured using the rotation of their polarization vector under an applied longitudinal magnetic field. This experiment can be considered as a resonant Faraday rotation and the rotation is proportional to the lifetime of polaritons due to their propagation time in the crystal ($\approx L \times v_g^{-1}$) and their

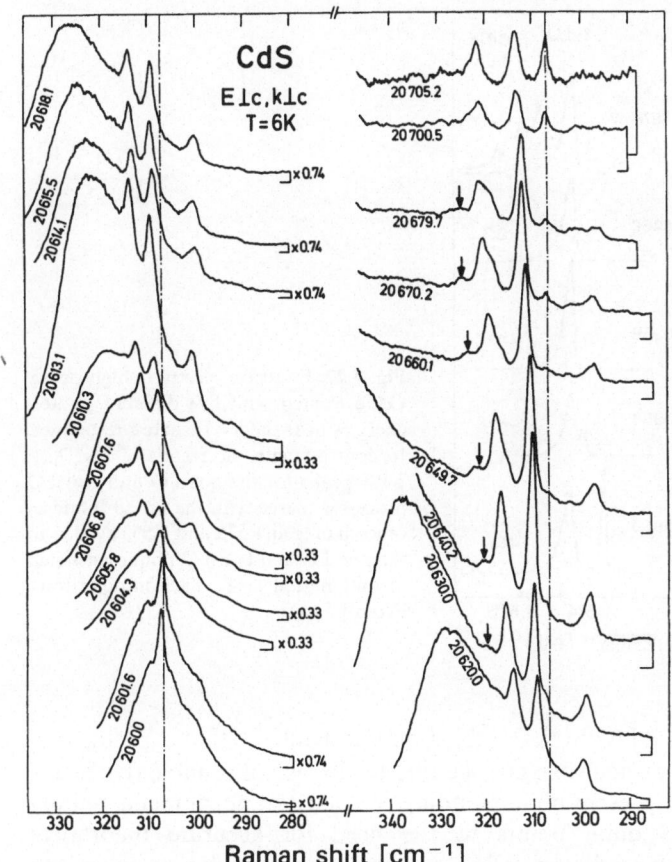

Fig. 7.21. Sharp RRS peaks superimposed on a broad background of luminescence in a light-scattering experiment in CdS. The three peaks observed are due, in order of increasing shift, to 1 LO, 1 LO + 1 TA and 1 LO + 1 LA phonon scatterings. From [7.148]

recombination time. The delay deduced from the rotation angle is well accounted for by this simple model. Due to the strong interactions of polaritons with other crystal excitations, one does not expect the scattered Raman peaks to be infinitely sharp either. Actually, this sharpness is determined by the balance between polariton lifetime and inelastic scatterings leading to a more or less complete thermalization. Very detailed studies of the thermalization effects in CdS have been performed by *Permogorov* and co-workers [7.160] and *Planel* et al. [7.161], and in CdSe by *Planel* et al. [7.162]. It is remarkable that sharp RRS peaks were always observed superimposed over broad phonon-assisted luminescence bands (Fig. 7.21). It is only in the case of CdTe that the complete transformation of sharp RRS peaks into broad luminescence bands was observed (Fig. 7.22). This is due to the high purity of the CdTe samples which leads to the most intrinsic character of observed effects. If a conclusion

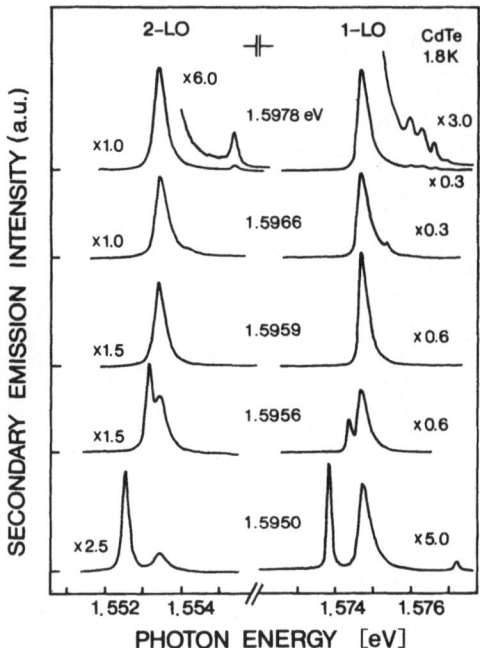

Fig. 7.22. Emission spectra of high-purity CdTe excited with five different photon energies near the $n=1$ exciton resonance. In contrast with the results of Fig. 7.21, RRS peaks for the one-LO and two-LO processes merge with the broad luminescence bands at 1.575 and 1.553 eV due to the one-LO and two-LO phonon assisted recombination of exciton-polaritons. From [7.149]

were to be drawn from RRS studies at this point, it would be that experimental results have shown a considerable change due to the use of tunable dye lasers and purer samples. Our present understanding of polariton phenomena, mainly originating from RBS data, points to the need for accurate theoretical treatments of polariton-mediated RRS, in particular in order to take into account specific polariton effects such as finite propagation times and multiple elastic and inelastic scattering events.

7.6 Multiphonon Scattering and Polarization Effects

7.6.1 Scattering by Two-Acoustic Phonons

From a simple inspection of two-acoustic phonon scattering (Fig. 7.23b), it seems improbable that such events would lead to well-defined peaks, as the various paths in K space correspond to different energies due to acoustic phonon dispersion. Nevertheless, in CdS, *Yu* and *Evangelisti* [7.163] were able to demonstrate that, due to (i) the large anisotropy of the polariton dispersion curve (Fig. 7.23c) and (ii) the q and directional dependence of the piezoelectric matrix element, sharp 2 LA, LA+TA or 2 TA phonon peaks could be observed. In this way, the "anomalous" RBS peaks previously reported by *Winterling* and *Koteles* [7.119, 164] were explained as two-phonon peaks.

TWO-PHONON BRILLOUIN SCATTERING

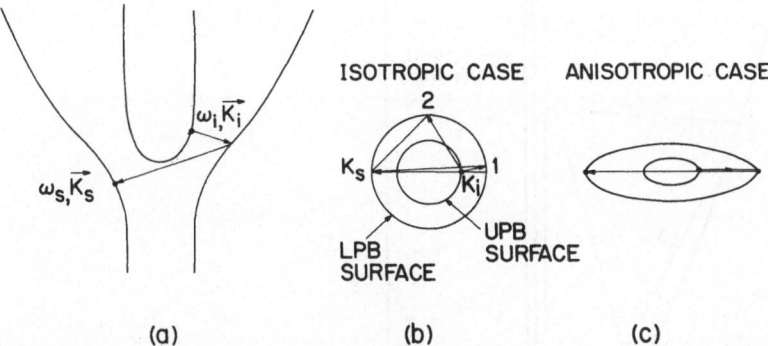

(a)   (b)   (c)

Fig. 7.23a–c. Kinematics of two-acoustic-phonon scattering in the vicinity of a polariton/exciton resonance: (a) dispersion curve $\omega(K)$; (b), (c) projections of K-space for initial and final energies ω_i, ω_s for isotropic and anisotropic $\omega(K)$ relations

7.6.2 Acoustic Phonon and LO Phonon

The combined $LO+LA$, $LO+TA$ phonon scattering has been observed in a number of semiconductors: CuCl [7.165], CdSe [7.166], ZnTe [7.167], Cu_2O [7.168], CdTe ([7.169], see [Ref. 7.149, Fig. 2] for spectra), CdS [7.148], AgBr [7.170]. The advantage of these events over one-LO phonon events is that they are *dispersive* due to the dispersion of the acoustic phonon (Fig. 7.24) and therefore bring essentially information similar to RBS about the polariton dispersion curve (with somewhat less precision, however) (Fig. 7.25). An interesting feature is the ordering of phonon emission, as noted by *Oka* and *Cardona* [7.167], originating in the density of intermediate states and q dependence of the matrix elements; on the in-going resonance ($\hbar\omega_i \approx E_T$, $\hbar\omega_s \approx E_T - \hbar\omega_{LO}$), the acoustic phonon is emitted first. On the out-going photon resonance ($\hbar\omega_i \approx E_T + \hbar\omega_{LO}$, $\hbar\omega_s \approx E_T$), the LO phonon is emitted first (the case of Fig. 7.25). Some recent calculations were made on the lineshape of such processes, but only in the exciton model and not taking into account polariton effects [7.146].

7.6.3 Two-LO Phonon and Multiphonons

As noted earlier for single-phonon scattering, most of the previous work was performed under conditions for which no polariton effect was expected (doped samples, high temperatures). A notable exception is to be found in the work by *Permogorov* and *Travnikov* on CdS [7.171]. They used the *ratio* of one to two-LO phonon scattering efficiencies to determine experimentally a portion of the LPB. For K's small enough ($K < a_B^{-1}$), (7.9) shows that the one-LO matrix

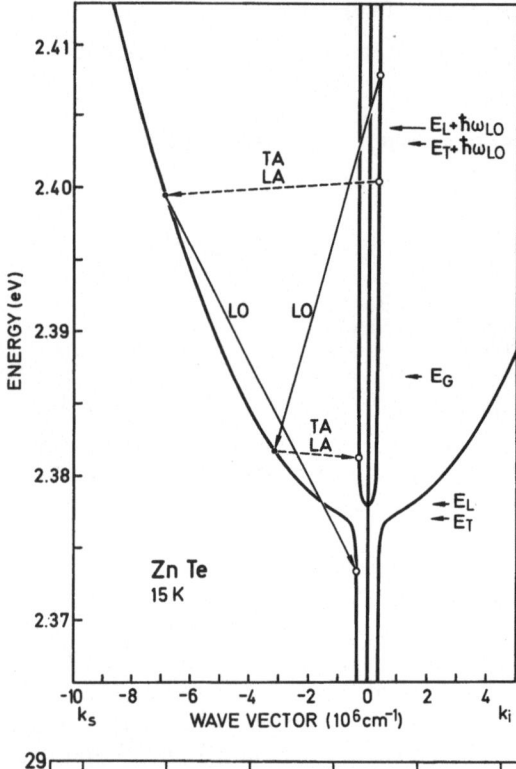

Fig. 7.24. Two-phonon (LO + LA, LO + TA) scattering processes of excitonic polaritons near the $n = 1$ exciton resonance in ZnTe. Note the dispersive character of the *acoustic* phonon contribution and also the difference in density-of-intermediate states depending on whether the LO phonon or the acoustic phonon is emitted first. From [7.167]

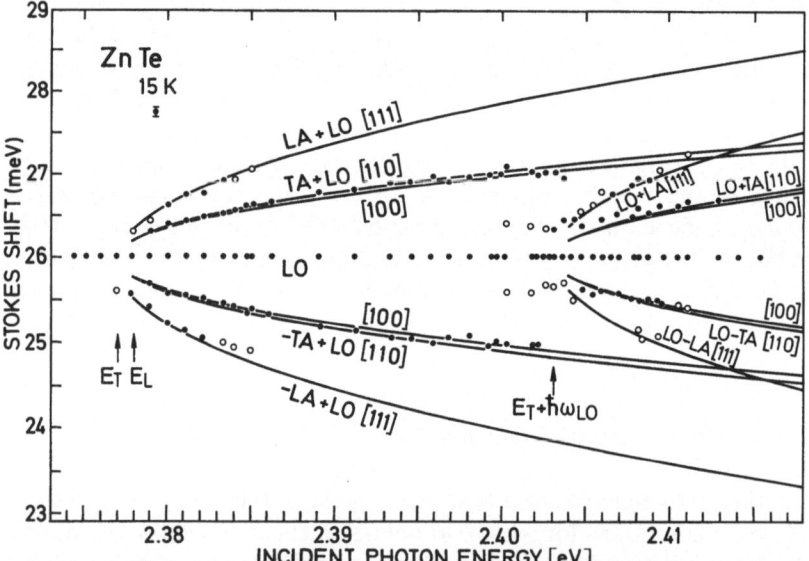

Fig. 7.25. Stokes shift of two-phonon (LO + TA, LO + LA) inelastic scattering peaks and the LO-Raman peak as a function of incident photon energy around and above the lowest exciton energy E_T in ZnTe. Full lines: theoretical fits for acoustic phonon propagation in the directions indicated in brackets. From [7.167]

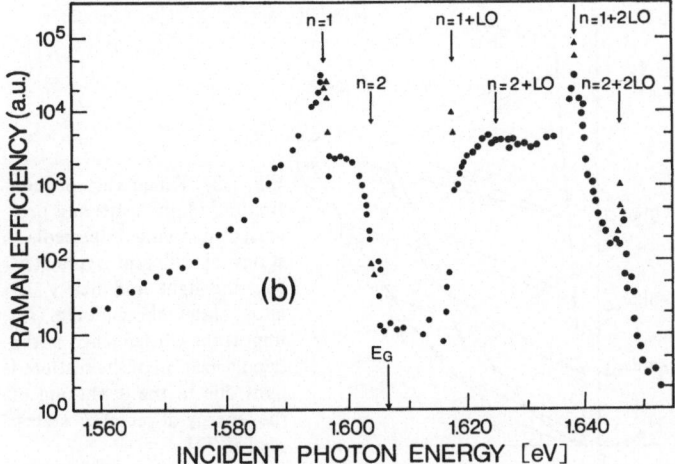

Fig. 7.26. Two-LO phonon scattering efficiency as a function of incident photon energy in CdTe. Distinct, sharp peaks are found for ingoing, intermediate and outgoing energies resonant with the $n = 1, 2, 3$ exciton levels. From [7.149]

element varies $\sim |K_i + K_s|^2$ whereas the two-LO phonon matrix element, representing a sum over a large K-intermediate state, is mostly independent of K_i and K_s. Assuming that other factors entering the scattering probabilities vary smoothly over the range of interest, the dispersion curve (K_i versus $\hbar\omega_i$) could be directly determined with fair precision.

Detailed measurements of the two-LO scattering efficiency were recently made on CdTe [7.149]. The data show that the efficiency is completely dominated by polariton effects, even when exciting light energy is well above the bandgap, as is the case in the out-going resonance and the intermediate state resonance (Fig. 7.26). Complete thermalization is also observed at every resonance.

7.6.4 Polarization Correlation in RRS

It has been known for a long time that standard selection rules break down in the case of resonant scattering [7.32, 75, 76] because of the involvement of real, long-lived, finite-K intermediate states. However, if no relaxation of the intermediate state occurs, the *polarization memory* of scattered polaritons (either linear of circular) should be total. Therefore, polarization memory measurements appear to be very sensitive probes of the strong interactions of these real intermediate states with other degrees of freedom of the crystal (such as impurities, phonons, interfaces).

As already mentioned in Sect. 7.5.2, the action of magnetic fields on linearly polarized polaritons allows the determination of their dynamics, as the Larmor precession can be used as an internal clock for polariton motion [7.159]. Polarization measurements can also be used to study multiple scatterings: due

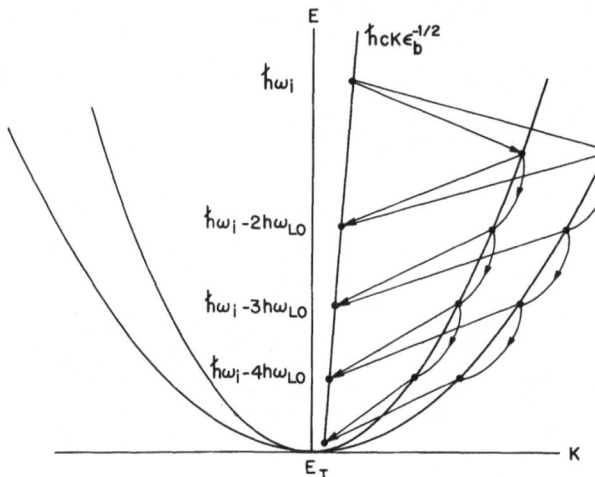

Fig. 7.27. Kinematics of multiphonon light scattering processes involving intermediate states of different symmetry, e.g., the light and heavy exciton states shown here. An important consequence is the depolarization of the scattered light due to the dephasing of the energy-degenerate states, see [7.173]

to the lack of dispersion of the LO phonon, most scattering events observed involve the combination of an LO scattering plus one or more elastic scatterings into the solid angle of observation. As shown by *Bonnot* et al. [7.172], most scattering events are depolarizing. By a careful study of the polarization as a function of exciting energy, *Planel* et al. [7.159, 162] were able to deduce elastic scattering times and wave function mixing in the resonance region of CdS and CdSe.

In multiphonon processes, depolarization occurs additionally because of the summation over intermediate states with different symmetries. Such intermediate states can exist, for instance, when exciting zinc-blende structure semiconductors above the bandgap (Fig. 7.27). In addition, elastic scattering can occur during the relaxation cascade in multiphonon RRS. Such effects were calculated by *Bir* et al. [7.173]. Experimental evidence of the depolarization was reported by *Permogorov* et al. in ZnTe [7.174] and by *Kwietnak* and co-workers in ZnSe and ZnTe [7.175].

7.7 Nonlinear Spectroscopy

We have been mainly concerned up to now with spontaneous effects in light scattering mediated by excitonic polaritons. A number of nonlinear processes are closely related to scattering processes. Although it would take us far astray to detail such effects, it is worth mentioning some of them, referring the reader to recent reviews by *Grun* et al. [7.176] for more details.

Two-photon processes (absorption, mixing, hyper-Raman scattering) can always be considered as scattering processes, and polariton effects can show up dramatically at resonance. An advantage of two-photon processes is the

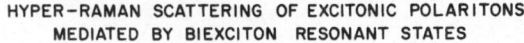

HYPER–RAMAN SCATTERING OF EXCITONIC POLARITONS
MEDIATED BY BIEXCITON RESONANT STATES

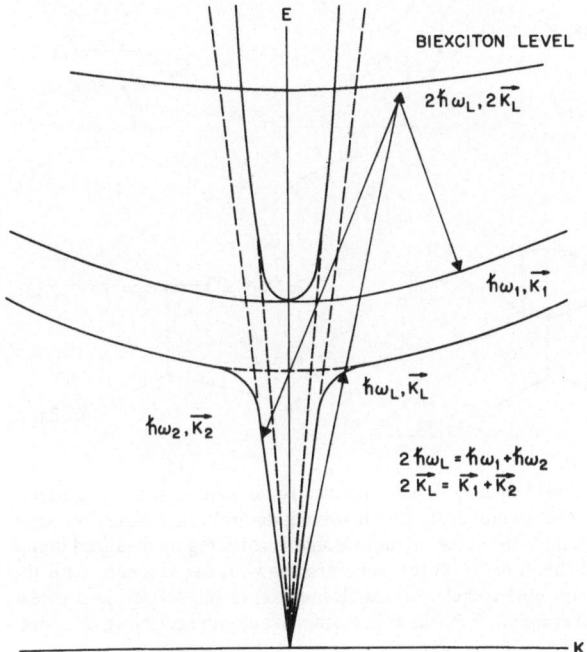

Fig. 7.28. Kinematics of hyper-Raman scattering involving the biexciton as an intermediate state (schematic, see text)

possibility to vary at will the directions of the two-light beams, allowing for *K-space spectroscopy*. The dispersion of the UPB in CuCl was determined by angular dependent two-photon absorption [7.177]. Two-photon hyper-Raman scattering has been successfully applied to CuCl [7.19], CuBr [7.178] and CdS [7.179]. The principle is the following (Fig. 7.28): two photons, with total energy near resonant with the biexciton energy E_B, are mixed at some angle in the crystal. When the virtual biexciton recombines, it emits two polaritons, one of which is usually observable in a given direction (ω_2, K_2 in Fig. 7.28). Working out the momentum and energy conservation equations, one can construct the whole dispersion curve from the initial and observed ω, K's of the photons [7.19]. Besides determining the kinematic properties of the scattering, polaritons play a role in the scattering efficiency through their density of states. These phenomena were discussed by *Vogt* in [Ref. 7.71, Chap. 4].

7.8 Electronic Raman Scattering (ERS)

The most recent development in the field of polariton-mediated Raman scattering concerns resonant impurity Raman scattering (RIRS) (reviews of nonresonant electronic Raman scattering are in [7.36, 180]). Whereas nonre-

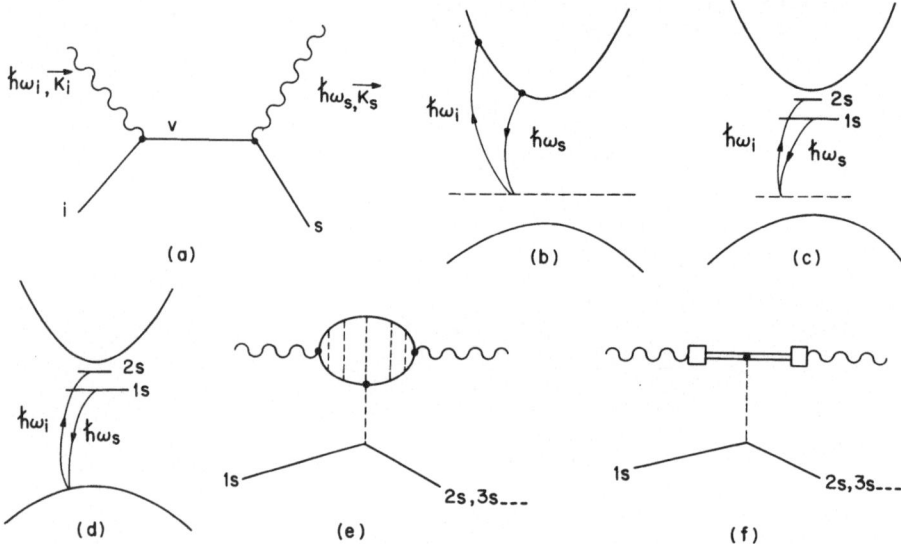

Fig. 7.29a–f. Diagrams and energy level schemes illustrating electronic Raman scattering (ERS) in *n*-doped semiconductors: **(a)**, **(b)** nonresonant scattering involving delocalized conduction band states and valence-band states as intermediate states: **(c)** nonresonant scattering on localized donor states (inelastic excitation $1s \rightarrow 2s$ of the donor); **(d)** the same process with $\hbar\omega_i$ resonant with the band gap; **(e)** inclusion of the electron-hole correlation (exciton effect) in the scattering process; **(f)** description of polariton-mediated resonant ERS in a first-order Coulomb scattering diagram

sonant cross sections are of the order of the square of the classical electron radius $r_0^2 = 6 \times 10^{-25} \, \text{cm}^2$ ([7.181], [Ref. 7.71, Sect. 2.3.2]), resonant cross sections [7.182] can reach values of $\sigma \approx 5 \times 10^{-14} \, \text{cm}^2$, i.e., a resonance enhancement of $\gtrsim 10$ orders of magnitude! Such a dramatic increase can be understood in terms of polariton-mediated scattering. The most striking feature of polariton-mediated ERS is that it is a third-order process, involving the electron–photon coupling twice and an electron–electron interaction once. As in the case of Raman scattering [7.22], this third-order process proves to be more efficient than the direct iterated electron–photon process [7.182]. The difference with Raman scattering is that the increased efficiency does not lie in the larger exciton–phonon or exciton–donor-electron interaction, but in the strong interaction of excitons with photons (leading to polaritons) as compared with the weaker interaction of delocalized band electrons with photons (Fig. 7.29).

Second-order perturbation theory between electron states yields for the scattering probability [7.180] (see also Sect. 2.3.2)

$$\frac{dP}{dt} \sim \left| \langle f | \frac{e^2 A^2}{2m_0} | i \rangle + \sum_I \left(\frac{\langle f | eA \cdot p/m_0 | I \rangle \langle I | eA \cdot p/m_0 | i \rangle}{E_i - E_I - \hbar\omega_i} \right. \right.$$

$$\left. \left. + \frac{\langle f | eA \cdot p/m_0 | I \rangle \langle I | eA \cdot p/m_0 | i \rangle}{E_f - E_I + \hbar\omega_s} \right) \right|^2 . \tag{7.24}$$

The first term represents first-order transitions due to the A^2 term of radiation–matter interaction. The last terms are the second-order terms due to the iterated $A \cdot p$ interaction. $|i\rangle$ and $|f\rangle$ represent the initial and final states of the crystal (band states or impurity states), I is any intermediate electronic state. In the context of electronic Raman scattering of electrons in a semiconductor (free in the conduction band or bound to donors), the dominant contribution to (7.24) is due to valence band states $|I\rangle$. The scattering event is then represented schematically by the virtual transitions shown in Fig. 7.29b, 7.29c.

As a convenient measure of cross sections, let us recall that the elastic cross section of free electrons *in vacuum*, due to the first term in (7.24), is given by the Thomson formula $\sigma = 8\pi r_0^2/3$ (r_0 is the classical electron radius $r_0 = e^2/4\pi\varepsilon_0 m_0 c^2$). For *conduction electrons* in a semiconductor, away from the bandgap resonance ($\hbar\omega_i, \hbar\omega_s \ll E_G$), this formula retains the same expression with the effective mass m^* replacing the free electron mass m_0 [7.183]. For $\hbar\omega$ not negligible compared to E_G, *Wolff* showed that the *conduction-electron* cross section becomes [7.183] (see [Ref. 7.71, Sect. 2.3.2])

$$\frac{d\sigma}{d\Omega} \approx \left(\frac{e^2}{4\pi\varepsilon_0 m^* c^2}\right)^2 \frac{E_G^2}{E_G^2 - (\hbar\omega)^2} \, . \tag{7.25}$$

This expression diverges at $\hbar\omega = E_G$ in which case perturbation theory breaks down.

For *donor-bound electrons* (the discussion can be straight forwardly transposed for acceptors), the initial and final wave functions are those of donor-bound electrons which can be written in the effective-mass approximation [7.184]

$$|i\rangle_{nl}, |f\rangle_{nl} = \sum_k \Phi_{nl}(k) u_{ck}(r) e^{ikr} , \tag{7.26}$$

where the $u_{ck}(r)$ are conduction-electron Bloch functions and $\Phi_{nl}(k)$ is the donor envelope wave function with the hydrogenic quantum numbers n, l (at low temperatures $|i\rangle_{nl} = |i\rangle_{1s}$). The matrix elements entering the predominant second-order terms in (7.24) are usually (nonresonant case) factorized, thanks to the slowly varying behavior of $\Phi_{nl}(k)$, as compared to $u_{ck}(r)$, into [7.185]

$$\langle u_v | A \cdot p | u_c \rangle \langle u_c | A \cdot p | u_v \rangle \int dr F_{nl}(r) F_{n'l'}(r') , \tag{7.27}$$

where the F's are the Fourier transforms of the Φ's. Due to the orthogonality of the F's, the last integral of (7.27) vanishes in the present approximation [7.185]. This explains the extremely small values of σ reported for nonresonant measurements. As can be evaluated from the data of *Wright* and *Mooradian* [7.36], $\sigma \approx r_0^2 \approx 10^{-25} \text{ cm}^2$ for donors and acceptors in GaAs and CdTe, significantly smaller than the calculated conduction electron σ given by (7.25).

It is only very recently that *resonant* impurity Raman scattering (RIRS) has been evaluated [7.182, 186, 187]. A first resonance behavior is obtained when

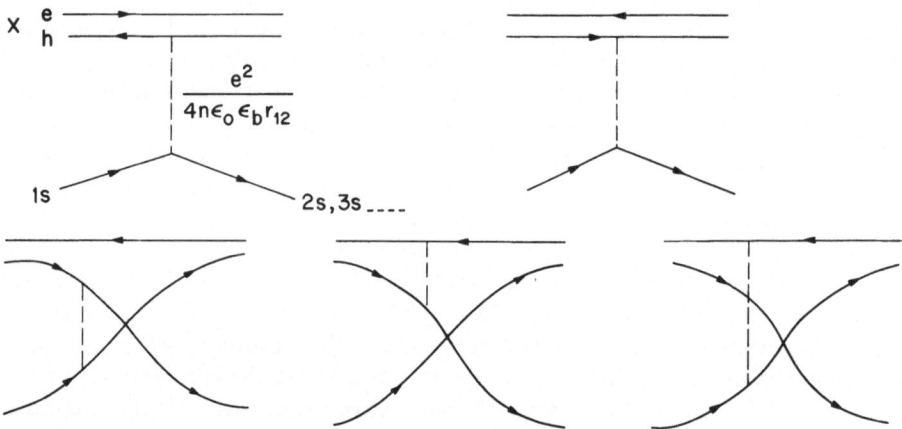

Fig. 7.30. Five diagrams illustrating direct and exchange part of the Coulomb interaction (---) between an exciton-polariton (e–h) pair (⇄) and a localized (donor bound) electron (→). From [7.187]

$\hbar\omega_i$, $\hbar\omega_f > E_G - E_{1s}$, E_{2s} (Fig. 7.29d). Then, real transitions to the valence band can occur and the denominator in (7.24) can vanish for some k values. The factorization leading to (7.27) cannot be performed. The complete calculation of σ in that case was made by *Van Hieu* and co-workers [7.186], both in the case of a simple valence band and degenerate valence bands. They obtained in the effective-mass approximation an expression for σ which can be put into the form

$$\sigma \approx a_B^2 \left(\frac{v_{el}}{v_{phot}}\right)^2, \qquad (7.28)$$

where v_{el} is the velocity of a conduction electron with the same K vector as the photon and v_{phot} is the light phase velocity in the crystal. This resonant cross section is ≈ 6 orders of magnitude larger than nonresonant conduction-electron cross sections.

However, a *third*-order process, polariton-mediated RIRS, can be shown to be still larger than this resonant second-order process [7.182, 187]. We start from (7.20) using as initial and final states direct products of polariton and donor electron states (denoted $|X, D_{nl}\rangle$), and as the interaction, the Coulomb interaction between the donor electron and the electron or hole of the exciton polariton (see Fig. 7.29f). The various direct and exchange interactions are shown in Fig. 7.30 [7.187]. It can be shown that direct interactions are negligibly small due to the orthogonality of $1s$ and $2s$ donor states, and only exchange terms contribute by an amount

$$\left|\langle X, D_{nl}| \frac{e^2}{4\pi\varepsilon_0 \varepsilon r_{12}} |X', D_{n'l'}\rangle\right|^2 \approx \left|\frac{e^2 a_B^2}{4\pi\varepsilon_0 \varepsilon}\right|^2, \qquad (7.29)$$

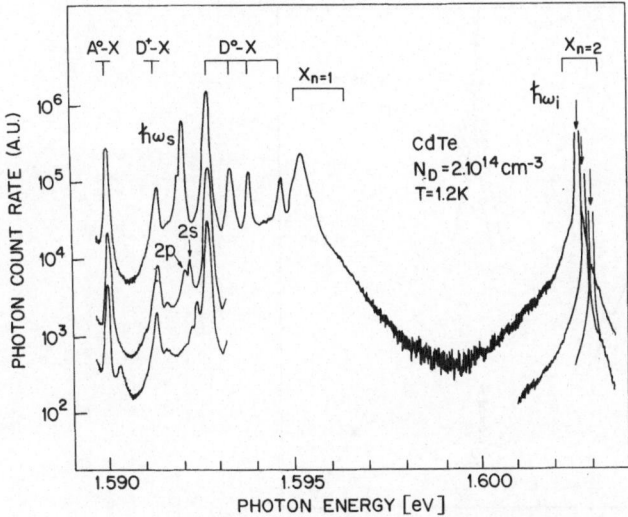

Fig. 7.31. Resonant electronic Raman scattering (ERS) spectra for three different incident photon energies $\hbar\omega_i$ (indicated by right-hand arrows) closely resonant with the $n=2$ exciton-polariton level $X_{n=2}$. The ERS peaks denoted $\hbar\omega_s$ are shifted by the donor excitation energy and show the $2s$, $2p$ donor level splitting. The other lines are due to bound exciton recombination processes and do not shift with excitation photon energy. From [7.182]

where $|X, D_{nl}\rangle$ and $|X', D_{n'l'}\rangle$ are the initial and final states of the exciton and donor electron, respectively. The scattering efficiency *inside the crystal and per donor* is then, from (7.20),

$$\frac{dP}{dt\,d\Omega_{in}} \approx \frac{k_f^2}{4\pi^2\hbar^2 v_{gf}}\left(\frac{e^2 a_B^2}{4\pi\varepsilon_0\varepsilon}\right)^2, \tag{7.30}$$

taking $T_i \approx T_f \approx 1$. This can be translated in a cross section per donor

$$\frac{d\sigma}{d\Omega_{in}} = \frac{1}{v_{gi}}\frac{dP}{dt\,d\Omega_{in}} = \frac{1}{4\pi\hbar^2}\frac{k_f^2}{v_{gi}v_{gf}}\left|\frac{e^2 a_B^2}{4\pi\varepsilon_0\varepsilon}\right|^2. \tag{7.31}$$

Using the exciton velocity expression $v_{ex}=\hbar K_f/M_X$ and (7.2), one finds for the total cross section

$$\sigma = a_B^2\frac{v_{ex}^2}{v_{gi}v_{gf}}. \tag{7.32}$$

A cross section of the order of a_B^2 $[\approx(100\,\text{Å})^2]$ represents ≈ 12 orders of magnitude increase as compared to the Thomson value. It was recently observed by tunable laser spectroscopy in GaAs and CdTe [7.182]. Figure 7.31 shows a typical spectrum when a CdTe crystal is excited in the region of the $n=2$

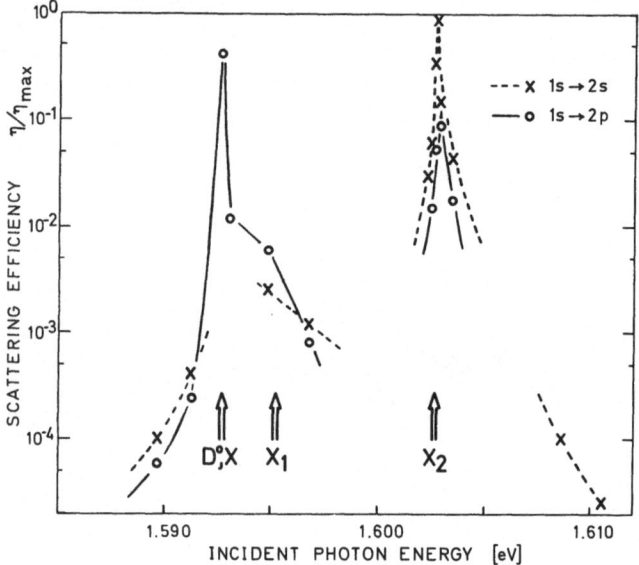

Fig. 7.32. Experimental electronic Raman scattering efficiency for the donor $1s \to 2s$ and $1s \to 2p$ excitation process as a function of incident photon energy $\hbar\omega_i$. The strongest resonance is observed when $\hbar\omega_i$ is at the $n=2$ exciton energy

exciton. The signals observed in that case are stronger than with excitation at $n=1$ (Fig. 7.32). We deal here with a situation of double resonance: v_{gi} and v_{gf} are both small, the former as it is the $n=2$ exciton-polariton resonance, the second as the scattered photon at energy $E_{X_{n=2}} - (E_{2s} - E_{1s})$ is just below the $n=1$ exciton-polariton resonance. The signals observed when exciting the $n=1$ exciton are significantly smaller as the out-going group velocity is ≈ 100 times smaller than in the former case. This is seen in Fig. 7.32 which also displays the resonance observed when incoming polaritons are resonant with the donor-bound exciton level (D^0, X) (a definitive account of impurity bound-exciton phenomena is [7.188]). A similar enhancement was also reported in ZnTe [7.189]. No theory of this (D^0, X) resonance has been made up to now because of the complexities of the (D^0, X) states.

The very reasonable agreement between the calculated efficiency at the $X_{n=2}$ resonance peak ($\eta_{calc} \approx 1.6 \times 10^{-5}$ and the measured one $\eta_{meas} \approx 8 \times 10^{-5}$), together with the well-explained predominance of the $X_{n=2}$ resonance provide very satisfactory support for the model of exciton-polariton-mediated RIRS. The very good sensitivity of the method (the spectra shown in Fig. 7.31 are obtained for impurity concentrations in the range $10^{14}\,cm^{-3}$) and the good precision in the measured $1s \to 2s, 2p$ energy should allow RIRS to be used as a powerful tool in analytic studies, both for the impurity concentration at low levels and the chemical nature of impurities (through their labeling, thanks to the $1s \to 2s, 2p$ chemical shifts).

7.9 Conclusion

The field of polariton-mediated light scattering is obviously a very rich one. It first gives a direct visualisation of those fundamental coupled-mode excitations, the excitonic polaritons. It helps to obtain evidence for the predictions made by Hopfield and others of some extraordinary properties of these objects. Let us mention, for instance, the dispersive, multiple Brillouin peaks, the AS/S ratio in RBS larger than unity, the disappearence of sharp RRS peaks into luminescence bands at resonance, the huge cross sections of RIRS and its double-resonance behavior on the $n=2$ exciton-polariton peak. In this latter case, we encounter once more the situation where a third-order process with two strong interactions overcomes a two-step process involving weak interactions. But polariton-mediated light scattering is much more than just another theoretical description of the light–matter interaction. It provides a huge amount of information on exciton parameters [masses, L-T splittings (i.e., polarizability), fine structure parameters (exchange energy)], and semiconductor parameters (refractive indices, sound velocities, phonon energies, K-linear energy terms, impurity energy levels and concentration). Exciton–phonon (acoustic as well as optical phonons) and exciton–impurity interactions can be studied in great detail with this type of resonance scattering. Due to this wealth, and also due to the relative ease with which such experiments can be performed, polariton-mediated light scattering still has a bright future. There is a number of unresolved and challenging problems. In the first place, of course, the ABC problem due to its generality and fundamental character. The extremely complicated, somewhat uncontrolled, experimental situation at the crystal interface has been discussed in various portions of the text. We want to stress again that only complete sets of measurements, i.e., transmission, reflectivity, RBS (efficiency, linewidth, lineshapes, high-resolution kinematics) can provide useful information to shed some light onto this delicate problem. Other remaining tasks concern realistic calculations of efficiencies and multiple scattering properties (in particular, thermalization of RRS events). Interesting experiments would be the observation of RBS for surface exciton polaritons, which would, as in the case of bulk polaritons, bring welcome information on these otherwise rather indirectly evidenced excitations.

References

7.1   E. Burstein, F. De Martini (eds.): Polaritons, Proceedings of the First Taormina Conference on the Structure of Matter, Taormina, Italy, Oct. 2–6, 1972, (Pergamon, New York 1974)
7.2   S. I. Pekar: Zh. Eksp. Teor. Fiz. **34**, 1176 (1958) [English transl.: Sov. Phys. JETP **7**, 813 (1958)]
7.3   J. J. Hopfield: Phys. Rev. **112**, 1555 (1958)
7.4   J. J. Hopfield, D. G. Thomas: Phys. Rev. Lett. **15**, 22 (1965)

7.5   V. A. Kiselev, B. S. Razbirin, I. L. Uraltsev: Pis'ma Zh. Eksp. Teor. Fiz. **18**, 504 (1973) [English transl.: Sov. Phys. JETP Lett. **18**, 296 (1973)]; Proc. XIIth Int. Conf. On the Physics of Semiconductors, Stuttgart, 1974, ed. by M. Pilkuhn (Teubner, Stuttgart 1974) p. 996; Phys. Status Solidi B **72**, 161 (1975)

7.6   J. Voigt: Phys. Status Solidi B **64**, 549 (1974)

7.7   C. Benoit a la Guillaume, A. Bonnot, J. M. Debever: Phys. Rev. Lett. **24**, 1235 (1970)

7.8   E. Gross, S. Permogorov, V. Travnikov, A. Selkin: Solid State Commun. **10**, 1071 (1972)

7.9   A. Bonnot, C. Benoit a la Guillaume: In Ref. 7.1, p. 197

7.10   a) D. D. Sell, S. E. Stokowski, R. Dingle, J. V. DiLorenzo: Phys. Rev. B **7**, 4568 (1973)
        b) C. Weisbuch, R. G. Ulbrich: Phys. Rev. Lett. **39**, 654 (1977)

7.11   H. Sumi: J. Phys. Soc. Jpn. **41**, 526 (1976)

7.12   L. N. Ovander: Usp. Fiz. Nauk. **86**, 3 (1965) [English transl.: Sov. Phys-Usp. **8**, 33 (1965)]

7.13   D. Fröhlich, E. Mohler, P. Wiesner: Phys. Rev. Lett. **26**, 554 (1971)

7.14   J. J. Hopfield, D. G. Thomas: Phys. Rev. **132**, 563 (1963)

7.15   F. Evangelisti, A. Frova, J. U. Fischbach: Phys. Rev. Lett. **29**, 1001 (1972)

7.16   F. Patella, F. Evangelisti, M. Capizzi: Solid State Commun. **20**, 23 (1976)

7.17   W. Brenig, R. Zeyher, J. L. Birman: Phys. Rev. B **6**, 4617 (1972)

7.18   R. G. Ulbrich, C. Weisbuch: Phys. Rev. Lett. **38**, 865 (1977)

7.19   B. Hönerlage, A. Bivas, Vu Duy Phach: Phys. Rev. Lett. **41**, 49 (1978)

7.20   B. Sermage, G. Fishman: Phys. Rev. Lett. **43**, 1043 (1979)

7.21   R. Sandercock: Phys. Rev. Lett. **28**, 237 (1972)

7.22   R. Loudon: Proc. R. Soc. London A **275**, 218 (1963)

7.23   R. Loudon: J. Phys. (Paris) **26**, 677 (1965)

7.24   A. K. Ganguly, J. L. Birman: Phys. Rev. **162**, 806 (1967)

7.25   E. Burstein, D. L. Mills, A. Pinczuk, S. Ushioda: Phys. Rev. Lett. **22**, 348 (1969)

7.26   D. L. Mills, E. Burstein: Phys. Rev. **188**, 1465 (1969)

7.27   B. Bendow, J. L. Birman: Phys. Rev. B **1**, 1678 (1970)
        B. Bendow: Phys. Rev. B **2**, 5051 (1970); B **4**, 552 (1971)

7.28   J. J. Hopfield: Phys. Rev. **182**, 945 (1969)

7.29   M. Sturge, E. Rashba (eds.): *Excitons* (North-Holland, Amsterdam 1982). See in particular the contributions by J. L. Birman, E. L. Ivchenko, E. Koteles, S. Permogorov, G. E. Pikus

7.30   B. Bendow: "Polariton Theory of Resonance Raman Scattering in Solids," in Springer Tracts in Modern Physics, Vol. 82, Electronic Structure of Noble Metals and Polariton-Mediated Light Scattering (Springer, Berlin, Heidelberg, New York 1978) p. 89

7.31   P. Y. Yu: "Studies of Excitons and Exciton-Phonon Interactions by Resonant Raman and Brillouin Spectroscopies," in *Excitons*, ed. by K. Cho, Topics in Current Physics, Vol. 14 (Springer, Berlin, Heidelberg, New York 1979) pp. 211–253

7.32   M. Cardona (ed.): *Light Scattering in Solids*, Topics in Applied Physics, Vol. 8 (Springer, Berlin, Heidelberg, New York 1975)

7.33   W. Hayes, R. Loudon: *Scattering of Light by Crystals* (Wiley, New York 1978)

7.34   R. Glauber (ed.): *Quantum Optics* (Academic Press, New York 1969). See p. 247 by H. Z. Cummins

7.35   C. T. Walker: In *Physics of Quantum Electronics*, Vol. 2, ed. by S. Jacobs, M. Sargent, M. O. Scully (Addison-Wesley, Reading 1975)

7.36   A. Mooradian: In *Laser Handbook*, Vol. 2, ed. by F. T. Arecchi, E. O. Schulz-Dubois (North-Holland, Amsterdam 1972) p. 1409

7.37   R. Knox: Solid State Phys. Suppl. **5** (1963)
        J. O. Dimmock: In *Semiconductors and Semimetals*, Vol. 3, ed. by R. K. Willardson, A. C. Beer (Academic Press, New York 1967) p. 259

7.38   K. Cho (ed.): *Excitons*, Topics in Current Physics, Vol. 14 (Springer, Berlin, Heidelberg, New York 1979)

7.39   B. Gerlach, J. Pollmann: Phys. Status Solidi B **67**, 93, 477 (1975)

7.40   G. D. Mahan, J. J. Hopfield: Phys. Rev. A **135**, 428 (1964)

7.41   A. Baldereschi, N. O. Lipari: Phys. Rev. B **3**, 439 (1971)

7.42   E. O. Kane: Phys. Rev. B **11**, 3850 (1975)

7.43 M. Altarelli, N.O. Lipari: Phys. Rev. B **15**, 4898 (1977)
7.44 C.G. Kuper, G.D. Whitfield (eds.): In *Polarons and Excitons* (Oliver and Boyd, Edinburgh 1962). See p. 295 by H. Haken
7.45 J. Pollmann, H. Büttner: Phys. Rev. B **16**, 4480 (1977)
      G. Behnke, H. Büttner: Phys. Status Solidi B **90**, 53 (1978)
      H. Trebin: Phys. Status Solidi B **92**, 601 (1979)
      H. Rössler, H.R. Trebin: Phys. Rev. B **23**, 1961 (1981)
7.46 E.O. Kane: Proc. 14th Int. Conf. on the Physics of Semiconductors, Edinburgh, 1978, ed. by B.L.H. Wilson (Inst. Phys. Conf. Ser. 43, London 1979) p. 1321
7.47 For the work pertaining up to 1963, see Ref. 7.37
7.48 Y. Onodera, Y. Toyozawa: J. Phys. Soc. Jpn. **22**, 833 (1967)
7.49 G.E. Bir, G.L. Pikus: Zh. Theor. Eksp. Fiz. **60**, 195 (1971) [English transl.: Sov. Phys.-JETP **33**, 108 (1971)]
7.50 M.M. Denisov, V.P. Makarov: Phys. Status Solidi B **59**, 9 (1973)
7.51 J. Treusch (ed.): In *Festkörperprobleme – Advances in Solid State Physics*, Vol. XIX, (Pergamon-Vieweg, Braunschweig 1979). See p. 77 by U. Rössler
7.52 G.E. Pikus, G.L. Bir: *Symmetry and Deformation Effects in Semiconductors* (Nauka, Moscow 1972) [English transl.: Wiley, New York 1974]
7.53 M. Cardona: Solid State Phys. Suppl. **11** (1969)
7.54 a) I. Broser, M. Rosenzweig: Phys. Rev. B **22**, 2000 (1980)
      b) H. Venghaus, B. Jusserand: Phys. Rev. B **22**, 932 (1980)
      c) C. Uihlein, S. Feierabend: Phys. Status Solidi B **94**, 153, 421 (1979)
7.55 D. Bimberg: In *Festkörperprobleme – Advances in Solid State Physics*, Vol. XVII, ed. by J. Treusch (Pergamon-Vieweg, Braunschweig 1977) p. 195
7.56 R.J. Elliot: Phys. Rev. **108**, 1384 (1957)
7.57 R.J. Elliott: In Ref. 7.44, p. 269
7.58 A.I. Ansel'm, Iu.A. Firsov: Zh. Eksp. Theor. Fig. **28**, 151 (1955); **30**, 719 (1956) [English transl.: Sov. Phys.-JETP **1**, 139 (1955); **3**, 564 (1956)]
7.59 Y. Toyozawa: Prog. Theor. Phys. **20**, 53 (1958); Prog. Theor. Phys. Suppl. **12**, 111 (1959)
7.60 S. Nakajima, Y. Toyozawa, R. Abe: *The Physics of Elementary Excitations*, Springer Series in Solid-State Sciences, Vol. 12 (Springer, Berlin, Heidelberg, New York 1980)
7.61 F. Urbach: Phys. Rev. **92**, 1324 (1953)
7.62 H. Sumi, Y. Toyozawa: J. Phys. Soc. Jpn. **31**, 342 (1971)
7.63 J.D. Dow, D. Redfield: Phys. Rev. B **5**, 594 (1972)
7.64 J.D. Dow: In *Optical Properties of Solids, New Developments*, ed. by B.O. Seraphin (North-Holland, Amsterdam 1976) p. 33
7.65 E. Mohler, B. Thomas: Phys. Rev. Lett. **44**, 543 (1980)
7.66 Y. Toyozawa: J. Phys. Chem. Solids **25**, 59 (1964)
7.67 K. Huang: Proc. R. Soc. London A **208**, 352 (1951) [Reprinted in Ref. 7.1]
7.68 M. Born, K. Huang: *Dynamical Theory of Crystal Lattices* (University Press, Oxford 1954)
7.69 C.P. Slichter: *Principles of Magnetic Resonance*, Springer Series in Solid-State Sciences, Vol. 1 (Springer, Berlin, Heidelberg, New York 1980)
7.70 A. Abragam: *The Principles of Nuclear Magnetism* (University Press, Cambridge 1961)
7.71 M. Cardona, G. Güntherodt (eds.): *Light Scattering in Solids II*, Topics in Applied Physics, Vol. 50 (Springer, Berlin, Heidelberg, New York 1982)
7.72 H.J.G. Meier, D. Polder: Physica **19**, 255 (1953)
7.73 A.R. Hutson: J. Appl. Phys. Suppl. **32**, 2287 (1961)
7.74 Duy-Phach Vu, Y. Oka, M. Cardona: Phys. Rev. B **24**, 765 (1981)
7.75 R.M. Martin, C.M. Varma: Phys. Rev. Lett. **26**, 124 (1971)
      R.M. Martin: Phys. Rev. B **4**, 3676 (1971)
7.76 W. Richter, R. Zeyher: *Festkörperprobleme – Advances in Solid State Physics*, Vol. XVI, ed. by J. Treusch (Pergamon-Vieweg, Braunschweig 1976) p. 15
      W. Richter: "Resonant Raman Scattering in Semiconductors", in Springer Tracts in Modern Physics, Vol. 78, Solid-State Physics (Springer, Berlin, Heidelberg, New York 1976)
7.77 C. Benoit a la Guillaume, J.M. Debever, F. Salvan: Phys. Rev. **177**, 567 (1969)

7.78 K.Matsuda, M.Hirooka, S.Sumakawa: Prog. Theor. Phys. **54**, 79 (1975)
7.79 H.Büttner: Phys. Status Solidi B**42**, 775 (1970)
7.80 E.Hanamura, H.Haug: Phys. Rep. **33**, 209 (1977)
7.81 O.S.Zinets, V.I.Sugakov: Fiz. Tver. Tela **11**, 243 (1969) [English transl.: Sov. Phys.-Solid State **11**, 197 (1969)]
7.82 W.C.Tait: Phys. Rev. B**5**, 648 (1972)
7.83 R.Le Toullec: Unpublished
7.84 G.Fishman: Solid State Commun. **27**, 1087 (1978)
7.85 W.Dreybrodt, K.Cho, S.Suga, F.Willmann, Y.Niji: Phys. Rev. B**21**, 4692 (1980)
7.86 J.Lagois: Phys. Rev. B**16**, 1699 (1977)
7.87 E.Ostertag: Phys. Rev. Lett. **45**, 372 (1980)
7.88 V.M.Agranovitch, V.L.Ginzburg: *Spatial Dispersion in Crystal Optics and the Theory of Excitons*, 1st and 2nd ed. (Nauka, Moscow 1963, 1979) [English transl.: 1st ed. (Wiley-Interscience, New York 1966), 2nd ed., Springer Series in Solid-State Physics, (Springer, Berlin, Heidelberg, New York 1982)
7.89 J.J.Hopfield: Proc. Int. Conf. on the Physics of Semiconductors, Kyoto 1966; J. Phys. Soc. Jpn. Suppl. **21**, 77 (1966)
7.90 J.J.Hopfield: Excitons and their Electromagnetic Interactions: In Ref. 7.34, p. 340
7.91 J.L.Birman, D.N.Pattanayak: In *Light Scattering in Solids*, ed. by J.L.Birman, H.Z.Cummins, K.K.Rebane (Plenum Press, New York 1979)
      J.L.Birman: In Ref. 7.29
7.92 J.D.Jackson: *Classical Electrodynamics*, 2nd ed. (Wiley, New York 1975), Chap. 7
7.93 M.F.Bishop, A.A.Maradudin: Phys. Rev. B**14**, 3384 (1976)
      A.V.Selkin: Fiz. Tverd. Tela **19**, 2433 (1977) [English transl.: Sov. Phys.-Solid State **19**, 1424 (1977)]
      F.Forstmann: Z. Phys. **32**, 385 (1979)
7.94 J.J.Sein: Phys. Lett. A**32**, 141 (1970)
      J.L.Birman, J.J.Sein: Phys. Rev. B**6**, 2482 (1972)
      G.S.Agarwal, D.N.Pattanayak, E.Wolf: Opt. Commun. **4**, 255 (1971); Phys. Rev. Lett. **27**, 1022 (1971); Phys. Rev. B**10**, 1447 (1974); B**11**, 1342 (1975)
7.95 R.Zeyher, J.L.Birman, W.Brenig: Phys. Rev. B**6**, 4613 (1972)
      R.Zeyher, J.L.Birman: In Ref. 7.1, p. 169
      C.S.Ting, M.J.Frankel, J.L.Birman: Solid State Commun. **17**, 1285 (1975)
7.96 A.Stahl, C.Uilhein: In Ref. 7.51, p. 159
7.97 F.Evangelisti, J.U.Fischbach, A.Frova: Phys. Rev. B**9**, 1516 (1974)
7.98 F.Evangelisti, A.Frova, F.Patella: Phys. Rev. B**10**, 4253 (1974)
7.99 J.Lagois, E.Wagner, W.Bludau, K.Lösch: Phys. Rev. B**18**, 4325 (1978)
7.100 M.F.Bishop: Solid State Commun. **20**, 779 (1976)
7.101 A.Bosacchi, B.Bosacchi, S.Franchi: Phys. Rev. Lett. **36**, 1086 (1976)
      J.M.Besson, R.Le Toullec, N.Piccioli: Phys. Rev. Lett. **39**, 671 (1977)
7.102 J.Voigt, F.Spiegelberg: Phys. Status Solidi **30**, 659 (1968)
      J.Voigt, M.Senoner, I.Ruckmann: Phys. Status Solidi B**75**, 213 (1976)
      F.Spiegelberg, E.Gutsche, J.Voigt: Phys. Status Solidi B**77**, 233 (1976)
      J.Voigt, F.Spiegelberg, M.Senoner: Phys. Status Solidi B**91**, 189 (1979)
      B.Dietrich, J.Voigt: Phys. Status Solidi B**93**, 669 (1979)
7.103 I.V.Makarenko, I.N.Uraltsev, V.A.Kiselev: Phys. Status Solidi B**98**, 773 (1980)
7.104 I.Broser, R.Broser, E.Beckmann, E.Birkicht: Proc. 15th Int. Conf. on the Physics of Semiconductors, Kyoto, 1980, J. Phys. Soc. Jpn. Suppl. A**69**, 401 (1980); Solid State Commun. **39**, 1209 (1981)
      I.Broser, M.Rosenzweig: Solid State Commun. **36**, 1027 (1980)
7.105 O.V.Konstantinov, Sh.R.Saifullaev: Fiz. Tverd. Tela **18**, 3433 (1976) [English transl.: Sov. Phys.-Solid State **18**, 1998 (1976)]
7.106 M.J.Frankel, J.L.Birman: Phys. Rev. A**15**, 200 (1977)
7.107 R.Loudon: J. Phys. A**3**, 233 (1970)
7.108 Y.Segawa, Y.Aoyagi, K.Azuma, S.Namba: Solid State Commun. **28**, 853 (1978)
      Y.Segawa, Y.Aoyagi, S.Namba: Solid State Commun. **32**, 229 (1979); Proc. 15th Int. Conf. on the Physics of Semiconductors, Kyoto, 1980, p. 389

7.109 Y. Masumoto, Y. Unuma, Y. Tanaka, S. Shionoya: J. Phys. Soc. Jpn. **47**, 1844 (1979); Proc. 15th Int. Conf. on the Physics of Semiconductors, Kyoto, 1980, p. 393

7.110 R. G. Ulbrich, G. W. Fehrenbach: Phys. Rev. Lett. **43**, 963 (1979)

7.111 L. N. Ovander: Fiz. Tverd. Tela **3**, 2394 (1961); **4**, 1466, 1471 (1962); **5**, 21 (1963) [English transl.: Sov. Phys.-Solid State **3**, 1377 (1962); **4**, 1078, 1081 (1962); **5**, 13 (1963)]
E. M. Verlan, L. N. Ovander: Fiz. Tverd. Tela **8**, 2435 (1967) [English transl.: Sov. Phys.-Solid State **8**, 1939 (1967)]

7.112 A. V. Selkin: Phys. Status Solidi B**83**, 47 (1977); Fiz. Tverd. Tela **19**, 2433 (1977) [English transl.: Sov. Phys.-Solid State **19**, 1424 (1977)]

7.113 M. Lax, D. F. Nelson: In *The Theory of Light Scattering in Condensed Matter*, ed. by V. M. Agranovitch, J. L. Birman (Plenum Press, New York 1976) p. 371; also in *Coherence and Quantum Optics.*, ed. by L. Mandel, E. Wolf (Plenum Press, New York 1973) p. 415

7.114 R. Zeyher, C. S. Ting, J. L. Birman: Phys. Rev. B**10**, 1725 (1974)

7.115 Duy-Phach Vu, Y. Oka, M. Cardona: To be published

7.116 R. Loudon: J. Raman Spectrosc. **7**, 10 (1978)

7.117 R. G. Ulbrich, C. Weisbuch: Verhandl. DPG (VI) **12**, 59 (1977)

7.118 R. H. Bruce, H. Z. Cummins, C. Ecolivet, F. H. Pollack: Bull. Am. Phys. Soc. **22**, 315 (1977)
R. H. Bruce, H. Z. Cummins: Phys. Rev. B**16**, 4462 (1977)

7.119 G. Winterling, E. Koteles: Solid State Commun. **23**, 95 (1977)

7.120 E. J. Flynn, S. Geschwind: Bull. Am. Phys. Soc. **26**, 488 (1981); to be published

7.121 J. Wicksted, M. Matsushita, H. Z. Cummins: To be published

7.122 A. S. Pine: Phys. Rev. B**5**, 3003 (1972)
D. K. Garrod, R. Bray: Phys. Rev. B**6**, 1314 (1972)

7.123 A. S. Pine: "Brillouin Scattering in Semiconductors", in Ref. 7.32, p. 253

7.124 S. Adachi, C. Hamaguchi: J. Phys. Soc. Jpn. **48**, 1981 (1980). This recent work contains references to studies of a number of semiconducting compounds

7.125 J. Sandercock: In *Festkörperprobleme – Advances in Solid State Physics*, Vol. XV, ed. by H. J. Queisser (Pergamon-Vieweg, Braunschweig 1975) p. 183

7.126 R. G. Ulbrich, C. Weisbuch: In *Festkörperprobleme – Advances in Solid State Physics*, Vol. XVIII, ed. by J. Treusch (Pergamon-Vieweg, Braunschweig 1978) pp. 217–240

7.127 P. Y. Yu: Comments Solid State Phys. **9**, 37 (1979): In *Light Scattering in Solids*, ed. by J. L. Birman, H. Z. Cummins, K. K. Rebane (Plenum Press, New York 1979) p. 143; Proc. 15th Int. Conf. on the Physics of Semiconductors, Kyoto, 1980, p. 533
J. L. Birman, E. Koteles: In Ref. 7.29

7.128 R. G. Ulbrich, C. Weisbuch: Unpublished

7.129 E. S. Koteles, G. Winterling: J. Lumin. **18/19**, 267 (1979); Phys. Rev. Lett. **44**, 948 (1980)

7.130 C. Hermann, P. Y. Yu: Solid State Commun. **28**, 313 (1978)

7.131 B. Sermage: Ph. D. Thesis, Université Paris VII, 1980
B. Sermage, G. Fishman: Phys. Rev. (to be published)

7.132 Y. Oka, M. Cardona: To be published

7.133 T. Goto, Y. Nishina: Solid State Commun. **31**, 751 (1979)

7.134 Y. Oka, Duy-Phach Vu: Proc. 15th Int. Conf. on the Physics of Semiconductors, Kyoto, 1980, p. 347

7.135 G. Winterling, E. S. Koteles, M. Cardona: Phys. Rev. Lett. **39**, 1286 (1977)

7.136 R. G. Ulbrich, C. Weisbuch: Unpublished

7.137 A. Dervisch, R. Loudon: J. Phys. C**9**, L669 (1976)
R. Loudon: J. Phys. C**11**, 403 (1978)

7.138 D. R. Tilley: J. Phys. C**13**, 781 (1980)

7.139 C. Weisbuch, R. G. Ulbrich: Proc. Int. Conf. on Lattice Dynamics, Paris 1977, ed. by M. Balkanski (Flammarion, Paris 1978) p. 167

7.140 C. Hermann, P. Y. Yu: Phys. Rev. B**21**, 3675 (1980)

7.141 D. F. Nelson, P. D. Lazay: J. Opt. Soc. Am. **67**, 1599 (1977)

7.142 V. A. Kiselev: Fiz. Tverd. Tela **20**, 2173 (1978) [English transl.: Sov. Phys.-Solid State **20**, 1255 (1978)]

7.143 R. G. Ulbrich, C. Weisbuch: Bull. Am. Phys. Soc. **24**, 442 (1979)

7.144 P. Y. Yu, F. Evangelisti: Phys. Rev. Lett. **42**, 1642 (1979); **43**, 474 (1979)

7.145   C.Trallero Giner, I.G.Lang, S.T.Pavlov: Phys. Status Solidi B**100**, 631 (1980) and references therein
7.146   H.Kurita, O.Sakai, A.Kotani: J. Phys. Soc. Jpn. **49**, 1920, 1929 (1980)
7.147   S.Permogorov: Phys. Status Solidi B**68**, 9 (1975)
        S.A.Permogorov: In Ref. 7.29
7.148   E.S.Koteles, G.Winterling: Proc. 14th Int. Conf. on the Physics of Semiconductors, ed. by B.L.H.Wilson (Inst. Physics Conf. Series 43, London 1979) p. 481; Phys. Rev. B**20**, 628 (1979)
7.149   A.Nakamura, C.Weisbuch: Solid State Commun. **32**, 301 (1979)
7.150   R.M.Martin: Phys. Rev. B**10**, 2620 (1974)
        R.Zeyher: Solid State Commun. **16**, 49 (1975)
        A.A.Klyuchikhin, S.A.Permogorov, A.N.Reznitskii: Zh. Eksp. Teor. Fiz. **71**, 2230 (1976) [English transl.: Sov. Phys.-JETP **44**, 1176 (1976)] and references therein
7.151   J.Reydellet, J.M.Besson: Solid State Commun. **17**, 23 (1975); J. Phys. Lett. **37**, L219 (1976)
        J.Reydellet, P.Y.Yu, J.M.Besson, M.Balkanski: Proc. 14th Int. Conf. on the Physics of Semiconductors, ed. by B.L.H.Wilson (Inst. Physics Conf. Series 43, London 1979) p. 1271
        J.Reydellet: Ph. D. Thesis, Université Paris VII (1979)
7.152   P.Y.Yu: In *Physics of Semiconductors*, Proc. 13th Int. Conf., Rome, 1976, ed. by F.Fumi (North-Holland, Amsterdam 1977) p. 235
7.153   B.Bendow, J.L.Birman: Phys. Rev. B**4**, 569 (1971)
        A.S.Barker, R.Loudon: Rev. Mod. Phys. **44**, 18 (1972)
7.154   P.J.Colwell, M.V.Klein: Solid State Commun. **8**, 2095 (1970)
        S.A.Permogorov, A.N.Reznitskii, Ya.V.Morozenko, B.A.Kazennov: Fiz. Tverd. Tela **16**, 2403 (1974) [English transl.: Sov. Phys.-Solid State **16**, 1562 (1975)]
        A.A.Klochikhin, S.A.Permogorov, A.N.Reznitskii: Fiz. Tverd. Tela **18**, 2239 (1976) [English transl.: Sov. Phys.-Solid State **18**, 1304 (1976)]
        A.A.Gogolin, E.I.Rashba: Solid State Commun. **19**, 1177 (1976)
        S.Permogorov, A.Reznitzky: Solid State Commun. **18**, 781 (1976)
7.155   M.V.Klein: Phys. Rev. B**8**, 919 (1973)
        Y.R.Shen: Phys. Rev. B**9**, 622 (1974); B**14**, 1772 (1976)
        J.R.Solin, H.Merkelo: Phys. Rev. B**12**, 624 (1975); B**14**, 1775 (1976)
7.156   P.F.Williams, D.L.Rousseau, S.H.Dworetsky: Phys. Rev. Lett. **32**, 196 (1974)
        D.L.Rousseau, J.M.Friedman, P.F.Williams: In *Raman Spectroscopy of Liquids and Gases*, ed. by A.Weber, Topics in Current Physics, Vol. 11 (Springer, Berlin, Heidelberg, New York 1979) p. 203
7.157   F.A.Novak, J.M.Friedman, R.M.Hochstrasser: In *Lasers and Coherence Spectroscopy*, ed. by J.I.Steinfeld (Plenum Press, New York 1978) p. 451. A simple description of transient effects in light-scattering
7.158   Y.Toyozawa: J. Phys. Soc. Jpn. **41**, 400 (1976)
        Y.Toyozawa, A.Kotani, A.Sumi: J. Phys. Soc. Jpn. **42**, 1495 (1977)
        T.Takagahara, E.Hanamura, R.Kubo: J. Phys. Soc. Jpn. **43**, 802, 811 (1977)
        T.Takagahara: "Resonant Raman Scattering and Luminescence" in *Relaxation of Elementary Excitations*, ed. by R.Kubo, E.Hanamura, Springer Series in Solid-State Sciences, Vol. 18 (Springer, Berlin, Heidelberg, New York 1980) p. 45
        H.Miyazaki, E.Hanamura: "First and Second Order Optical Responses in Exciton-Phonon System", in *Relaxation of Elementary Excitations*, ed. by R.Kubo, E.Hanamura, Springer Series in Solid-State Sciences, Vol. 18 (Springer, Berlin, Heidelberg, New York 1980) p. 71
7.159   R.Planel: Ph. D. Thesis, Université Paris VII (1977)
        M.Nawrocki, R.Planel, C.Benoit a la Guillaume: Phys. Rev. Lett. **36**, 1343 (1976)
        R.Planel, M.Nawrocki, C.Benoit a la Guillaume: Il Nuovo Cim. B**39**, 519 (1977)
7.160   S.A.Permogorov, V.Travnikov: Solid State Commun. **29**, 615 (1979) and references therein
7.161   R.Planel, A.Bonnot, C.Benoit a la Guillaume: Phys. Status Solidi B**58**, 251 (1973)
7.162   R.Planel, C.Benoit a la Guillaume: Proc. 14th Int. Conf. on the Physics of Semiconductors, Edinburgh, 1978, ed. by B.L.H.Wilson (Inst. Phys. Conf. Series 43, London 1979) p. 1263
7.163   P.Y.Yu, F.Evangelisti: Solid State Commun. **27**, 87 (1978)

7.164 G. Winterling, E. Koteles: In Ref. 7.139, p. 170
7.165 S. Suga, P. Hiesinger, M. Cardona: Unpublished
7.166 P. Y. Yu: Solid State Commun. **19**, 1087 (1976)
7.167 Y. Oka, M. Cardona: Solid State Commun. **30**, 447 (1979)
7.168 P. Y. Yu, Y. R. Shen: Phys. Rev. B**12**, 1377 (1975)
7.169 A. Nakamura, C. Weisbuch: Unpublished
7.170 J. Windscheif, H. Stolz, W. Von der Osten: Solid State Commun. **24**, 607 (1977)
7.171 S. A. Permogorov, V. V. Travnikov: Fiz. Tverd. Tela **13**, 709 (1971) [English transl.: Sov. Phys.-Solid State **13**, 586 (1971)]; Phys. Status Solidi B**78**, 389 (1976)
7.172 A. Bonnot, R. Planel, C. Benoit a la Guillaume: Phys. Rev. B**9**, 690 (1974)
7.173 G. L. Bir, E. L. Ivchenko, G. E. Pikus: Izv. Akad. Nauk SSSR Ser. Fiz. **49**, 1866 (1976)
1.174 S. A. Permogorov, Ya. V. Morozenko: Fiz. Tvead. Tela **21**, 784 (1979 [English transl.: Sov. Phys.-Solid State **21**, 458 (1979)]
7.175 M. Kwietnak, Y. Oka, T. Kushida: J. Phys. Soc. Jpn. **44**, 558 (1978)
7.176 J. B. Grun: Proc. 15th Int. Conf. on the Physics of Semiconductors, Kyoto, 1980, p. 563; In Ref. 7.29
7.177 D. Fröhlich, E. Mohler, P. Wiesner: Phys. Rev. Lett. **26**, 554 (1971)
7.178 B. Hönerlage, U. Rössler, Vu Duy Phach, A. Bivas, J. B. Grun: Phys. Rev. B**22**, 797 (1980)
7.179 H. Schrey, V. G. Lyssenko, C. Klingshirn, B. Hönerlage: Phys. Rev. B**20**, 5267 (1979)
7.180 M. V. Klein: In Ref. 7.32, p. 147
       P. M. Platzman, P. A. Wolff: Solid State Phys. Suppl. **13**, (1973)
7.181 Estimated from the data of Wright and Mooradian in [7.36]
7.182 R. G. Ulbrich, C. Weisbuch: Verhandl. DPG (VI) 14, 81 (1979)
       R. G. Ulbrich, Nguyen Van Hieu, C. Weisbuch: Phys. Rev. Lett. **46**, 53 (1981)
7.183 P. A. Wolff: Phys. Rev. Lett. **16**, 225 (1966); *Proc. Scottish Universities School on Non-Linear Optics*, ed. by S. M. Kay, A. Maitland (Academic, New York 1970) p. 169
7.184 W. Kohn: In *Solid State Physics*, Vol. 5, ed by F. Seitz, D. Turnbull (Academic Press, New York 1957)
7.185 P. J. Colwell, M. V. Klein: Phys. Rev. B**6**, 498 (1972)
7.186 Nguyen Van Hieu, Nguyen Ai Viet: Phys. Status Solidi B**92**, 537 (1979)
       Nguyen Ba An, Nguyen Van Hieu, Nguyen Toan Thang, Nguyen Ai Viet: Phys. Status Solidi B**99**, 635 (1980)
7.187 Nguyen Ba An, Nguyen Van Hieu, Nguyen Toan Thang, Nguyen Ai Viet: J. Phys. (Paris) **41**, 1067 (1980)
7.188 P. J. Dean, D. C. Herbert: "Bound Excitons in Semiconductors", in Ref. 7.38, p. 55
7.189 R. Romestain, N. Magnea: Solid State Commun. **32**, 1201 (1979)
       P. J. Dean, D. C. Herbert, A. M. Labee: Proc. 15th Int. Conf. on the Physics of Semiconductors, Kyoto, 1980, p. 185

Subject Index

Light Scattering in Solids I

Editor: **M. Cardona**

With contributions by numerous experts
1975. 111 figures, 3 tables. XIII, 339 pages
(Topics in Applied Physics, Volume 8). ISBN 3-540-07354-X

"...The present book should find its audience in the solid-state optics community among those investigators who want a good and comprehensive introduction to laser light scattering in solids. The book is warmly recommended to them."

Journal of the Optical Society of America

Light Scattering in Solids II

Basic Concepts and Instrumentation

Editors: **M. Cardona, G. Güntherodt**

With contributions by numerous experts
1982. 88 figures. XIII, 251 pages
(Topics in Applied Physics, Volume 50). ISBN 3-540-11380-0

This Topics volume, the second of a 4-part treatment, is concerned with the basic principles of light scattering in solids and the instrumentation for its measurement. The basic theoretical principles are described, with emphasis on absolute scattering efficiencies and resonance phenomena near interband critical points. Topics and highlights of light scattering in solids during the past five years are briefly summarized. An updated review of light scattering in amorphous and disordered materials is given. Instrumentation and novel techniques are treated, with particular emphasis on optical multichannel detection and its applications to time-resolved and spatially resolved measurements and on coherent and hyper-Raman techniques such as hyper-Raman effect, coherent anti-Stokes Raman scattering, Raman-induced Kerr effect and multiwave mixing.

Physics of Intercalation Compounds

Proceedings of an International Conference Trieste, Italy, July 6–10, 1981

Editors: **L. Pietronero, E. Tosatti**

1981. 167 figures. IX, 323 pages
(Springer Series in Solid-State Sciences, Volume 38)
ISBN 3-540-11283-9

This book containing the papers delivered at the Trieste International Conference on the **Physics of Intercalation Compounds** in July 1981, is meant to represent an up-to-date reference point for all workers in the field, as well as an easy source of consultation for anybody wishing to be informed.

The contents deal largely – but not exclusively – with intercalated graphite. Structural, electronic, transport and other important properties are discussed by leading scientists, covering both experimental and theoretical aspects. Relevant side-issues, such as the status of 2-dimensional melting and stability and distortions of chain and layered compounds are also touched.

Springer-Verlag
Berlin
Heidelberg
NewYork

Coherent Nonlinear Optics

Recent Advances

Editors: **M. S. Feld, V. S. Letokhov**

1980. 2 portraits, 134 figures, 18 tables.
XVIII, 377 pages
(Topics in Current Physics, Volume 21)
ISBN 3-540-10172-1

Contents: *M. S. Feld, V. S. Letokhov:* Advances in Coherent Nonlinear Optics. – *M. S. Feld, J. C. MacGillivray:* Superradiance. – *V. P. Chebotayev:* Coherence in High Resolution Spectroscopy. – *G. Grynberg, B. Cagnac, F. Biraben:* Multiphoton Resonant Processes in Atoms. – *C. D. Cantrell, V. S. Letokhov, A. A. Makarov:* Coherent Excitation of Multilevel Systems by Laser Light. – *A. Laubereau, W. Kaiser:* Coherent Picosecond Interactions. – *M. D. Levenson, J. J. Song:* Coherent Raman Spectroscopy.

M. Lannoo, J. Bourgoin

Point Defects in Semiconductors I

Theoretical Aspects

With a Foreword by J. Friedel
1981. 87 figures. XVII, 265 pages
(Springer Series in Solid-State Sciences, Volume 22)
ISBN 3-540-10518-2

Contents: Atomic Configuration of Point Defects. – Effective Mass Theory. – Simple Theory of Deep Levels in Semiconductors. – Many-Electron Effects and Sophisticated Theories of Deep Levels. – Vibrational Properties and Entropy. – Thermodynamics of Defects. – Defect Migration and Diffusion. – References. – Subject Index.

Laser Spectroscopy of Solids

Editors: **W. M. Yen, P. M. Selzer**

1981. 117 figures. XI, 310 pages
(Topics in Applied Physics, Volume 49)
ISBN 3-540-10638-3

Contents: *G. F. Imbusch, R. Kopelman:* Optical Spectroscopy of Electronic Centers in Solids. – *T. Holstein, S. K. Lyo, R. Orbach:* Excitation Transfer in Disordered Systems. – *D. L. Huber:* Dynamics of Incoherent Transfer. – *P. M. Selzer:* General Techniques and Experimental Methods in Laser Spectroscopy of Solids. – *W. M. Yen, P. M. Selzer:* High Resolution Laser Spectroscopy of Ions in Crystals. – *M. J. Weber:* Laser Excited Fluorescence Spectroscopy in Glass. – *A. H. Francis, R. Kopelman:* Excitation Dynamics in Molecular Solids.

Photoemission in Solids I

General Principles

Editors: **M. Cardona, L. Ley**

1978. 90 figures, 17 tables. XI, 290 pages
(Topics in Applied Physics, Volume 26)
ISBN 3-540-08685-4

Contents: *M. Cardona, L. Ley:* Introduction. – *W. L. Schaich:* Theory of Photoemission: Independent Particle Model. – *S. T. Manson:* The Calculation of Photoionization Cross Sections: An Atomic View. – *D. A. Shirley:* Many-Electron and Final-State Effects: Beyond the One-Electron Picture. – *G. K. Wertheim, P. H. Citrin:* Fermi Surface Excitations in X-Ray Photoemission Line Shapes from Metals. – *N. V. Smith:* Angular Dependent Photoemisson. – Appendix.

Photoemission in Solids II

Case Studies

Editors: **L. Ley, M. Cardona**

1979. 214 figures, 26 tables. XVIII, 401 pages
(Topics in Applied Physics, Volume 27)
ISBN 3-540-09202-1

Contents: *L. Ley, M. Cardona:* Introduction. – *L. Ley, M. Cardona, R. A. Pollak:* Photoemission in Semiconductors. – *S. Hüfner:* Unfilled Inner Shells: Transition Metals and Compounds. – *M. Campagna, G. K. Wertheim, Y. Baer:* Unfilled Inner Shells: Rare Earths and Their Compounds. – *W. D. Grobman, E. E. Koch:* Photoemission from Organic Molecular Crystals. – *C. Kunz:* Synchrotron Radiation: Overview. – *P. Steiner, H. Höchst, S. Hüfner:* Simple Metals. – Appendix: Table of Core-Level Binding Energies. – Additional References with Titles. – Subject Index.

Springer-Verlag
Berlin
Heidelberg
New York